The Elements of
Astronomy for Surveyors

J. B. MACKIE, M.Sc., B.E.

A.O.S.M., F.G.S., F.N.Z.I.S., F.R.A.S.N.Z., M.I.P.E.N.Z.

Professor Emeritus, University of Otago, New Zealand
Past President of the New Zealand Institute of Surveyors
Past President of the Royal Astronomical Society of New Zealand
Formerly Chairman of the Carter Observatory Board, New Zealand

NINTH EDITION

CHARLES GRIFFIN & COMPANY LTD
High Wycombe

CHARLES GRIFFIN & COMPANY LIMITED

Registered Office:
Charles Griffin House, Crendon Street
High Wycombe, Bucks, HP13 6LE
England

Copyright © 1985

All rights reserved. No part of this book may be reproduced or transmitted in any form or by any means, electronic or mechanical, including photocopying, recording, or by any information storage and retrieval system, without permission in writing from the publisher.

Author: Sir Robert Chapman

First published	1918
Fourth edition	1933
Five impressions	

Author: J. B. Mackie

Fifth edition	1953
Sixth edition	1964
Seventh edition (reset)	1971
Eighth edition (reset)	1978
Reprinted	1982
Ninth edition (reset)	1985

Mackie, J. B.
   The elements of astronomy for surveyors.—
  9th ed.
  1. Surveying   2. Astronomy, Spherical and
  Practical
  I. Title
  526'.6   TA597
  ISBN 0-85264-244-X

Set by Joshua Associates, Oxford
Printed in Great Britain by
The Bath Press, Avon

PREFACE TO NINTH EDITION

The ordinary surveyor's greatest use of field astronomy is in the determination of azimuth to control the orientation of his survey in areas where the density of control, or its reliability, is low. In the process, he may well need to find his position by astronomical observation, but 1 arc-second of accuracy here is often much more than adequate. There are still many parts of the earth where such azimuth control is needed and where it must be obtained quickly and at reasonable cost, without the need for expensive satellite positioning equipment.

It has been said that modern survey techniques do not require traditional astronomy, and that the need for a book like this would hardly exist were it not for teaching institutions and licensing authorities continuing to include the subject in their curricula. The comment was made in the light of developments such as satellite positioning of survey control points, and combination or "total station" instruments that turn horizontal and vertical angles and read the distance by electromagnetic means.

The ordinary surveyor is not in the business of trying to fix the position of his ground mark *for cadastral survey purposes* by astronomical methods; even if he could guarantee to observe to 1 arc-second, he could still be 30 metres in error in latitude, which greatly exceeds the usual cadastral requirements. Thus, with present-day satellite positioning (by Doppler) giving results to the order of 0.8 metre, the comment is perhaps fair enough. It is essential to remember however, that this 0.8 metre accuracy is obtained only after several days of "observations" with sophisticated and expensive equipment not available to most surveyors. There is little doubt that satellite techniques will improve in the not-too-distant future to allow position to within 10 centimetres to be achieved—a figure which is coming closer to the requirements of cadastre in developed countries and of scientific inquiry into problems like those of plate tectonics and earth deformation.

This revision has been prompted by several factors: the need to update some of the text, particularly that relating to the timing of observations; the desirability of including examples of almanac tables, especially of the *Star Almanac for Land Surveyors* to which so much reference is made; and the necessity for correcting a few minor errors discovered after the 8th edition went to press. In the revision the opportunity has also been taken to add two further appendices, one to cater for surveyors in the U.S.A. who use a set of sun and star tables published by the U.S. Department of the Interior, Bureau of Land Management; and the other to provide some calculator and microcomputer programs relating specifically to the methods of field astronomy.

Preface

Sample pages from The *Star Almanac for Land Surveyors* have been reproduced with the permission of the Controller of Her Majesty's Stationery Office, Norwich, England; from *Apparent Places of Fundamental Stars* with the permission of the publishers, G. Braun of Karlsruhe, Federal Republic of Germany; from the catalogue *Positions and Proper Motions of 258,997 Stars for the Epoch and Equinox of 1950.0* with the permission of the Smithsonian Astrophysical Observatory, Cambridge, Massachusetts, U.S.A.; and from *Ephemeris of the Sun, Polaris, and other Selected Stars* with permission from the Divison of Cadastral Survey, Bureau of Land Management, Washington, D.C., U.S.A. I am indebted to the organizations concerned for their ready approval of my request to include this material in *Astronomy for Surveyors*.

Chapter 6 has been purged of references to mechanical time-keepers, and largely rewritten around digital clocks and watches and more modern electronic technology.

The two new Appendices are quite substantial. Appendix I has a good selection of examples worked out for locations in U.S.A., using the *Ephemeris* produced by the Division of Cadastral Survey. A number of these examples have also been worked out using the *Star Almanac for Land Surveyors*, so that readers who want more experience with this set of tables now have additional northern hemisphere illustrations for their guidance. The content of Appendix I was kept separate from the main text because of the considerable differences between the two sets of tables and the need to avoid confusion in the minds of users. However, those who have mastered the problems of time transformations and the use of the *Star Almanac for Land Surveyors* may find it interesting to see the U.S.A. approach.

Appendix II contains listings of a number of computer programs which users of this book may find useful in reducing the observations made in field astronomy. They are divided into two categories: those for use with pocket and desk-top calculators employing the RPN logic system (e.g. HP-67/97), and those for use with a 16K Level II BASIC microcomputer (Z-80 CPU). These programs have been checked and appear to work satisfactorily. However, one seldom finds the perfect computer program written by someone else and I expect some users will modify these to their own ends.

The worked examples illustrating the methods of field astronomy explained in the text have not been changed; although the observations on which they were based were done in 1974/75 (except for later ones inserted in Appendix I), they still demonstrate the reductions which would be made of data collected at the time of this revision. The only "dated" matter in those early examples is the reference to chronometers, but if viewed in the proper context this is no detraction; a modern digital clock or good wrist-watch controlled by a quartz crystal oscillator is probably as much entitled to the name chronometer as its mechanical forerunners.

Before bringing these remarks to an end I would like to thank the many

Preface

reviewers of the Eighth Edition for their helpful comments. Constructive reviews are most welcome, although it is not always possible to incorporate every suggestion made. One writer, for example, commented that no gesture had been made to the use of satellites in determining position. Artificial earth satellites may certainly be regarded as astronomical objects, but as stated earlier in this Preface, observing them for position fixing requires very specialized equipment. The subject is a complex one that would require explanation in a separate treatise. All I can do for the reviewer in question and any other interested readers is to refer them by way of introduction to two relatively simple papers in *The New Zealand Surveyor*: Rowe, G. H., 1981, vol. 29(6), No. 258, pp. 608–24; and Mackie, J. B., 1983, vol. 30(4), No. 262, pp. 412–20. Both these papers contain references to wider reading.

I am also indebted to the Surveyor-General of New Zealand, Mr W. N. Hawkey, for permission to publish the photographs in Fig. 6.2, 6.3, 8.2 and 8.3 and to Mr Glen Rowe of his staff for taking them in the field.

Finally, I must say it is seldom that an author has such a forbearing and helpful publisher and editor, and I would be most remiss if I did not record my sincere thanks to Mr James R. Griffin and Mr Edmund V. Burke, respectively, in this regard.

1984   J.B.M.

CONTENTS

		Page
1	**The solution of spherical triangles**	1

Basic formulae—Great circles and spherical triangles—Solution of right-angled spherical triangles and of oblique spherical triangles—Derivation of formulae—Solution by calculator or microcomputer

| 2 | **The celestial sphere and astronomical coordinates** | 17 |

Apparent motion of the stars—Celestial equator—Systems of astronomical coordinates—Altitude and azimuth—Right ascension and declination–Sidereal time—Hour angle—The prime vertical—Synopsis of astronomical terms—Exercises

| 3 | **The earth** | 28 |

Terrestrial latitude and longitude—Measurement of great circle arcs along the earth's surface—The figure of the earth—Astronomic, geodetic, geographic and geocentric latitude—Deviation of the vertical—Exercises

| 4 | **The sun and the stars** | 42 |

The sun's apparent motion among the stars—The earth's orbit round the sun—The sun's motion in right ascension and declination—The sun's apparent path on the celestial sphere—The slight variations in the right ascensions and declinations of the stars—Star catalogues—Mean and Apparent Places of stars—Reduction to Apparent Place from catalogue epoch—Sample extracts from star catalogues—Exercises

| 5 | **Time** | 64 |

BASIC KINDS OF TIME—Sidereal time—Atomic time and Coordinated Universal Time (U.T.C.)—Solar time—The Equation of Time—Systems of time measurement—Local sidereal time—Local mean time—Greenwich mean time (Universal Time), U.T.C., U.T.0., U.T.1—Local standard time. TIME TRANSFORMATIONS—Local mean time from local standard time—Sidereal interval from mean solar interval and vice versa—Local

Contents

standard time of upper transit of the First Point of Aries on a given date—True sun's local hour angle from local standard time and vice versa—Local standard time of sun's upper transit across the meridian on a given date—Coefficients for computing R, E, sun's declination and semi-diameter—Sample extracts from the *Star Almanac*—Exercises

6 The timing of observations   109

Timepieces—Methods of time-keeping—Personal equation—The Chronocord—Impersonal telescope eyepieces—Radio time-signals—DUT1 correction—Clock error and rate by comparison with radio time-signals—Special time-comparison equipment—Accuracy of timing observations

7 Astronomical and instrumental corrections to observations of altitude and azimuth   119

Corrections to altitude observations—(A) Astronomical: parallax; celestial refraction; semi-diameter (sun observations); curvature of path—(B) Instrumental: vertical collimation error; alidade bubble off centre. Corrections to azimuth observations—(A) Astronomical: semi-diameter (sun observations); curvature of path—(B) Instrumental: horizontal collimation error; dislevelment of trunnion axis—Summary

8 Preparation for observations   147

Set-up of theodolite—Illuminating gear—Reference mark and its approximate azimuth—Time-keeping equipment—Assistant—Observing programme—Computer predictions—Field equipment and accessories required

9 The location of objects on the celestial sphere   157

Given the right ascension and declination of a star at a certain time and place of observation, to find its altitude and azimuth at that time, and some short interval of time later—Identification of stars by determination of right ascension and declination—from rate of change of altitude and azimuth. Worked examples throughout—Exercise

10 The determination of azimuth   167

The need for azimuth—Star observations in daylight—Azimuth by (1) Observation of a star at equal altitudes—(2) Observation of

Contents

a circumpolar star near elongation—Circum-elongation observations of stars—(3) Ex-meridian observations of the sun or a star by measuring altitudes and horizontal angles—(4) Ex-meridian observations of the sun or a star by measuring hour angles and horizontal angles—(5) Time observations of a close circumpolar star at any hour angle. Procedures and worked examples throughout—Choice of method—References—Exercises

11 **The determination of latitude**   233

The need for astronomical latitude—Latitude by (1) Observation of the sun or a star at meridian transit—(2) Zenith pair observations of stars—(3) Circum-meridian observations of the sun or a star—(4) Measuring altitude of a close circumpolar star at any hour angle—(5) Observing two stars at the same altitude. Procedures and worked examples throughout—Choice of method—References—Exercises

12 **The determination of local time by observation**   271

The need for astronomically determined time—Determination of local time by (1) Meridian transits on either or both sides of the zenith—(2) Ex-meridian altitudes of the sun or stars—(3) Noting the instants when stars are at the same altitude. Procedures and worked examples throughout—Prismatic astrolabe, astrolabe attachment—Sundials; types and construction. Choice of method—Exercises

13 **The determination of longitude**   300

Relative and absolute methods—Longitude by comparison of local time and radio time-signals—Worked example—References—Exercises

14 **Simultaneous determination of latitude and longitude from position lines**   306

Principle of the method—Application: observations and computations—Prismatic astrolabe observations—Observing procedure for theodolite work—Star tables for prediction and reduction of observations—Worked example—Least squares computation of position from multiple observations—References—Exercises

A Short General Bibliography   322

Contents

Appendices

I Explanation and examples of the use of the U.S. *Ephemeris*   325

Identification of stars and prediction of star positions—Determination of azimuth—Determination of latitude—Determination of local time—Determination of longitude—Simultaneous determination of latitude and longitude from position lines—Sample extracts from *Ephemeris* and *SALS*

II Some desk-top and microcomputer programs for use in field astronomy   366

Introduction—The Reverse Polish Notation programs: Azimuth by altitude of the sun—Azimuth by altitude of a star—Azimuth by hour angle of the sun—Azimuth by hour angle of a star. The BASIC programs: Calculating azimuth, latitude, local time (watch error) and longitude—Finding latitude and longitude using the position line method—Computing the solar ephemeris

III The convergence of meridians   396

Method of computation of convergence—Sense of application of adjustment for convergence in Northern and Southern Hemispheres —Correction for deviation of the vertical

Index   401

NOTATION

Earlier editions of this book followed the notation used in the UK War Office's *Text Book of Field Astronomy* and in *Star Almanac for Land Surveyors*. However, in some respects this notation was at variance with the recommendations of the International Astronomical Union, *vide* Lee in "Astronomical Notation", *Survey Review*, **20**, No. 156, April 1970, pp. 290-2. The eighth edition was altered to follow the IAU recommendations. The main changes were from h = hour angle and H = altitude to t = hour angle and h = altitude.

P	the elevated pole	
Z	the observer's zenith; also the zenith angle of the astronomical triangle	
S	the celestial body observed; also the parallactic angle of the astronomical triangle	
t, H.A.	the hour angle of a celestial body	
p	the polar distance of a celestial body	
δ	the declination of a celestial body	
ζ	the zenith distance of a celestial body	
h	the altitude of a celestial body above the horizon	
ψ	celestial refraction	
ω	the co-latitude of a terrestial station	
ϕ	the latitude of a terrestrial station	
λ	the longitude of a terrestrial station	
A	azimuth clockwise from north	
R.A., α	the right ascension of a celestial body	
♈	the First Point of Aries	
P.V.	Prime Vertical	
R.O.	Reference object ⎫	synonymous
R.M.	Reference mark ⎭	
L.A.S.T.	Local Apparent Sidereal Time	
L.S.T.	Local Sidereal Time (= L.A.S.T. unless qualified)	
L.M.S.T.	Local Mean Sidereal Time	
G.A.S.T.	Greenwich Apparent Sidereal Time (= L.A.S.T. on the Greenwich meridian)	
G.S.T.	Greenwich Sidereal Time (= G.A.S.T. unless qualified)	
G.M.S.T.	Greenwich Mean Sidereal Time (= L.M.S.T. on the Greenwich meridian)	
L.M.T.	Local Mean Time	
U.T.C.	Coordinated Universal Time (from most radio time-signals)	

Notation

U.T.1	U.T.C. + DUT1
DUT1	U.T.1 − U.T.C. (DUT1 is coded on time-signals)
U.T.	Universal Time (= U.T.1 in this book unless qualified)
G.M.T.	Greenwich Mean Time (probably = U.T.C. unless qualified)
L.Std.T.	Local Standard Time (= U.T.C., ± Longitude of Zone meridian)
P.S.T.	Pacific Standard Time (U.S.A.)
E.S.T.	Eastern Standard Time (U.S.A.)
L.H.A.	Local Hour Angle of a celestial body
G.H.A.	Greenwich Hour Angle of a celestial body
E	difference between the Greenwich Hour Angle of the Sun and Universal Time
R	difference between the Greenwich Hour Angle of ♈ and Universal Time
S.A.	*Star Almanac for Land Surveyors* (H.M.S.O.); also *SALS*
SAO	Smithsonian Astrophysical Observatory (Catalogue)
APFS	*Apparent Places of Fundamental Stars* (Catalogue)

Greek alphabet

A	α	alpha	N	ν	nu
B	β	beta	Ξ	ξ	xi
Γ	γ	gamma	O	o	omicron
Δ	δ	delta	Π	π	pi
E	ϵ	epsilon	P	ρ	rho
Z	ζ	zeta	Σ	σ	sigma
H	η	eta	T	τ	tau
Θ	θ	theta	Υ	υ	upsilon
I	ι	iota	Φ	ϕ	phi
K	κ	kappa	X	χ	chi
Λ	λ	lambda	Ψ	ψ	psi
M	μ	mu	Ω	ω	omega

Atmospheric Pressure Conversion Table

Inches of mercury	Millimetres of mercury	Millibars mb	Inches of mercury	Millimetres of mercury	Millibars mb
32·0	812·8	1084	29·0	736·6	982
31·9	810·3	1080	28·9	734·1	979
31·8	807·7	1077	28·8	731·5	975
31·7	805·2	1073	28·7	729·0	972
31·6	802·6	1070	28·6	726·4	969
31·5	800·1	1067	28·5	723·9	965
31·4	797·6	1063	28·4	721·4	962
31·3	795·0	1060	28·3	718·8	958
31·2	792·5	1057	28·2	716·3	955
31·1	789·9	1053	28·1	713·7	952
31·0	787·4	1050	28·0	711·2	948
30·9	784·9	1046	27·9	708·7	945
30·8	782·3	1043	27·8	706·1	941
30·7	779·8	1040	27·7	703·6	938
30·6	777·2	1036	27·6	701·0	935
30·5	774·7	1033	27·5	698·5	931
30·4	772·2	1029	27·4	696·0	928
30·3	769·6	1026	27·3	693·4	924
30·2	767·1	1023	27·2	690·9	921
30·1	764·5	1019	27·1	688·3	918
30·0	762·0	1016	27·0	685·8	914
29·9	759·5	1013	26·0	660·4	880
29·8	756·9	1009	25·0	635·0	847
29·7	754·4	1006	24·0	609·6	813
29·6	751·8	1002	23·0	584·2	779
29·5	749·3	999	22·0	555·8	745
29·4	746·8	996	21·0	533·4	711
29·3	744·2	992	20·0	508·0	677
29·2	741·7	989	19·0	482·6	643
29·1	739·1	985	18·0	457·2	610

1 inch = 25·4 mm = 33·863 886 mb

Temperature Conversion Table

Fahrenheit degrees F	Celsius degrees C	Fahrenheit degrees F	Celsius degrees C	Fahrenheit degrees F	Celsius degrees C
0	−17·8	51	10·6	84	28·9
10	−12·2	52	11·1	85	29·4
20	−6·7	53	11·7	86	30·0
21	−6·1	54	12·2	87	30·6
22	−5·6	55	12·8	88	31·1
23	−5·0	56	13·3	89	31·7
24	−4·4	57	13·9	90	32·2
25	−3·9	58	14·4	91	32·8
26	−3·3	59	15·0	92	33·3
27	−2·8	60	15·6	93	33·9
28	−2·2	61	16·1	94	34·4
29	−1·7	62	16·7	95	35·0
30	−1·1	63	17·2	96	35·6
31	−0·6	64	17·8	97	36·1
32	0	65	18·3	98	36·7
33	0·6	66	18·9	99	37·2
34	1·1	67	19·4	100	37·8
35	1·7	68	20·0	101	38·3
36	2·2	69	20·6	102	38·9
37	2·8	70	21·1	103	39·4
38	3·3	71	21·7	104	40·0
39	3·9	72	22·2	105	40·6
40	4·4	73	22·8	106	41·1
41	5·0	74	23·3	107	41·7
42	5·6	75	23·9	108	42·2
43	6·1	76	24·4	109	42·8
44	6·7	77	25·0	110	43·3
45	7·2	78	25·6		
46	7·8	79	26·1		
47	8·3	80	26·7		
48	8·9	81	27·2		
49	9·4	82	27·8		
50	10·0	83	28·3		

°F to °C = (°F − 32°) × $\frac{5}{9}$

°C to °F = (°C × $\frac{9}{5}$) + 32°

For degrees Kelvin, add 273°·16 to °C

CHAPTER 1
The solution of spherical triangles

1.01 Introduction

In this chapter the principal formulae of spherical trigonometry, such as will be afterwards applied to calculations on the celestial sphere, are brought together for convenient reference. A brief synopsis will be given of the usual methods for the solution of spherical triangles under different conditions, and to conclude the chapter some of the formulae will be established.

1.02 Great circles and spherical triangles

The line of intersection made with the surface of a sphere by a plane passing through the centre of the sphere is known as a *great circle*. If this circle passes through two points A and B on the surface of the sphere, then the shortest distance between A and B, measured along the sphere's surface, is that measured along the arc of the great circle joining them. Only one great circle can be drawn to pass through two given points on the surface of a sphere, unless they happen to be at the opposite extremities of a diameter, and the length of the shorter arc of this great circle between the two points is the shortest distance between them. Meridians of longitude on the earth's surface are great circles, assuming the earth to be a perfect sphere.

The line which passes through the centre of the sphere at right angles to the plane of any great circle is the *axis* of that circle, while the two points where the axis cuts the surface of the sphere are the *poles* of the circle.

The three great circles shown in Fig. 1.1, namely *EABD, GACF* and *HBCJ*, cut one another to form a number of spherical triangles of which *ABC* is one. In spherical trigonometry it is always assumed that the arcs representing the sides of the triangles considered are arcs of great circles. In the figure we know that the arcs *AB, AC* and *BC* of the great circles are each equal in length to the product of the radius of the sphere and the angle subtended by the arc at the centre of the sphere, O, but as we are dealing in astronomy with the celestial sphere which has an enormous and arbitrary radius, we cannot conveniently concern ourselves with the lengths of these arcs.

However, since all our triangles will be on one single sphere, we can assume the radius of that sphere to be unity, even though the unit is an extremely large one, and then the arc becomes equal to the angle it subtends at the centre.

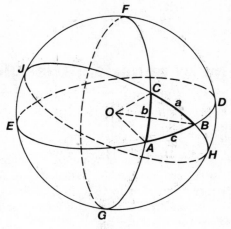

Fig. 1.1

We now discover that our spherical triangle has for its elements *six angles*. Looking at triangle *ABC* in Fig. 1.1 we see that these are:

A (or *BAC*), the angle between the planes of the great circles *GACF* and *EABD*, part of whose line of junction is the line *OA*.
B (or *ABC*), the angle between the planes of the great circles *EABD* and *HBCJ*, part of whose line of junction is the line *OB*.
C (or *ACB*), the angle between the planes of the great circles *GACF* and *HBCJ*, part of whose line of junction is the line *OC*.
a (or the side *CB*) which is equal to angle *COB* at the centre of the sphere.
b (or the side *AC*) which is equal to angle *AOC* at the centre of the sphere.
c (or the side *AB*) which is equal to angle *AOB* at the centre of the sphere.

Three properties of spherical triangles should be noted:

(i) the sum of the three apical angles is not, as in a plane triangle, equal to 180°, but is always greater by an amount called the spherical excess, E; E may be calculated from the formula:

$$\tan^2 \tfrac{1}{2}E = \tan \tfrac{1}{2}s \cdot \tan \tfrac{1}{2}(s-a) \cdot \tan \tfrac{1}{2}(s-b) \cdot \tan \tfrac{1}{2}(s-c),$$

where

$$s = \tfrac{1}{2}(a+b+c).$$

(Sometimes the spherical excess is denoted by ϵ.)

(ii) if two of the apical angles are equal, the sides opposite them are equal; and
(iii) if one apical angle is greater than another, the side opposite the larger one is greater than that opposite the smaller one.

Solution of spherical triangles

1.03 Small circles

The line of intersection made with the surface of a sphere by a plane that does not pass through the centre is known as a *small circle*. The ordinary formulae of spherical trigonometry do not apply to triangles having one or more sides that are arcs of small circles. A parallel of latitude on the earth's surface is a small circle, assuming the earth to be a true sphere. It follows that the shortest distance between two points in the same latitude is not that measured along the parallel of latitude, but is that measured along the *arc of the great circle* joining them.

1.04 Basic formulae

If, as in Fig. 1.1, we denote the angles of a spherical triangle by A, B and C and the sides opposite to these angles by a, b and c respectively, the sides being measured by the angles which they subtend at the centre of the sphere, we have the following fundamental relations:

(a) The sines of the angles are proportional to the sines of the opposite sides:

$$\frac{\sin A}{\sin a} = \frac{\sin B}{\sin b} = \frac{\sin C}{\sin c}. \tag{1.1}$$

(b) One side of a triangle is expressed in terms of the two other sides and the angle included between them by the formula:

$$\cos b = \cos a . \cos c + \sin a . \sin c . \cos B. \tag{1.2}$$

(c) From this may be derived another set of useful relationships of which the following two are types:

$$\cot c . \sin a = \cos a . \cos B + \sin B . \cot C \tag{1.3a}$$

$$\cot c . \sin b = \cos b . \cos A + \sin A . \cot C. \tag{1.3b}$$

(d) Again the cosine of the side of a triangle may be expressed in terms of the sines and cosines of the apical angles by the formula:

$$\cos A = -\cos B . \cos C + \sin B . \sin C . \cos a. \tag{1.4}$$

All the above basic formulae are in a form appropriate for use with calculating machines, but (1.2), (1.3a), (1.3b) and (1.4) must be transformed into different modes (as given later) if they are to be employed in logarithmic computation. All the above equations are *cyclic*, so that (1.2), (1.3a), (1.3b) and (1.4) each represents only one of the three in the set (see **1.06**).

Astronomy for surveyors

1.05 The solution of right-angled spherical triangles

If one of the apical angles of the spherical triangle is a right angle, the basic formulae in the preceding section become much simpler. For example, with $C = 90°$,

(1.1) becomes $\dfrac{\sin A}{\sin a} = \dfrac{\sin B}{\sin b} = \dfrac{1}{\sin c}$

(1.2) becomes $\cos c = \cos a . \cos b$, using the cyclic equation containing $\cos C$
(1.3a) becomes $\cot c . \sin a = \cos a . \cos B$
(1.3b) becomes $\cot c . \sin b = \cos b . \cos A$
(1.4) becomes $\cos A = \sin B . \cos a$.

To help us find easily the relationships between the sides and angles of a right-angled spherical triangle there are two mnemonic rules devised by Napier, the inventor of logarithms. These are known as Napier's Rules of Circular Parts.

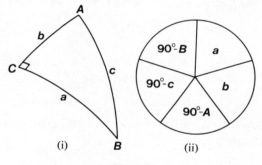

Fig. 1.2

In Fig. 1.2(i) we have a spherical triangle which is right-angled at C. The apices of the triangle are lettered alphabetically in a clockwise direction. Figure 1.2(ii) depicts a circle divided into five equal segments; in one segment is placed a, denoting one of the sides of the spherical triangle next to the right angle C. To establish the figure which is the key to the mnemonic we must now range the "circular" parts of the triangle around the circle in Fig. 1.2(ii) clockwise in the order in which they stand in the triangle. After a would come C, but since this is the right angle it is omitted. Next comes b, followed by $(90° - A)$, $(90° - c)$ and $(90° - B)$. Note that the elements of the triangle in the three brackets are in their correct clockwise order round the triangle, but they are represented by "circular" parts $(90° - A)$, $(90° - c)$ and $(90° - B)$.

Now if any one of these five parts is selected and called the *middle* part, the two parts on each side of it are called the *adjacent* parts, and the remaining two are called the *opposite* parts. For instance, if a is chosen as the middle part,

Solution of spherical triangles

$(90° - B)$ and b are the adjacent parts, and $(90° - c)$ and $(90° - A)$ are the opposite parts. Then Napier's Rules are:

Sine of *middle* part = product of *tangents* of *adjacent* parts,
Sine of *middle part* = product of *cosines* of *opposite* parts.

Thus
$$\sin a = \tan(90° - B) \cdot \tan b = \cot B \cdot \tan b$$

and
$$\sin a = \cos(90° - c) \cdot \cos(90° - A) = \sin c \cdot \sin A.$$

As an aid to memory, it may be noticed that the vowels in the words *sine* and *middle* are the same, so with *tangent* and *adjacent, cosine* and *opposite*.

By choosing different parts in turn as the middle parts, we obtain all the possible relationships between the sides and angles, and with a little practice it is easy to choose the particular ones wanted. For example, if we know a and b, and wish to find c, we must select $(90° - c)$ as the middle part; using a and b then as opposite parts:

$$\sin(90° - c) = \cos a \cdot \cos b$$

i.e.
$$\cos c = \cos a \cdot \cos b.$$

If a and A are known, and B is required, we must choose $(90° - A)$ as the middle part:

$$\sin(90° - A) = \cos(90° - B) \cdot \cos a$$

or
$$\cos A = \sin B \cdot \cos a.$$

There are six cases to consider in the solution of right-angled triangles, and the formulae required, which are readily obtained from Napier's Rules, are as follows:

(a) Given the hypotenuse c and an angle A:

$$\tan b = \tan c \cdot \cos A \tag{1.5a}$$

$$\cot B = \cos c \cdot \tan A \tag{1.5b}$$

$$\sin a = \sin c \cdot \sin A. \tag{1.5c}$$

(b) Given a side b and the adjacent angle A:

$$\tan c = \frac{\tan b}{\cos A} \tag{1.6a}$$

Astronomy for surveyors

$$\tan a = \tan A . \sin b \qquad (1.6b)$$

$$\cos B = \cos b . \sin A . \qquad (1.6c)$$

(c) Given the two sides a and b, then

$$\cos c = \cos a . \cos b \qquad (1.7a)$$

$$\cot A = \cot a . \sin b \qquad (1.7b)$$

$$\cot B = \cot b . \sin a . \qquad (1.7c)$$

(d) Given the hypotenuse c and side a:

$$\cos b = \frac{\cos c}{\cos a} \qquad (1.8a)$$

$$\cos B = \frac{\tan a}{\tan c} \qquad (1.8b)$$

$$\sin A = \frac{\sin a}{\sin c}. \qquad (1.8c)$$

(e) Given the two angles A and B:

$$\cos c = \cot A . \cot B \qquad (1.9a)$$

$$\cos a = \frac{\cos A}{\sin B} \qquad (1.9b)$$

$$\cos b = \frac{\cos B}{\sin A}. \qquad (1.9c)$$

(f) Given a side a and opposite angle A:

$$\sin c = \frac{\sin a}{\sin A} \qquad (1.10a)$$

$$\sin b = \tan a . \cot A \qquad (1.10b)$$

$$\sin B = \frac{\cos A}{\cos a}. \qquad (1.10c)$$

1.06   The solution of oblique spherical triangles

In this section many of the formulae are cyclic, that is, by putting b for a, c for b, a for c, and B for A, C for B and A for C, as appropriate, another formula is set up for computing the next quantity in order. For example (1.2) of **1.04** is

$$\cos b = \cos a . \cos c + \sin a . \sin c . \cos B.$$

Solution of spherical triangles

This can be cyclically changed to

$$\cos c = \cos b . \cos a + \sin b . \sin a . \cos C$$

and again to

$$\cos a = \cos c . \cos b + \sin c . \sin b . \cos A.$$

(a) Given the three sides a, b and c, to find A, B, C.

(i) $$\cos A = \frac{\cos a - \cos b . \cos c}{\sin b . \sin c} \quad \text{(cyclic)} \qquad (1.11)$$

When A, B and C are all found they may be checked by (1.1):

$$\frac{\sin A}{\sin a} = \frac{\sin B}{\sin b} = \frac{\sin C}{\sin c}. \qquad (1.1)$$

(ii) Let $s = \tfrac{1}{2}(a + b + c)$; then

$$\tan \tfrac{1}{2}A = \sqrt{\left(\frac{\sin(s-b).\sin(s-c)}{\sin s.\sin(s-a)}\right)} \quad \text{(cyclic)} \qquad (1.12a)$$

$$= \sqrt{\left(\frac{\cos(b-c) - \cos a}{\cos a - \cos(b+c)}\right)} \quad \text{(cyclic)} \qquad (1.12b)$$

$$= \frac{\tan r}{\sin(s-a)} \quad \text{(cyclic)} \qquad (1.12c)$$

where

$$\tan r = \sqrt{\left(\frac{\sin(s-a).\sin(s-b).\sin(s-c)}{\sin s}\right)}$$

and r is the radius of the inscribed circle.

(iii) $$\sin \tfrac{1}{2}A = \sqrt{\left(\frac{\sin(s-b).\sin(s-c)}{\sin b.\sin c}\right)} \quad \text{(cyclic)} \qquad (1.13a)$$

$$= \sqrt{\left(\frac{\cos(b-c) - \cos a}{2 \sin b.\sin c}\right)} \quad \text{(cyclic)} \qquad (1.13b)$$

$$\cos \tfrac{1}{2}A = \sqrt{\left(\frac{\sin s.\sin(s-a)}{\sin b.\sin c}\right)} \quad \text{(cyclic)} \qquad (1.14a)$$

$$= \sqrt{\left(\frac{\cos a - \cos(b+c)}{2 \sin b.\sin c}\right)} \quad \text{(cyclic)} \qquad (1.14b)$$

(b) Given two sides a, b and the included angle C, to find A, B, c.

(i)
$$\cot A = \frac{\cot a . \sin b - \cos b . \cos C}{\sin C} \qquad (1.15)$$

$$\cot B = \frac{\sin a . \cot b - \cos a . \cos C}{\sin C} \qquad (1.16)$$

$$\cos c = \cos a . \cos b + \sin a . \sin b . \cos C. \qquad (1.17)$$

Check: $\dfrac{\sin a}{\sin A} = \dfrac{\sin b}{\sin B} = \dfrac{\sin c}{\sin C}.$ (1.1)

(ii) Let $\tan m = \tan a . \cos C$, $\tan n = \tan b . \cos C$; then

$$\cot A = \frac{\sin(b - m) . \cot C}{\sin m} \qquad (1.18)$$

$$\cot B = \frac{\sin(a - n) . \cot C}{\sin n} \qquad (1.19)$$

$$\cos c = \frac{\cos a . \cos(b - m)}{\cos m} = \frac{\cos b . \cos(a - n)}{\cos n} \qquad (1.20)$$

(iii)
$$\tan \tfrac{1}{2}(A + B) = \frac{\cos \tfrac{1}{2}(a - b)}{\cos \tfrac{1}{2}(a + b)} . \cot \tfrac{1}{2}C \qquad (1.21)$$

$$\tan \tfrac{1}{2}(A - B) = \frac{\sin \tfrac{1}{2}(a - b)}{\sin \tfrac{1}{2}(a + b)} . \cot \tfrac{1}{2}C \qquad (1.22)$$

and
$$A = \tfrac{1}{2}(A + B) + \tfrac{1}{2}(A - B)$$
$$B = \tfrac{1}{2}(A + B) - \tfrac{1}{2}(A - B).$$

$$\tan \tfrac{1}{2}c = \frac{\sin \tfrac{1}{2}(A + B)}{\sin \tfrac{1}{2}(A - B)} . \tan \tfrac{1}{2}(a - b) \qquad (1.23a)$$

$$= \frac{\cos \tfrac{1}{2}(A + B)}{\cos \tfrac{1}{2}(A - B)} . \tan \tfrac{1}{2}(a + b) \qquad (1.23b)$$

$\left(\text{or } \sin c = \dfrac{\sin a . \sin C}{\sin A} = \dfrac{\sin b . \sin C}{\sin B} \text{ but these are ambiguous} \right).$

(1.17) can also be used to find c without ambiguity.

(c) Given two sides a and b and the angle opposite one of them A, to find B, C, c. There will be only one solution if $|90° - a| < |90° - b|$, but two solutions (i.e. ambiguity) if $|90° - a| > |90° - b|$.

Solution of spherical triangles

(i) Let $\tan m = \tan b \cdot \cos A$, $\cos n = \dfrac{\cos m \cdot \cos a}{\cos b}$,

$\cot M = \cos b \cdot \tan A$, $\cos N = \cot a \cdot \tan b \cdot \cos M$;

then
$$\sin B = \frac{\sin b \cdot \sin A}{\sin a} \quad \text{(two values of } B\text{)}$$

or
$$\cot B = \frac{\sin n \cdot \cot A}{\sin m} = \cos a \cdot \tan N \qquad (1.24)$$

or
$$\cos B = \tan n \cdot \cot A = \frac{\sin N \cdot \cos A}{\sin M}. \qquad (1.25)$$

$$C = M \pm N, \quad c = m \pm n.$$

(ii)
$$\sin B = \frac{\sin b \cdot \sin A}{\sin a} \quad \text{(two values of } B\text{)}$$

$$\tan \tfrac{1}{2} C = \frac{\cos \tfrac{1}{2}(a-b)}{\cos \tfrac{1}{2}(a+b)} \cdot \cot \tfrac{1}{2}(A+B) \qquad (1.26a)$$

$$= \frac{\sin \tfrac{1}{2}(a-b)}{\sin \tfrac{1}{2}(a+b)} \cdot \cot \tfrac{1}{2}(A-B); \qquad (1.26b)$$

then use (1.23a) or (1.23b) to find c.

(d) Given two angles A, B and the included side c, to find a, b, C.

(i) $\cos C = \cos c \cdot \sin A \cdot \sin B - \cos A \cdot \cos B \qquad (1.27)$ cf. (1.4)

$$\cot a = \frac{\cot A \cdot \sin B + \cos c \cdot \cos B}{\sin c} \qquad (1.28)$$

$$\cot b = \frac{\sin A \cdot \cot B + \cos c \cdot \cos A}{\sin c} \qquad (1.29)$$

Check: $\dfrac{\sin a}{\sin A} = \dfrac{\sin b}{\sin B} = \dfrac{\sin c}{\sin C} \qquad (1.1)$

(ii) Let $\cot M = \cos c \cdot \tan A$, $\cot N = \cos c \cdot \tan B$; then

$$\cos C = \frac{\cos A \cdot \sin(B-M)}{\sin M} = \frac{\cos B \cdot \sin(A-N)}{\sin N} \qquad (1.30)$$

$$\cot a = \frac{\cos(B-M).\cot c}{\cos M} \qquad (1.31)$$

$$\cot b = \frac{\cos(A-N).\cot c}{\cos N}. \qquad (1.32)$$

(iii)
$$\tan\tfrac{1}{2}(a+b) = \frac{\cos\tfrac{1}{2}(A-B)}{\cos\tfrac{1}{2}(A+B)} . \tan\tfrac{1}{2}c \qquad (1.33)$$

$$\tan\tfrac{1}{2}(a-b) = \frac{\sin\tfrac{1}{2}(A-B)}{\sin\tfrac{1}{2}(A+B)} . \tan\tfrac{1}{2}c \qquad (1.34)$$

and $\quad a = \tfrac{1}{2}(a+b) + \tfrac{1}{2}(a-b), \quad b = \tfrac{1}{2}(a+b) - \tfrac{1}{2}(a-b).$

$$\cot\tfrac{1}{2}C = \frac{\cos\tfrac{1}{2}(a+b)}{\cos\tfrac{1}{2}(a-b)} . \tan\tfrac{1}{2}(A+B) \qquad (1.35a)$$

$$= \frac{\sin\tfrac{1}{2}(a+b)}{\sin\tfrac{1}{2}(a-b)} . \tan\tfrac{1}{2}(A-B) \qquad (1.35b)$$

$$\left(\text{or } \sin C = \frac{\sin c.\sin A}{\sin a} = \frac{\sin c.\sin B}{\sin b}, \text{ but these are ambiguous}\right).$$

(e) Given two angles A, B and the side a opposite one of them, to find b, c, C. There will be only one solution if $|90° - A| < |90° - B|$, but two solutions if $|90° - A| > |90° - B|$.

(i) Let $\tan m = \tan a . \cos B, \quad \sin n = \sin m . \cot A . \tan B$ (two values of n)

$$\cot M = \cos a.\tan B, \quad \sin N = \frac{\sin M.\cos A}{\cos B} \text{ (two values of } N\text{)} \qquad (1.36)$$

then
$$C = M + N, \quad c = m + n. \qquad (1.37)$$

$$\cos b = \frac{\cos n.\cos a}{\cos m} = \cot N.\cot A, \qquad (1.38)$$

or
$$\cot b = \cot n.\cos A = \frac{\cot a.\cos N}{\cos M} \qquad (1.39)$$

Solution of spherical triangles

(ii) $$\sin b = \frac{\sin a \cdot \sin B}{\sin A} \quad \text{(two values of } b),$$

then use (1.23a) or (1.23b) to find c and (1.26a) or (1.26b) to find C.

(f) Given the three angles A, B, C, to find a, b, c.

(i) $$\cos a = \frac{\cos A + \cos B \cdot \cos C}{\sin B \cdot \sin C} \quad \text{(cyclic)} \qquad (1.40)$$

When a, b and c are all found, a check can be made with (1.1).

(ii) $$\tan \tfrac{1}{2} a = \sqrt{\left(-\frac{\cos A + \cos(B + C)}{\cos A + \cos(B - C)} \right)} \text{ (cyclic)} \qquad (1.41\text{a})$$

$$= \sqrt{\left(\frac{-\cos S \cdot \cos(S - A)}{\cos(S - B) \cdot \cos(S - C)} \right)} \text{ (cyclic)} \qquad (1.41\text{b})$$

$$= \tan R \cdot \cos(S - A) \text{ (cyclic)} \qquad (1.41\text{c})$$

where $$2S = A + B + C$$

and $$\tan R = \sqrt{\left(\frac{-\cos S}{\cos(S - A) \cdot \cos(S - B) \cdot \cos(S - C)} \right)}$$

R being the radius of the circumcircle.

Note. The student may wonder at the negative signs within large parentheses on the right-hand sides of equations (1.41a) and (1.41b); however, since ($A + B + C$) is always greater than 180° for a spherical triangle, $S = \tfrac{1}{2}(A + B + C)$ will always be greater than 90°, hence $\cos S$ is negative. Thus the expression within large parentheses in (1.41b) will be positive. (1.41a) is just another way of writing (1.41b) without using S.

1.07 Solution by calculator or microcomputer

Appendix II contains a selection of programs for the solution of spherical triangles, in two categories: (1) those suitable for a programmable Hewlett-Packard pocket 67 or desk-top 97 calculator using Reverse Polish Notation (RPN); and (2) those written in BASIC language for a microcomputer such as a Model I TRS-80 using a Z-80 CPU (TRS-80 is a trade mark of the Tandy Corporation, U.S.A.).

The selection of programs is not extensive since HP calculators and the TRS-80 microcomputer are but two of many such devices now available, which use a variety of digital computing techniques, central processing units (CPUs) and languages. However, Appendix II should be useful to readers with access to those particular computers or to similar ones.

Astronomy for surveyors

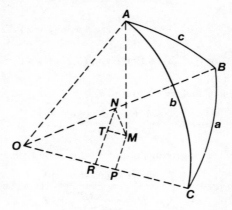

Fig. 1.3

1.08 Derivation of some of the formulae set out above

In Fig. 1.3, ABC is a spherical triangle;
AM is the perpendicular from A to the plane BOC;
MN is the perpendicular from M to BO;
MP is the perpendicular from M to CO;
NR is the perpendicular from N to CO;
MT is the perpendicular from M to NR;
Angle $AOB = c$; $AO = BO = CO = R$, the radius of the sphere;
Angle $AOC = b$;
Angle $BOC = a$; O is the centre of sphere.

To avoid confusion in the figure, the lines AN and AP, perpendicular to OB and OC respectively, are not drawn. However, since AMN is a plane perpendicular to OB (the line of intersection of planes AOB and BOC)

$$\text{angle } ANM = B;$$
similarly,
$$\text{angle } APM = C.$$

Since MN and NR are respectively perpendicular to BO and CO, angle $MNT = BOC = a$.

(a) *Sine formulae*:
In Fig. 1.3,

$$\sin b = \sin AOP = \frac{AP}{AO} = \frac{AM}{R \sin C} \tag{1.42}$$

$$\sin c = \sin AON = \frac{AN}{AO} = \frac{AM}{R \sin B} \tag{1.43}$$

12

Solution of spherical triangles

Dividing (1.43) by (1.42), we have

$$\frac{\sin c}{\sin b} = \frac{AM}{R \sin B} \times \frac{R \sin C}{AM} = \frac{\sin C}{\sin B}$$

that is,
$$\frac{\sin B}{\sin b} = \frac{\sin C}{\sin c}.$$

Similarly,
$$\frac{\sin A}{\sin a} = \frac{\sin B}{\sin b} = \frac{\sin C}{\sin c}. \tag{1.1}$$

(b) *Cosine formulae*:
In Fig. 1.3,

$$\cos a = \cos BOC = \frac{OR}{ON} = \frac{OP - PR}{ON} = \frac{OP - MT}{ON} \tag{1.44}$$

$OP = OA \cos AOP = R \cos b;$ $ON = OA \cos AON = R \cos c;$
$MT = MN \sin MNT = MN \sin a = AN \cos ANM . \sin a$
$ = AN \cos B . \sin a = AO \sin AON . \cos B . \sin a$
$ = R \sin c . \cos B . \sin a.$

Substituting, then, in (1.44) we get:

$$\cos a = \frac{R \cos b - R \sin c . \cos B . \sin a}{R \cos c}$$

$$= \frac{\cos b - \sin c . \sin a . \cos B}{\cos c}$$

$$\therefore \cos a . \cos c = \cos b - \sin c . \sin a . \cos B$$

or
$$\cos b = \cos a . \cos c + \sin a . \sin c . \cos B. \tag{1.2}$$

Similarly, it can be shown that
$$\cos a = \cos b . \cos c + \sin b . \sin c . \cos A$$
$$\cos c = \cos a . \cos b + \sin a . \sin b . \cos C.$$

(c) *Cotangent formulae*:
In Fig. 1.3,
$$NR = NT + TR = NT + MP$$

i.e.
$$ON \sin a = NM \sin NMT + AP \cos C$$
$$= NM \sin(90° - a) + AP \cos C$$
$$= NM \cos a + AP \cos C. \tag{1.45}$$

Astronomy for surveyors

But $\quad NM = AN\cos B = R\sin c . \cos B$

and $\quad AP = R\sin b$

and $\quad ON = R\cos c.$

So that, substituting in (1.45), we get:

$$R\cos c . \sin a = R\sin c . \cos B . \cos a + R\sin b . \cos C$$

i.e.

$$\cos c . \sin a = \sin c . \cos B . \cos a + \sin b . \cos C.$$

Dividing through by $\sin c$,

$$\cot c . \sin a = \cos B . \cos a + \frac{\sin b}{\sin c} . \cos C$$

$$= \cos B . \cos a + \frac{\sin B}{\sin C} . \cos C \text{ (using (1.1))}$$

$$\therefore \cot c . \sin a = \cos a . \cos B + \sin B . \cot C. \qquad (1.3a)$$

If in Fig. 1.3 a perpendicular is dropped from P on to OB, and another from M on to this, we get a further expression by similar working:

$$\cot b . \sin a = \cos a . \cos C + \sin C . \cot B.$$

Now, working cyclically from this and (1.3a) we get the other four formulae:

$$\cot a . \sin b = \cos b . \cos C + \sin C . \cot A$$

$$\cot b . \sin c = \cos c . \cos A + \sin A . \cot B$$

$$\cot c . \sin b = \cos b . \cos A + \sin A . \cot C \qquad (1.3b)$$

$$\cot a . \sin c = \cos c . \cos B + \sin B . \cot A.$$

(d) *Formulae involving half-angles and half sum of sides*:

(1.2) gives $\quad\cos B = \dfrac{\cos b - \cos a . \cos c}{\sin a . \sin c}$

$$\therefore 1 + \cos B = 1 + \frac{\cos b - \cos a . \cos c}{\sin a . \sin c}$$

$$= \frac{\sin a . \sin c + \cos b - \cos a . \cos c}{\sin a . \sin c}$$

Solution of spherical triangles

$$= \frac{\cos b - \cos(a+c)}{\sin a . \sin c}.$$

$$\therefore 2\cos^2 \tfrac{1}{2}B = \frac{\cos b - \cos(a+c)}{\sin a . \sin c}$$

$$= \frac{2\sin\tfrac{1}{2}(a+b+c).\sin\tfrac{1}{2}(a+c-b)}{\sin a . \sin c}$$

If $(a+b+c) = 2s$, then $(a+c-b) = 2s - 2b$

$$\therefore 2\cos^2 \tfrac{1}{2}B = \frac{2\sin s . \sin(s-b)}{\sin a . \sin c}$$

or

$$\cos\tfrac{1}{2}B = \sqrt{\left(\frac{\sin s . \sin(s-b)}{\sin a . \sin c}\right)}.$$

Cyclically, from this, we get

$$\cos\tfrac{1}{2}A = \sqrt{\left(\frac{\sin s . \sin(s-a)}{\sin c . \sin b}\right)} \qquad (1.14a)$$

$$\cos\tfrac{1}{2}C = \sqrt{\left(\frac{\sin s . \sin(s-c)}{\sin a . \sin b}\right)}.$$

Again, from (1.2)

$$1 - \cos B = 1 - \frac{\cos b - \cos a . \cos c}{\sin a . \sin c}$$

$$= \frac{\sin a . \sin c - \cos b + \cos a . \cos c}{\sin a . \sin c}$$

$$= \frac{\cos(a-c) - \cos b}{\sin a . \sin c}$$

i.e.

$$2\sin^2 \tfrac{1}{2}B = \frac{2\sin\tfrac{1}{2}(b+a-c).\sin\tfrac{1}{2}(b-a+c)}{\sin a . \sin c}$$

$$\sin^2 \tfrac{1}{2}B = \frac{\sin(s-c).\sin(s-a)}{\sin a . \sin c} \quad \text{where } s = \tfrac{1}{2}(a+b+c)$$

or

$$\sin\tfrac{1}{2}B = \sqrt{\left(\frac{\sin(s-a).\sin(s-c)}{\sin a . \sin c}\right)}$$

and cyclically,

$$\sin\tfrac{1}{2}A = \sqrt{\left(\frac{\sin(s-b).\sin(s-c)}{\sin b.\sin c}\right)} \qquad (1.13a)$$

$$\sin\tfrac{1}{2}C = \sqrt{\left(\frac{\sin(s-a).\sin(s-b)}{\sin a.\sin b}\right)}$$

(1.13a) ÷ (1.14a) gives

$$\tan\tfrac{1}{2}A = \sqrt{\left(\frac{\sin(s-b).\sin(s-c)}{\sin s.\sin(s-a)}\right)} \qquad (1.12a)$$

and cyclically:

$$\tan\tfrac{1}{2}B = \sqrt{\left(\frac{\sin(s-a).\sin(s-c)}{\sin s.\sin(s-b)}\right)}$$

$$\tan\tfrac{1}{2}C = \sqrt{\left(\frac{\sin(s-a).\sin(s-b)}{\sin s.\sin(s-c)}\right)}.$$

Note. The formulae which are derived cyclically in the above proofs may be found by "tipping over" the diagram, Fig. 1.3, so that *AOC* is the horizontal plane, dropping a perpendicular from *B* on to *AOC,* and duplicating the construction. A final tipping-over to make *AOB* the horizontal plane, and dropping a perpendicular from *C*, etc., will complete the series.

CHAPTER 2

The celestial sphere and astronomical coordinates

2.01 The celestial sphere

Though we may easily imagine, looking up to the heavens on a cloudless night, that the stars are distributed over the surface of the spherical vault of sky above us, we know that in fact the distances of the stars differ tremendously, as refined measurements have proved. For practical purposes we are never concerned with the distances of the stars, but only with their directions, and in order to record these it is convenient to picture the stars as distributed over the surface of an imaginary spherical sky having its centre at the position of the observer. Thus has arisen the conception of the *celestial sphere*, which we may consider as a geometrical device to enable us to record and measure the directions of the stars.

In Fig. 2.1, suppose that O represents the position of the observer. With O as centre, imagine a spherical surface described with a radius of any length

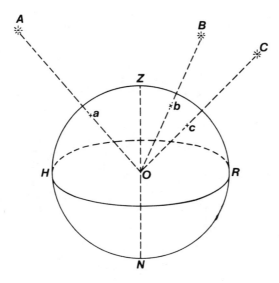

Fig. 2.1

Astronomy for surveyors

we please; we may make it ten thousand or ten billion metres, it makes no difference. Now let *A, B* and *C* be three of these immensely distant stars, and let the lines *OA, OB* and *OC* cut our imaginary sphere in *a, b* and *c* respectively. Then, if we are only concerned with the directions of the stars, we may just as well picture them as occupying the positions *a, b* and *c* as their actual places *A, B* and *C*. In fact, to the observer at *O* their appearance would be unaltered. So, proceeding in this way, we may picture all the stars in the sky as occupying places on this imaginary surface, which is then known as the *celestial sphere*. It may be considered as the spherical surface upon which the stars *appear* to lie, but, of course, in reality they are not all equally distant from us, and they are only represented in this way in order conveniently to measure their directions.

If through the point *O* a vertical line be drawn to intersect the celestial sphere over the observer's head in *Z*, and to cut it vertically below his feet at *N*, the point *Z* is called the *zenith* and the point *N* the *nadir*. The zenith is thus the point in the celestial sphere directly over the observer.

If a horizontal plane *HR* be drawn through *O*—a plane, that is to say, at right angles to the vertical *ZO*, the direction which gravity acts—it will cut the celestial sphere in a great circle, which is called the *celestial horizon*. To an observer whose eye was close to the surface of a calm ocean, the celestial horizon would form the boundary of the visible part of the celestial sphere.

2.02 The apparent motion of the stars

Continued observation shows that, leaving the few planets out of account, the stars always maintain the same relative positions, and hence they are

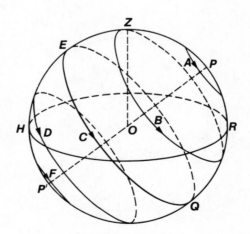

Fig. 2.2

commonly referred to as the *fixed* stars. Whilst, however, there is no motion relative to one another, they all appear to revolve from east to west in a period slightly less than twenty-four hours round a point in the sky that is known as the *celestial pole*. The motion is just as though the whole celestial sphere, carrying the stars, revolved about an axis passing through this point and its own centre. The ancients thought that this was really what occurred, but we know now that this motion is apparent only, and is due to the fact that we view the stars from a revolving earth. Thus, referring to Fig. 2.2, the whole of the stars appear to describe circles slowly about a point P in the celestial sphere, just as though the whole sphere revolved about the axis OP, so that every star completes its circle in the same time. Some stars, such as A, which are comparatively near to the point P, describe only a small circle, which never takes them below the horizon, so that such stars are always visible. Thus, in the northern celestial hemisphere, the Pole star (Polaris), which is only about 1° from the north pole, is always visible on clear nights from places in the northern terrestrial hemisphere; likewise the Southern Cross in the latitudes of southern Australia and New Zealand can be seen throughout all cloudless nights and never sets. Other stars, such as B and C, which are further away from P, describe much larger circles, which take them, as is shown in the figure, below the horizon for a portion of their revolution, so that such stars rise in the east and set in the west. This diurnal motion of the stars may be prettily demonstrated by fixing a camera so that it is directed towards the celestial pole on a clear night, and leaving the film exposed for an hour or two. The images of the brightest stars will leave trails upon the film which are all seen to be arcs of circles having a common centre at the celestial pole.

Now, the stars are so distant that their apparent direction in space is unaltered by any movement of the observer over the earth's surface. The direction of any particular star is precisely the same, even when determined by our most refined instruments, whether viewed from Melbourne, London or Perth. More than this, we know that the earth, in the course of a year, describes a path round the sun that is approximately a circle whose diameter is about 299 million kilometres, yet even this great shift of the point of observation produces no appreciable change in the directions of the fixed stars. At six-monthly intervals, when the points of observation are at opposite ends of this enormous diameter, a slight difference in direction, amounting in no case to more than 1.5 seconds of arc, may be detected in a few "near" stars with the refined observations possible at fixed observatories. Even so, this parallax of 1.5 arc seconds shows the "near" star to be distant 4.1×10^{13} km, or about 4.4 light-years. Most stars have parallaxes much smaller than 1.5, so that we may regard the position of the observer on the earth's surface as of absolutely no importance when measuring the direction of the stars in space. Considering Fig. 2.2, we may regard the earth as a tiny speck at O, the centre of the great celestial sphere, and no matter where we take the point O on this tiny speck, the direction of the line *OP* remains the same within the possibilities of our means of measurement, so that the lines joining any one of the fixed stars to different points on the earth's surface may all be considered as parallel.

It follows from this that the portion of the sky visible to an observer at any point on the earth's surface presents exactly the same appearance as it would do if it were possible for him to view it from the earth's centre. This statement refers only to the fixed stars.

Therefore, if we imagine an observer anywhere on a small spherical earth at the centre of a great celestial sphere of dimensions infinitely great compared to the earth, and suppose the earth to rotate about an axis through its centre, the successive pictures of the sky presented to the observer during a revolution will be precisely the same as they would be if the earth remained stationary and the great celestial sphere itself were to rotate about the same axis.

Thus, looking again at Fig. 2.2, if we produce the line *PO* backwards to cut the celestial sphere below the plane of the horizon in P', the fixed stars appear to the observer at *O* to revolve on the celestial sphere about the axis PP'. In reality it is the earth that is revolving, and it is the earth's axis that lies in the direction PP', so that the celestial poles *P* and P' are the points in which the axis of the earth, if infinitely produced, would cut the celestial sphere. If the observer is in the southern hemisphere, the elevated pole *P* visible to him will be that to which the earth's south pole is directed. If he is in the northern hemisphere the visible elevated celestial pole is that towards which the earth's north pole points.

2.03 Celestial equator

If we take a plane through *O* perpendicular to the line PP', it will cut the celestial sphere in a great circle *EQ*, which is known as the *celestial equator*. Its plane clearly is coincident with the plane of the equator of the earth. Since two great circles of a sphere always intersect at opposite extremities of a diameter, it follows that a star revolving in the celestial equator has its path divided into two equal parts by the circle of the celestial horizon *HR*, so that the time during which it is visible above the horizon will be equal to the time it is out of sight below.

To an observer in southern latitudes, the celestial pole *P* lies to the south, and, since the line PP' (Fig. 2.2) marks also the direction of the earth's axis, the celestial pole will be in the direction of the true geographical south. Any star such as *B*, lying to the south of the celestial equator, will trace the greater part of its circular path above the plane of the horizon. On the other hand, a star such as *D*, to the north of the celestial equator, will trace out the smaller portion of its path only above the horizon, so that it will be visible for less than half of its time of revolution. Stars such as *F*, sufficiently far to the north, will not be visible at all to a person in this latitude, but will complete the whole of their revolution below the plane of the horizon, as shown in the figure.

Celestial sphere and astronomical coordinates

2.04 Astronomical coordinates

If we wish to mark the position of a point on a plane, we may do so by measuring its distances from two fixed straight lines in the plane at right angles to each other. A knowledge of these two distances is sufficient to enable us to fix the position of the point, but one distance only would not be enough. Measured in this way, these two distances are spoken of as the "coordinates" of the point. Now, in astronomical observation, we commonly require to determine the position of a star on the celestial sphere, and so it is necessary to have some system of coordinate measurement applicable to the purpose. Either one of two sets of coordinates is commonly employed. In the first set, the coordinates are *altitude* and *azimuth*; in the second they are *right ascension* and *declination*. We will consider the two sets in turn.

2.05 Altitude (Alt. or h) and azimuth (A)

In Fig. 2.3, let O be the position of the observer, Z the zenith and P the celestial pole. Then the plane ZOP will cut the plane of the horizon through O in the north and south points M_N and M_S respectively. $M_S Z M_N$ is known as the plane of the *meridian*.

Suppose that S is a star describing its circular path ASC round the pole P.

The plane ZSO cuts the plane of the horizon in the line DO. Then it is clear that if we know the angle DOM_N, which is the angle that the plane ZOD makes with the plane of the meridian, our knowledge is sufficient to fix the position of the plane ZOD.

If, in addition, we know the angle DOS, the position of the star S may be fixed on the celestial sphere.

The angle DOM_N, which the plane passing through the zenith and the star makes with the meridian, measures what is known as the *azimuth* of the star. It is generally measured from the north towards the right.

The angle DOS, measuring the angular altitude of the star in a vertical plane above the horizon, is spoken of as the *altitude* of the star.

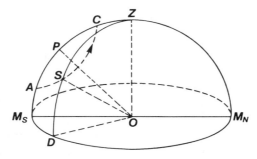

Fig. 2.3

Astronomy for surveyors

Instead of the altitude (h) we may measure the angle ZOS, which is known as the *zenith distance* (ζ), and is clearly the complement of the altitude.

If we know both the altitude and azimuth of a star at any time we can mark its position on the celestial sphere. The ordinary theodolite is adapted for measurement in this system of coordinates. It should be noted that, because the star is always moving, its azimuth and altitude are always changing; S is merely its position at any one instant.

2.06 Right ascension (R.A. or α) and declination (Dec. or δ)

In Fig. 2.4 let O be the position of the observer, Z the zenith, P the celestial pole and $M_S PZM_N$ the plane of the meridian.

Suppose that S is a star travelling round the pole in the direction of the arrow in a circle of which only half is shown.

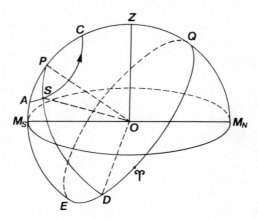

Fig. 2.4

EDQ is the plane of the celestial equator drawn through O at right angles to OP.

PSD is the arc of a *great* circle of the celestial sphere intersecting the celestial equator in D. The plane of this great circle must pass through O, and the angle POD is a right angle.

Then clearly if we know the position of the point D on the celestial equator, and also know either the angle POS or the complementary angle DOS, we shall be able to fix the position of the star S on the celestial sphere.

The position of the point D on the equator may be determined if we know its angular distance from some known fixed point also on the equator. The fixed point selected for the purpose is known as the *First Point of Aries*. It is usually indicated by the symbol ♈, denoting a pair of ram's horns (Aries, the Ram, being one of the signs of the Zodiac). The exact nature of this point

Celestial sphere and astronomical coordinates

we shall discuss a little later on, but for the present all that we want to know is that it is a point whose position can always be accurately determined.

If we know, then, the angular measure of the arc ♈D—that is to say, the angle which the arc subtends at the centre O, and also the direction in which it is measured from ♈—this is sufficient to determine D.

To avoid any confusion as to the direction in which the arc ♈D should be measured, it is always measured from ♈ in the direction opposite to that in which ♈ travels round the celestial equator EQ (♈ moves round with the rest of the fixed stars from east to west).

Measured in this way, the angular measure of the arc ♈D is known as the *right ascension* of the star S. It may have any value from 0° to 360°. It is commonly denoted by the letters R.A., or by α.

The right ascension of the star being known, its position may be fixed if we know either the angle *POS*, the angular measure of the arc *PS*, or the angle *DOS*, the angular measure of the arc *DS*. The angular measure of the arc *PS* is known as the *polar distance* (p) of the star S. It is often denoted by the letters N.P.D. or S.P.D., according as it is measured from the north or the south pole.

The angular measure of the arc *DS* is called the *declination* of the star S, and the circle *PSD* is known as the *hour circle* of the star, while the small circle *ASC* is called a *parallel of declination*. Some star tables adopt the convention of regarding declinations north of the celestial equator as positive (+) and declinations south of the celestial equator as negative (−). In this connection, however, it should be noted that in the *Star Almanac for Land Surveyors* northerly declinations are given as "N", and southerly declinations as "S".

Polar distance (p) and declination (δ) are complementary to each other, their sum being 90°, so that if one is known the other is found by simple subtraction from 90° (but note that a star with a southerly declination has a polar distance = 90° + δ from the north celestial pole, and a star with a northerly declination has a polar distance = 90° + δ from the south celestial pole).

2.07 Comparative advantages of the two coordinate systems

The altitude and azimuth of a star are readily measured with a theodolite, and serve to fix the position of a star at any particular instant, but owing to the diurnal motion of the stars these coordinates are continually changing.

On the other hand, the right ascension and declination of a star are constant, for the reference point, the First Point of Aries, partakes of the diurnal motion of the stars. These coordinates are in consequence the most convenient for recording the relative positions of the stars on the celestial sphere, and it is in this way that the *Star Almanac* catalogues them (pages 26-51), in the order of their right ascensions. See pages 56-57 herein.

2.08 The sidereal day and sidereal time

As the revolution of the whole system of stars about the polar axis takes place uniformly from east to west, the period of revolution serves as a convenient unit of time for astronomical purposes. All the stars complete their circles of revolution in the same period, which is known as the *sidereal day*. This day is about 4 minutes shorter than the ordinary mean solar day. Sidereal clocks, adjusted to keep sidereal time—the sidereal day being divided into 24 hours—are used in fixed observatories. Such clocks are arranged to mark $0^h\ 0^m\ 0^s$ when the First Point of Aries, the point on the celestial equator from which right ascensions are measured, crosses the meridian of the observer, at upper transit. Thus the *sidereal time* at any instant is the interval that has elapsed, measured in sidereal hours, minutes and seconds, since the last upper meridian transit of the First Point of Aries. (Note that the celestial objects cross the observer's meridian *twice* during the course of a complete revolution—once above the pole at upper transit (culmination) and once below the pole at lower transit.)

Referring to Fig. 2.4, it is clear that all stars on the same hour circle, such as *PSD*—that is to say, all stars having the same right ascension— will cross the meridian at the same instant. A star whose right ascension is 180° will cross the meridian 12 sidereal hours after the First Point of Aries, and one whose right ascension is 15° will cross the meridian at 1^h sidereal time. Thus we deduce the important result that the *right ascension of a star, when reduced to time at the rate of* 24 *hours for* 360° *or* 1 *hour for* 15°, *gives the sidereal time at the moment when it crosses the meridian at upper transit.*

2.09 Hour angle (H.A. or t)

In Fig. 2.4, the angle *SPZ*, which is the angle that the plane of the hour circle *PSD* makes with the plane of the meridian, is known as the *hour angle* (t) of the star *S*. The usual convention is that hour angles are measured from the upper meridian towards the west, and when the figure for an hour angle is quoted without qualification, this is what is meant. However, when a celestial body is approaching the meridian from the east, its hour angle may be very large, say 21 hours, but in 3 hours it will be on the meridian. We would then say that its *east hour angle* was 3 hours. In this case the spherical triangle we use in any computation lies to the east of the meridian, and it is the east hour angle which is either used, or found, during the calculation.

If we know the hour angle of a star, and also its polar distance, we can clearly mark the position of the star on the celestial globe at any instant, so that these two may be used as another system of coordinates. The hour angle of a star is continually changing, but owing to the uniform character of the star's motion, it varies at a constant rate. A knowledge of the hour angle at once gives us the time that has elapsed since the star last crossed the meridian at upper transit,

Celestial sphere and astronomical coordinates

or, subtracting the hour angle from 24 hours, the time that the star will take to reach the meridian at its next transit (i.e. the east hour angle, E.H.A.).

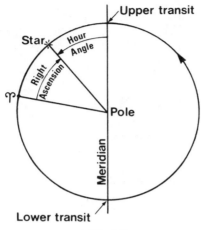

Fig. 2.5

From the definitions of right ascension, hour angle and sidereal time, it follows that

$$\text{local sidereal time} = \text{hour angle} + \text{right ascension},$$

i.e.
$$\text{L.S.T.} = t + \text{R.A.}$$

Further, it should be obvious that the local sidereal time can be called the right ascension of the meridian (see Fig. 2.5).

2.10 Prime vertical (P.V.)

The plane through the zenith at right angles to the meridian—that is, the vertical plane running east and west—is known as the *prime vertical*. The east-and-west line, which is the line of intersection of the prime vertical with the plane of the horizon, is also the line of intersection of the plane of the celestial equator with the horizon, as will be evident from Fig. 2.6a and 2.6b.

2.11 Synopsis of astronomical terms

For purposes of reference, the principal quantities dealt with in this chapter are illustrated in Fig. 2.6a and 2.6b.

Figure 2.6a is drawn for an observer in the northern hemisphere and Fig. 2.6b for the southern hemisphere.

In these figures, *PZS* is known as the astronomical triangle.

Astronomy for surveyors

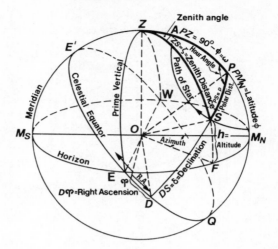

Fig. 2.6a

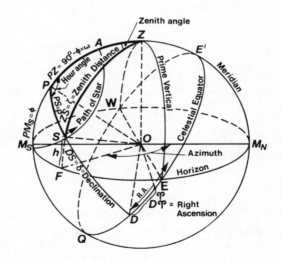

Fig. 2.6b

O is the observer; *S*, any star
$M_N E M_S W$, the plane of the horizon
Z, the zenith
P, the celestial pole; *OP*, the polar axis
$M_N P Z M_S$, the plane of the meridan
$E'WQE$, the celestial equator
WZE the prime vertical
M_N, M_S, *W*, *E*, the north, south, west and east points
ZSP, the parallactic angle

ZPS, the hour angle of *S*
PSD, the hour circle of *S*
PS, the polar distance of *S*
SD, the declination of *S*
ΥD, the right ascension of *S*
ZSF, the vertical plane through *S*
SF, the altitude of *S*
SZ, the zenith distance of *S*
$M_N F$, the azimuth of *S*

The unlabelled arrow in each diagram shows the direction of rotation of celestial bodies.

Celestial sphere and astronomical coordinates

EXERCISES

1. The R.A. of a star being $35° 20'$, what is the local sidereal time when the star is on the meridian?

Ans. $2^h 21^m 20^s$.

2. If the RA of a star is $295°$ and the sidereal time is 15 hours, is the star to the east or west of the meridian?

Ans. To the east.

3. What is the declination of a star that rises exactly in the east?

Ans. $0°$.

4. What is the east hour angle of the star in Question 2?

Ans. $70°$.

5. The declination of a star is $-35°$; determine its S.P.D. and its N.P.D.

Ans. $55°$ and $125°$.

6. If the First Point of Aries crosses the meridian exactly 2 hours, as measured by a sidereal clock, after a certain star, what is the R.A. of the star?

Ans. $330°$ and 22^h.

7. The declination of the Pole Star is $+89° 05'$. What is the difference between its greatest and least zenith distances?

Ans. $1° 50'$.

8. At the time of the year when the R.A. of the sun is zero, determine approximately the time of rising of a star with declination $0°$ and R.A. $150°$.

Ans. 4 p.m.

9. Where is the point on the celestial sphere whose declination is equal to the latitude of the observer (on the same side of the equator as the elevated pole), and whose hour angle is zero?

Ans. At the zenith.

CHAPTER 3

The earth

3.01 The earth a globe

That the earth is a globe is no longer a matter of dispute. It is circumnavigated by sea and by air, and has been mapped and measured. We see its round shadow cast upon the moon during a partial eclipse. We see the planets as great balls of comparable dimensions revolving at different distances around the central sun; the law of gravitation explains the form of their orbits and enables their movements to be predicted with the greatest exactitude. That our earth is a globe like the other planets, revolving in a similar way around the sun, explains their apparently involved movements in the heavens; and in recent years practical proof has been obtained from observations by astronauts orbiting the earth and the moon, in spacecraft.

In the case of some of the other planets our telescopes show them to be in rotation in a similar manner to that in which our own earth rotates, producing the phenomena of night and day and of the diurnal rotation of the stars. In the planet Mars we see the poles or extremities of the axis of rotation surrounded by white caps analogous to the caps of ice and snow that surround the poles of our own earth. Man has landed on the moon, soft-landed instrumented capsules on Venus and Mars, and sent others to investigate and photograph Jupiter and Saturn from close quarters.

3.02 Terrestrial latitude (Lat. or ϕ) and longitude (Long. or λ)

The extremities of the axis of rotation of the earth are called the *poles*, and are distinguished as the *north* and *south* poles.

A plane through the earth's centre at right angles to the axis cuts the earth's surface in a circle known as the *equator*. Every point on the terrestrial equator is thus equidistant from the north and south poles.

In order to mark the position of a point on the earth's surface, it is necessary to have a system of coordinates similar to those we have already discussed in connection with the celestial sphere.

Suppose that X (Fig. 3.1) is a point on the earth's surface, the position of which it is desired to locate. A plane passing through X and the earth's axis PP' will cut the earth's surface in a great circle $PXMP'$ which is known as a *meridian*. Suppose this meridian cuts the equator EQ at the point M. Then

The earth

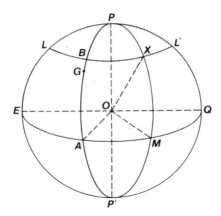

Fig. 3.1

clearly, if we know the position of the point *M* on the equator, and also the length of the arc *XM* or the angle which it subtends at the earth's centre, we shall be able to fix the point *X*.

The position of *M* on the equator is determined by the *longitude* of *X*. To measure this, some arbitrary place *A* must be selected on the equator as a starting-point. The point actually chosen is the point of intersection of the meridian passing through Greenwich, shown as *PGAP'* in the figure, and the equator. The angular measure of the arc *AM*—that is to say, the angle *AOM*—is known as the *longitude* (λ) of *X*. Thus, all points on the meridian *PGAP'*, passing through Greenwich, have zero longitude. The longitude of other places is reckoned as so many degrees east or west of Greenwich until we come to 180°, which is the longitude of the meridian exactly opposite to the Greenwich meridian.

The angle *XOM*, subtended at the centre between *X* and the plane of the equator, measures what is known as the *latitude* (ϕ) of *X*. If we draw a plane through *X* at right angles to the earth's axis, it will intersect the earth in a small circle *LXL'* parallel to the equator. Such a circle is known as a *parallel of latitude*, and all points on the same parallel clearly have the same latitude.

Latitude is measured as so many degrees north (+) or south (−) of the equator. The latitudes of the poles are 90° N. and 90° S.

Thus, if we know the position of the meridian of zero longitude, the latitude and longitude of a place are sufficient to enable us to mark its position on the globe.

3.03 The length of a degree of longitude

If the parallel of latitude through *X* intersects the meridian through Greenwich in *B*, it is clear that the arc *BX* will be much smaller than the arc *AM*. It will

Astronomy for surveyors

have the same angular measurement on a much smaller circle. If X were very near to the north pole, the arc BX would be very small indeed. Thus two places in the same latitude but differing by, say, ten degrees of longitude, will be very much closer together if they are in a "high" latitude—that is to say, a latitude approaching 90°—than they will be if both are on or near the equator. Thus a degree of longitude has its greatest value when measured in distance along the earth's surface at the equator, its value becoming less and less as we approach the poles. At the equator a degree of longitude is equivalent to a distance of about 111 kilometres. At latitude ϕ, the length of a degree of longitude is approximately equal to $111 \cos \phi$ kilometres.

A degree of latitude, on the other hand, is always of approximately the same value, about 111 kilometres, whether it is measured near the poles or near the equator, because it is measured along meridians which are all great circles of the same diameter (but see **3.07**).

3.04 The zones of the earth (Fig. 3.2)

Certain parallels of latitude divide the earth's surface into five belts or divisions, termed *zones*. These mark in a general way a natural division of the earth's surface according to climate. The parallel of latitude 23° 26½' north of the equator is termed the *tropic of Cancer*, and the corresponding parallel south of the equator is termed the *tropic of Capricorn*. As we shall presently see, at all places between these parallels at some part of the year the sun shines directly overhead at midday. As a consequence, the belt included between these is the hottest portion of the earth's surface, and it is known as the *torrid zone*.

The parallel of latitude 66° 33½' north of the equator is called the *Arctic circle*, and the corresponding parallel south of the equator the *Antarctic circle*.

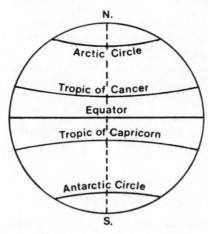

Fig. 3.2

The earth

The belt between the Arctic circle and the tropic of Cancer is known as the *north temperate zone*, and that between the Antarctic circle and the tropic of Capricorn as the *south temperate zone*. The regions around the two poles bounded by the Arctic and Antarctic circles respectively are termed the *frigid zones*. At all places within the frigid zones the sun remains completely below the horizon for some portion of the year.

3.05 The latitude of a place of observation is equal to the altitude of the celestial pole there

In Fig. 3.3, let O be the position of the observer and C the earth's centre. Then the direction of the pull of gravity at O is in the direction OC. This, then, will mark the direction of the vertical at O, and the zenith Z of the observer will be in CO produced.

HOR, at right angles to OZ and tangent to the earth, marks the plane of the horizon.

If CP, the earth's axis, be produced to cut the celestial sphere in P_1, then P_1 will be the celestial pole.

Draw OP_2 parallel to CP_1.

Then the celestial pole being, as we have seen, at a distance from the earth that is practically infinite in comparison with the earth's radius, OP_2 will mark the direction in which the celestial pole is seen by the observer at O.

Draw the plane of the equator ECQ at right angles to the earth's axis. Then, from our definition, the latitude of O is measured by the angle ECO.

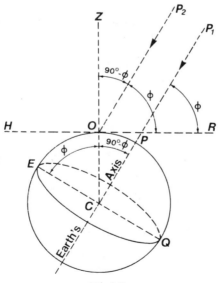

Fig. 3.3

Now the angle ZOP_2 = angle OCP_1, and the complements of these angles are therefore equal.

Therefore, the angle P_2OR = angle ECO; i.e. the altitude of the pole = the latitude of the observer.

It follows from this that if the observer travels equal distances north and south from O, since his latitude will change by equal amounts, the altitude of the celestial pole will also be increased or decreased by equal amounts. As this is actually the case from observation, the fact forms a strong proof of the sphericity of the earth.

3.06 To find the shortest distance, measured along the earth's surface, between two places whose latitude and longitude are given, assuming the earth to be a true sphere

In Fig. 3.4 let Q and R be two places whose latitudes and longitudes are known. The shortest distance between Q and R, measured along the earth's surface, will be the length of the arc of the great circle joining them.

Draw the meridians passing through Q and R. Then, if we know the latitudes, we know the angular measure of the meridian arcs PQ and PR, P being the north pole.

If Q is in north latitude, the arc PQ is the complement of the latitude. If R is in south latitude, the arc PR is $90°$ + the latitude.

The angle QPR (= angle AOB) is the difference of the longitudes of Q and R if both are measured in the same direction, or the sum of the longitudes, if one is east and the other west.

Thus in the spherical triangle PQR, we know the sides PQ and PR and the

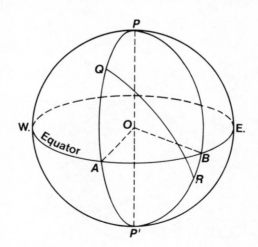

Fig. 3.4

The earth

included angle *QPR*. Then by the ordinary methods of spherical trigonometry (e.g. **1.06(b)**) we can compute the angular measurement of the great circle arc *QR*, and consequently its lineal measurement, since we know the radius of the earth to be appoximately 6373 kilometres.

EXAMPLE. Find the shortest distance measured along the earth's surface between Perth (Long. 115° 50′ E., Lat. 31° 57′ S.) and Brisbane (Long. 153° 01′ E., Lat. 27° 28′ S.), assuming that the earth is a sphere of radius 6373 kilometres.

In this case, both places being in the southern hemisphere, it will be preferable to solve the triangle *QP′R* (Fig. 3.5a) rather than *QPR*.

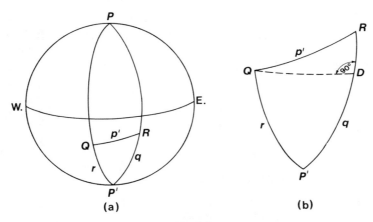

Fig. 3.5

If *R* denotes the position of Brisbane, *Q* of Perth and *P′* the south pole, we shall have, in the spherical triangle *QP′R* (Fig. 3.5b),

$$P'R = q = 90° - 27° \ 28' = 62° \ 32'$$
$$P'Q = r = 90° - 31° \ 57' = 58° \ 03'$$
$$P' = 153° \ 01' - 115° \ 50' = 37° \ 11'.$$

We may solve this triangle by using formula (1.2):

$$\cos p' = \cos q . \cos r + \sin q . \sin r . \cos P'$$
$$= \cos 62° \ 32' . \cos 58° \ 03'$$
$$+ \sin 62° \ 32' . \sin 58° \ 03' . \cos 37° \ 11'$$
$$= 0.843 \ 89$$
$$\therefore p' = 32°.447 \ 13 = 0.566 \ 31 \text{ radian.}$$

Thus the distance required = 0.56631 × 6373
= 3609.1 km.

Another way of solving the triangle of this problem is to divide it into two right-angled triangles by drawing a great circle arc QD to cut $P'R$ at right angles, as in Fig. 3.5b.

Then we have, from the right-angled triangle QDP',

$$\tan P'D = \cos P' . \tan r$$

so that

$$P'D = 51°.94637 \text{ or } 51° 56' 47''.$$

Thus

$$RD = P'R - P'D = 10°.58696 \text{ or } 10° 35' 13''$$

and in the right-angled triangle QRD,

$$\cos p' = \cos RD . \cos QD$$
$$= \cos RD . \cos r / \cos P'D$$
$$= 0.84389$$

and

$$p' = 0.56631 \text{ radian, as in the first method.}$$

This second method is not as short as the first, but it illustrates another approach to the solution and gives a check.

It should be pointed out at this stage that the earth is not a perfect sphere as is assumed in the above example, but rather resembles a slightly flattened sphere.

3.07 The figure of the earth

If, as in Fig. 3.6, F and G are two points on the same meridian, their difference of latitude will be measured by the angle FOG. If we know this angle, and also the length of the arc FG, we shall then be able to calculate the length of the earth's radius FO. The difference of latitude between F and G may be determined by astronomical observation, measuring the altitude of the celestial pole at each place. The length of the arc FG may be either directly measured or it may be computed by means of a triangulation survey from a measured base-line on some suitable adjacent part of the earth's surface. Determinations of the radius of the earth on these simple principles were made by the Greeks 2000 years ago.

If the earth were a true sphere, measurements of the radius of the earth made

The earth

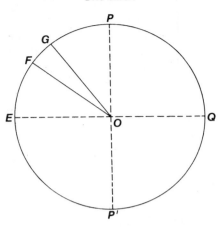

Fig. 3.6

in this way at different parts of its surface would be all the same. But when it became possible to make the necessary observations with sufficient precision it was found that such was not the case. When Newton discovered and investigated the results of the law of gravitation in the seventeenth century, he proved that one consequence was that if the earth is a plastic body, revolving on an axis and acted on by its own attraction, it must take the form of a slightly flattened sphere with its polar diameter less than its equatorial diameter. Measurements of two arcs made by the Cassinis in France seemed, on the other hand, to indicate that the length of a degree of latitude decreased towards the north, which would imply that the shape of the earth was such that its polar diameter was greater than its equatorial diameter, contrary to Newton's gravitational theory. The French Academy equipped two expeditions in order to settle the problem. One of these measured an arc in the equatorial regions of Peru (1735-1741), and the other an arc in the polar regions of Lapland (1736-1737). The results showed that a degree of latitude was longer in the polar regions than in parts near the equator, and corroborated Newton's theory. Since then many arcs have been measured in different parts of the world, and the observations have conclusively established the fact that the shape of the earth is not a true sphere, but is very approximately an *oblate spheroid*, the figure formed by revolving an ellipse about its minor axis.

The shape of the earth is thus like that of a sphere slightly flattened at the poles. The amount of flattening is not, however, very great, for if a model were made 10 metres in diameter the polar diameter would be shorter than the equatorial by a trifle over $33\frac{1}{2}$ millimetres.

The actual surface of the earth is, of course, far from being smooth and regular; it is certainly flat and featureless in some parts but in others there are rugged mountains and precipitous ocean deeps. This irregular surface obviously departs considerably in many places from the true spheroidal form. If we were to cut a network of tiny channels through the continents so that they kept below sea-level, and then allowed the sea to run into them under frictionless conditions, the surface coinciding with mean sea-level in the oceans and with the level of the water in the channels would be much more nearly spheroidal than the actual surface of the earth. This imaginary, complete extension of the ocean surface beneath the land gives us what is called the "geoid". However, because of irregular density distribution within the body of the earth, and other factors, the geoid has a very gently undulating surface departing from the spheroidal form by amounts of up to about 100 metres, at inclinations which may be as great as 1 minute. When we set up a surveying instrument and level it, the long axis of the level bubble is set tangential to the equipotential surface through it, and the vertical axis is thus made to lie in the perpendicular to the surface at the station, and hence in the perpendicular to the *geoid* below the station if the slight curvature of the vertical is ignored.

Sections of the geoid through its north-south axis approximate very closely to an ellipse whose semi-major axis (equatorial semi-diameter) is equal to 6 378 137 metres and whose semi-minor (polar) axis is equal to 6 356 752.3 metres. These figures were recommended in 1979 by the International Association of Geodesy for general use in computing national control surveys; the order of accuracy of the lengths is ±2 metres. The figure of the earth to be used for the computations, then, is obtained by revolving this ellipse about its minor axis.

Such a spheroid is thus a regular, curved, mathematically definable reference surface from which the uneven topographical surface—the province of the land surveyor—never departs more than a few kilometres, and upon which observed first-order directions and lengths are projected and adjusted for control of lower-order surveys. The resulting spheroidal, geodetic coordinates of points may, in turn, be transformed into the appropriate plane coordinates for control of mapping.

It is of interest to note that the change in the length of a degree of latitude which takes place as we proceed along a meridian is not precisely the same along all meridians. It seems that the equatorial section of the earth is not exactly circular, but is very slightly elliptical. The true shape would thus appear to be more nearly a triaxial *ellipsoid*. For practical purposes, however, computations in geodetic work are usually based upon the assumption that the figure of the earth is an oblate spheroid.

3.08 Astronomic, geodetic, geographic and geocentric latitude

In Fig. 3.7, the ellipse $PQP'E$ drawn in the full line represents the intersection of the plane of the meridian of a point R with the spheroid. The broken curve which is everywhere quite close to the ellipse, and which cuts it in places, represents the intersection of the plane of the meridian of R with the geoid;

The earth

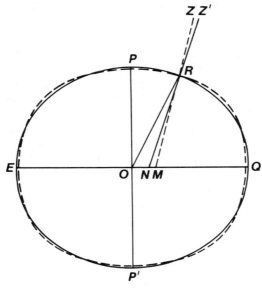

Fig. 3.7

its differences from the ellipse are exaggerated, of course, in order that they may be seen on the diagram. A theodolite set up and levelled at R will have its vertical axis in the broken line ZRM, at right angles to the surface of the geoid at R. The angle RMQ gives the *astronomical latitude* of R, since this is the latitude which would be yielded from an astronomical observation at R, with Z at the zenith. The full line Z'RN is perpendicular to the surface of the spheroid at R, and the angle RNQ measures the *geodetic* or *geographic latitude* of R—the latitude of R computed through a geodetic triangulation. The angle ROQ at the centre of the spheroid is called the *geocentric latitude* of R, and the difference between it and the astronomic and geodetic latitudes is never very great. There is no difference between geographic and geocentric latitudes at the poles and at the equator, and the maximum difference is in latitude 45° where it amounts to about 11' 44" of arc. Geocentric latitude is computed from the geodetic value by the formula:

$$\tan ROQ = \frac{OP^2}{OQ^2} \tan RNQ.$$

3.09 Deviation of the vertical

In Fig. 3.7 we have seen that ZRM is the normal to the geoid in the plane of the meridian, while Z'RN is the normal to the spheroid in that plane. The small angle ZRZ' is called the *deviation of the vertical* in the plane of the meridian at R. It is thus the difference between the geodetic and astronomic latitudes in this plane.

In the figure the geoid is shown intersecting the spheroid at the point R. The two surfaces will, of course, really intersect along a line if they are viewed three-dimensionally (unless they are parallel or touch each other). Generally the plane through the normals to the two surfaces at a point will not coincide with the plane of either the meridian or the prime vertical, but will fall in one of the quadrants formed by these, and the deviation of the vertical (also called the deflection of the vertical) would therefore be measured in this plane if it were possible to do so. However, the only positional information the surveyor can obtain from his astronomical observations at a point is astronomical latitude and longitude. By comparing the observed, i.e. astronomical, latitude with the geodetic latitude of the point which can only be found by computation (from values assumed at an origin on the reference spheroid), he may find the component of the deviation in the meridian; the difference between the respective longitudes multiplied by the cosine of the latitude will give him the component of the deviation in the prime vertical.

Thus, if ξ is the component of the deviation of the vertical in the meridian at a point, and η is the component in the prime vertical, and the suffixes A and G refer to the astronomical and geodetic values respectively of latitude ϕ and longitude λ,

$$\xi = \phi_A - \phi_G; \text{ and}$$

$$\eta = (\lambda_A - \lambda_G)\cos\phi.$$

The deviation of the vertical $= (\xi^2 + \eta^2)^{\frac{1}{2}}$. Its azimuth may be found from $\tan^{-1}(\eta/\xi)$. North latitudes are positive, south negative; east longitudes are positive and west negative.

Since the surveyor measures his horizontal angles about an axis which, from levelling the instrument, must lie in the perpendicular to the geoid, an astronomically determined azimuth will not have the same value as the geodetic azimuth if the vertical at the station does not coincide with the normal to the spheroid at the station, i.e. if there is any deviation there. Since the geodetic value of the azimuth is the one used for surveying and mapping control of horizontal orientation, but is something which cannot be directly observed, it is derived from the astronomical value via Laplace's equation, where A (not a suffix) stands for azimuth:

$$(\lambda_A - \lambda_G)\cos\phi = (A_A - A_G)\cot\phi$$

which leads to

$$A_G = A_A - \eta.\tan\phi,$$

where ϕ may be either ϕ_A or ϕ_G since $\tan\phi$ is required to two places of decimals only.

The earth

This is a very important relationship, since it permits control of the computed geodetic azimuths of lines of a network at properly distributed stations (Laplace stations) by precise observation of astronomical azimuth, latitude and longitude there.

It will be appreciated that, up to the present time, the geodetic values of ϕ, λ and A in the various national networks are based upon assumed values at unrelated national origins (of datums), and that measures of ξ and η are also relative to those assumed values. Because of the extent of the oceans, it is not possible to extend a complete network of conventional survey control over the whole surface of our planet and hence set up a single datum referred to the centre of the earth. However, the science of satellite geodesy has overcome this difficulty and the use of Doppler techniques now routinely allows position anywhere on the earth's surface to be determined to an accuracy of the order of ± 0.8m, both horizontally and vertically, relative to the geocentre, with still higher accuracy in prospect from VLBI (very long base interferometry), lunar laser ranging and artificial satellite laser ranging. This will allow the datum shifts from the various national datums to a single world datum to be computed, and hence values of ξ and η close to what might be called their absolute values to be determined. It remains to be seen, however, how many national geodetic organizations will immediately revise their control networks to put them in terms of an accepted world datum. Nevertheless, a long-awaited goal has been reached, and its scientific significance, if not its surveying and mapping importance, is very considerable.

The surveyor will realize from the above brief discussion that his carefully measured astronomical latitudes, longitudes and azimuths at a point will seldom agree exactly with the geodetic (geographical) values derived from a large-scale map, or from ground survey connections to horizontal control points. However, the relative deviations as defined earlier provide interesting information on geoid shape in the area covered by a geodetic survey. It is the departure of the shape of the geoid from that of the idealized spheroid surface, of course, which gives rise to the generally small disagreement between astronomical and geodetic values at a point. For further information on this subject the reader should consult works such as Bomford's *Geodesy* (Clarendon Press, Oxford, 1980).

EXERCISES

(Except where otherwise stated, take the earth to be a sphere of radius 6373 km.)

1. Find the shortest distance measured along the earth's surface between Mount Gambier (Long. 140° 45' E., Lat. 37° 50' S.) and Palmerston (Long. 130° 50' E., Lat. 12° 28' S.).

Ans. 2989 km.

2. Find the shortest distance measured along the earth's surface between Baltimore (Lat. 39° 17′ N., Long. 76° 37′ W.) and Cape Town (Lat. 33° 56′ S., Long. 18° 26′ E.).

Ans. 12 703 km

3. How far would a place be due south from the equator if the altitude of the S. celestial pole was exactly 20°?

Ans. 2225 km.

4. Two places are in S. Latitude 30°, one Longitude 115° E., and the other 35° E. Find the difference in the paths of the two ships sailing from one port to the other, one along the parallel of latitude and the other along the arc of the great circle joining the places.

Ans. 181 km.

5. What is the declination of a star that passes through the zenith at a place in Latitude 35° N.?

Ans. N. 35°.

6. A ship sails along the great circle joining two places, each of Latitude 45° N., the difference between their longitudes being $2a$. Show that the highest latitude reached during the passage is given by the formula $\cot \phi = \cos a$.

7. A ship from Latitude 8° 25′ N. sails south for 600 nautical miles. What latitude is she in? (1 nautical mile = 01′ of arc of a great circle.)

Ans. 1° 35′ S.

8. At a place in Latitude ϕ north, a star with declination δ rises 60° E. of north. Show that $\cos \phi = 2 \sin \delta$.

9. At a place in Latitude 42° 00′ S. a line is run from a point A on a bearing of 220° 00′ for a distance of 48 280.24 m to a point B. Find the bearing from B to A.

Ans. 40° 15′ 12″.

10. Given that the latitude of London is 51° 32′ N., latitude of Jerusalem 32° 44′ N., bearing of Jerusalem from London, 110° 04′, find the longitude of Jerusalem, its distance from London, and the bearing of London from Jerusalem.

Ans. Longitude, 37° 25′ E.
Distance 3667 km.
Bearing of London from Jerusalem, 316° 00′.

11. The latitude of a Trig. Station A is 33° 51′ S., and its longitude is 151° 12′ 42″ E. The bearing and distance to another Trig. Station B are 284° 08′ 44″, 32 187 m.

Compute the latitude and longitude of B, and the bearing of B to A, on the assumption that the earth is a sphere with radius 6 367 502 m.

Ans. Latitude, 33° 46′ 44″.
Longitude, 150° 52′ 26″ E.
Bearing, 104° 20′ 01″.

The earth

12. Find the great-circle distance in kilometres from Wellington, N.Z., to Panama, treating the earth as a sphere, and one degree as equal to 111.125 km.
Wellington, Lat. 41° 17′ S., Long. 174° 47′ E.
Panama, 9° 00′ N., Long. 70° 31′ W.

Ans. 12 714 km.

13. Two places are each in Latitude 50° N., and their difference of longitude is 47° 36′. Find their distance apart.

Ans. 3344 km.

14. On a perfectly level plain three poles of equal height are set up in a straight line at distances of 603.5 m, and a man looks from the top of the first pole at the top of the third. Show that, owing to the rotundity of the earth, the top of the second pole will be above his line of sight, and calculate the difference approximately. Ignore atmospheric refraction.

Ans. 2.9 cm.

15. The following astronomical observations were made at a control station A whose geodetic coordinates are:
Latitude: N. 41° 22′ 25″.
Longitude: W. 123° 06′ 10″.
Observed astronomical values:
Latitude: N. 41° 22′ 10″.
Longitude: W. 123° 06′ 35″.
Azimuth to station B: 305° 41′ 16″.
What are the amount and direction of the deviation of the vertical at A, and what is the geodetic azimuth to station B?

Ans. Amount of deviation is 24.0″;
Azimuth of deviation is 51° 21′.
Geodetic azimuth to B is 305° 41′ 32.5″.

(*Note*: the deviation components are reckoned positive when the downward vertical is deviated to the south and west of the inward normal to the spheroid.)

CHAPTER 4

The sun and the stars

4.01 The sun's apparent motion among the stars

Like the fixed stars, the sun shares in the apparent general daily rotation of the heavens, but unlike them it does not always maintain the same position relative to other objects on the celestial sphere. In addition to its daily circling of the sky, it appears to shift its position gradually with respect to the stars. Neither its declination nor its right ascension remains constant. Very little consideration will show that its declination must alter during the year, for, if it did not, the sun would always describe the same circle in the heavens. If this were the case, then, like the fixed stars, it would always rise and set at the same points on the horizon, and it would always attain the same altitude when on the meridian. Since it does not do this, it is clear that the declination of the sun must change during the year. That the sun also has a movement in right ascension among the stars is not quite so obvious, but the fact may be readily inferred if we watch the stars that are visible in the east on succeeding mornings just before sunrise, or in the west just after sunset. Stars in the east that rise just before the sun, so that in a very short time after rising they are masked by the sun's rays, will on each succeeding morning be seen for a longer time. Similarly stars in the west, setting just after the sun, will be visible for shorter and shorter periods as we watch them on successive evenings until finally they are lost altogether in the strong sunlight, other stars further east taking their places. Hence we infer that the sun has a progressive movement among the stars from west to east.

The problem of determining the sun's place on the celestial sphere with regard to the fixed stars was a difficult one to early astronomers, because as soon as the sun becomes visible its strong light prevents the stars from being observed at the same time. Some used the moon, and Tycho Brahe (1546-1601) used the bright planet Venus in order to get the connection, observing the relative positions of the sun and moon or of the sun and Venus when both were visible, and afterwards measuring the position of the moon or Venus with regard to the stars when the sun had set. But as both the moon and Venus also move amongst the stars, the movement that had taken place in the interval had to be allowed for, and the method was thus not particularly simple. The sun's position is nowadays determined by much more accurate methods.

4.02 The earth's orbit round the sun

All of these movements of the sun are apparent only and not real. Just as its apparent daily rotation in the heavens is due to the rotation of the earth on its axis, so the sun's apparent movements in right ascension and declination are really due to the fact that the earth moves in a great orbit round the sun once a year.

Actually the earth moves round the sun in a path that is very nearly a circle with a radius of about 149.5 million kilometres. More accurately, the path is described as an ellipse, one focus of the ellipse being occupied by the sun. The curve traced out by the centre of the earth lies in a plane that passes through the centre of the sun; for the practical purposes of surveyors the plane may be regarded as fixed. The earth traces out its complete orbit once a year, and all the time it is spinning on its own axis once a day, the direction of the spin on its axis being the same as that in which it moves round the sun. The earth's axis is not at right angles to the plane of its orbit, but it makes with this plane an angle of approximately 66° $33\frac{1}{2}'$. That the direction of the earth's axis is practically constant we know from the fact that the position of the celestial pole amongst the fixed stars shows no appreciable shift throughout the year, except to refined observation. Thus, as is illustrated in Fig. 4.1, the earth moves round the sun, spinning on its axis, which is inclined to the plane of the orbit,

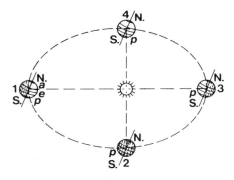

Fig. 4.1

and the axis always remains parallel to itself, pointing in the same direction amongst the fixed stars, whose distances, it must be remembered, are practically infinitely great even in comparison with the immense distance of the earth from the sun.

When the earth is in the position marked 1, the sun will be shining directly overhead at a place such as a north of the equator. If e is a point on the earth's equator on the same meridian of longitude as a, O being the earth's centre, the angle aOe will be the complement of 66° $33\frac{1}{2}'$ or 23° $26\frac{1}{2}'$—that is to say, a will

Astronomy for surveyors

be a point on the tropic of Cancer. In this position, then, the sun at midday will be vertically overhead at all points on the tropic of Cancer. This statement is not quite accurate, because the earth does not remain in the one position in its orbit while its makes a complete revolution on its axis; it is moving forward in its orbit all the time, but as it takes a whole year to go round the sun, its relative movement is not very great in one day.

As the earth moves from position 1 to position 2, its axis always remaining parallel to its original direction, it will be seen that the sun will appear to shine directly overhead at points successively nearer and nearer to the equator, until in position 2 the sun's rays fall vertically at the equator.

Similarly, as the earth moves on to position 3, the sun's rays will fall vertically at points further and further south of the equator, until at position 3 the sun will appear at midday to be overhead at a point on the tropic of Capricorn. From there on to position 4 the sun will shine vertically at points successively nearer to the equator, until at 4 the sun is once more overhead at the equator.

The earth is in the position marked 1 on 22 June, in that marked 2 on 23 September, at 3 on 22 December and at 4 on 21 March. Thus, if we consider the appearance of the sun to an observer at some point p to the south of the tropic of Capricorn, on 22 June the sun will appear to be further from the zenith and lower down in the sky than at any other period of the year. On 22 December, when the earth is in position 3, the sun at midday will be nearer the zenith of p than at any other time of the year.

The earth's orbit being an ellipse, its distance from the sun is not constant. It is furthest from the sun when about in the position 1, and nearest to the sun about in the position 3.

4.03 The equinoxes

On 21 March and 23 September the sun, being vertically overhead at the equator, will appear to an observer at any part of the earth to be in the celestial equator. Now we have seen that when any heavenly body is in the celestial equator its path is bisected by the horizon, so that the time during which it can be seen in the sky is equal to the time during which it is not visible. Thus, when the earth is in either of these positions the days and nights are of equal length all over the world. These points are consequently called the *equinoxes*.

4.04 Motion in right ascension and declination

It thus appears that on 21 March and 23 September the sun's declination is zero, as it lies on the celestial equator. From 21 March to 23 September it will appear in the sky to the north of the equator, so that its declination will be north with a maximum value of $23° \ 26\frac{1}{2}'$ on 22 June. From 23 September

The sun and the stars

to 21 March its declination will be south with a similar maximum value on 22 December.

It is also evident that the sun's right ascension changes throughout the year, because as the earth revolves round it the apparent position of the sun among the fixed stars must obviously change. The stars that would be seen by an observer on the earth when in position 1, looking in the direction of the sun, would be seen by an observer at 3 when looking in the direction opposite to that of the sun. Clearly, in the course of the year the sun will appear to trace out a complete circle among the fixed stars.

The declination of the sun is given at 6-hourly intervals of Universal Time (U.T.) throughout the year in the columns headed "Dec". on pages 2-25 of the *Star Almanac for Land Surveyors*. If the declination is required at intermediate times (as will usually be the case), interpolation from the table is necessary (see **10.20**). See also pages 98-101 herein.

The right ascension of the sun is not required in surveying work, but as a matter of interest it may be found by subtracting the quantity E from the quantity R for any given value of U.T. on pages 2-25 of the *Star Almanac*; if $R < E$, use $(24^h + R)$ instead of R. After reading Chapter 5 the student should reason out why this is so.

4.05 The sun's semi-diameter

The disc of the sun subtends at the eye of an observer an angle of about half a degree. By accurately measuring the angle subtended by diameters taken in different directions, we find that these are all equal, so that the disc is circular in form. In order to mark the position that the sun occupies on the celestial sphere at any time, we require to determine the position of the centre of the circular disc. But there is no mark at the centre that we can recognize, and so in practice we must observe a point on the edge of the sun and then make an allowance for the distance of this point from the sun's centre.

From what we have just seen of the nature of the earth's motion round the sun, it is clear that the sun is not at all times of the year at the same distance from us, and consequently we should not expect its angular diameter to remain constant. As the earth completes its orbit round the sun in a year and then goes over the same path again, we might anticipate that the variations in the value of the sun's apparent diameter would follow a yearly cycle. This is found to be the case, a slow decrease taking place from about 3 January to about 4 July, and a slow increase during the second half of the year.

As the semi-diameter is sometimes required in reducing sun observations, its average value (correct to one-tenth of a minute of arc) for each half-month is given on pages 2-25 of the *Star Almanac* in the same lines as the headings "Sunrise" and "Sunset". See pages 98-101 herein.

4.06 To plot the position of the sun's centre on the celestial sphere

Supposing that we know the direction of true north and south, and also the latitude of the place of observation, we may readily measure the declination of the sun at midday. With a telescope pointed in the direction of the meridian we may observe the altitude of the sun's upper or lower edge (*limb*, as it is usually called) at the moment when it crosses the meridian.

Making due allowance for the sun's semi-diameter, we shall thus obtain the meridian altitude of the sun's centre. Thus, as in Fig. 4.2, where P represents the pole and Z the zenith, we measure either $S_1 M_N$ or $S_2 M_S$, according as the sun is in a position such as S_1 or as S_2. Now, we have previously shown that the altitude of the celestial pole, PM_N, is equal to the latitude of the place. Thus, if the sun is situated as at S_1, on the same side of the zenith as the pole, the difference between the observed altitude $S_1 M_N$ and the latitude PM_N gives the sun's polar distance PS_1. If the sun is at S_2, on the opposite side of the zenith to the pole, then arc $S_2 M_N$ is equal to $180°$ minus the observed altitude $S_2 M_S$. The difference between $S_2 M_N$ and the latitude PM_N gives the sun's polar distance as before. The declination of the sun is the complement of its polar distance, or when the polar distance is greater than $90°$, it is equal to P.D. $-90°$.

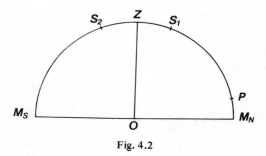

Fig. 4.2

Having measured the declination of the sun in this way, in order to fix its position on the celestial sphere it only remains to determine the difference between its right ascension and that of some star whose coordinates are known. But we have seen that the difference of right ascension of any two stars is measured by the interval in time between their transits across the meridian, as given by the sidereal clock. If, with the sidereal clock, the times be measured when the first and second limbs of the sun cross the meridian, the mean of the two times will give the instant when the centre crosses the meridian. If, therefore, the time of passage across the meridian of some selected known star is also observed, the interval between the two times, reduced to degrees, will give the difference between the right ascension of the sun and of the star.

These observations give us the elements necessary to plot the position of the sun.

4.07 The sun's apparent path on the celestial sphere

In Fig. 4.3 let A represent the position of the selected fixed reference star as plotted on a globe representing the celestial sphere, P being the pole, QR the great circle of the equator, and $M_S W M_N E$ the horizon. Then, if we set out the angle APS equal to the observed difference of right ascension and measure off the arc PS equal to the observed polar distance of the sun, the point S will represent the position of the sun's centre on the star globe.

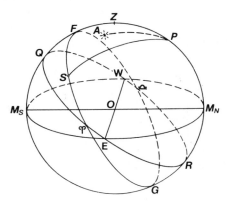

Fig. 4.3

When observations similar to those just described are made day after day, and the corresponding positions of the sun plotted on the globe, those positions are all found to lie on a great circle which cuts the equator at two opposite points ♈ and ♎ in the figure, and is inclined to it at an angle of about 23° 26½'.

The great circle, the plane of which contains the sun's yearly path, is called the *ecliptic*, and the angle this makes with the equator is spoken of as the *obliquity of the ecliptic*.

Its points of intersection with the equator are called the *equinoctial points*, one (♈) being known as the *First Point of Aries*, and the other (♎) as the *First Point of Libra*.

The sun is at the first of these points on or about 21 March (the *vernal equinox* in the north hemisphere), and at the second about 23 September (the *autumnal equinox* in the northern hemisphere), its declination being then 0° and its polar distance 90°.

As we have already seen, ♈ is the point selected on the equator as that from which right ascensions are measured, so that the right ascension of ♈ is 0° and that of ♎ 180°.

At the two points on the ecliptic whose right ascensions are respectively 90° and 270°, the sun will have its greatest declination north and south of the equator. These are known as the *solstitial points*. The sun reaches them on

Astronomy for surveyors

or about 22 June and 22 December. On 22 June the sun has its greatest declination of about 23° 26½' north of the equator, and on 22 December its greatest declination south.

4.08 Tabular information about the sun

Because of the sun's changing right ascension and declination and the fact that it is a convenient celestial object for the surveyor to use for obtaining azimuth and position, daily information about it in tabular form is provided in the *Star Almanac for Land Surveyors*. This material is given* from pages 2 through 25 each year, the double page at each opening covering one month, the quantities R and E (explained in Chapter 5) and the declination being given at six-hourly intervals. R changes at a constant rate, and a special interpolation table, "Interpolation Table for R", is given† near the end of the almanac. The declination and E do not change at a constant rate, so tabular differences are given in the six-hourly tables and another interpolation table, "Interpolation Table for Sun", is given‡ immediately after that for R.

Also on pages 2 through 25 are given the sun's semi-diameter (S.D.), and sunrise and sunset times for latitudes 0°-60° in each hemisphere; the phases of the moon are also shown, so that observation times can be planned to avoid the bright moonlight which may make the star constellations difficult to identify and fainter stars hard to see in the telescope.

The student should read the Introduction in the *Star Almanac* for more detailed information about the tabular material presented.

4.09 The slight variations in the right ascensions and declinations of the stars

In this and the foregoing two chapters, we have assumed that the direction of the earth's axis of rotation is fixed in space and that the so-called fixed stars always have exactly the same positions on the celestial sphere relative to one another. While we shall retain these ideas for the purposes of explanation, it is as well to note that they are not exactly true. A study of pages 26-51[§] of the *Star Almanac*, for example, will show that the right ascensions and declinations of the "fixed" stars are changing slightly all the time.

Because the earth is not quite a true sphere, the mutual attractions between it and the sun, moon and planets cause a slow secular movement of the line of the equinoxes (i.e. the line joining ♈ and ♎) around the plane of the ecliptic. The direction of this motion is opposite to the movement of the earth in its orbit, and amounts to about 50.2" per year. The phenomenon is called the

---

* See extract from *S.A.* on pages 98-101.
† See extract from *S.A.* on pages 102-104.
‡ See extract from *S.A.* on page 105.
§ See extract from *S.A.* on pages 56-7.

The sun and the stars

precession of the equinoxes, or simply *precession*. Further, because of this precession and the obliquity of the ecliptic, the pole of the equator (the celestial pole) will, in about 25 000 years, describe a circle around the pole of the ecliptic, if the plane of the ecliptic be accepted as the plane of reference. Thus, through the centuries, ♈ will migrate away from its present position in the constellation of Pisces, and, in about 100 years, the north celestial pole will begin to move away from the present Pole Star.

The attractive forces causing precession do not remain constant because of the changing distances between earth, sun and moon and, more particularly, because the plane of the moon's orbit does not lie in the ecliptic. As a result, a small periodic oscillation of the earth's axis, called *nutation*, is produced.

Since the earth travels in its orbit at a speed which is an appreciable fraction of the velocity of light, there results a displacement of each star from its mean position by a constant amount of about $20.4''$. This displacement is known as *annual aberration*. A similar but much smaller effect, called *diurnal aberration*, results from the earth's diurnal motion; it is nil at the poles where the velocity of rotation is nil, and a maximum at the equator.

Both the solar system and the "fixed" stars are in motion, and are therefore gradually changing their absolute positions in space. The amount of this *proper motion* is different for every star but is so small that it does not perceptibly affect their *relative* positions as far as we are concerned.

It is mainly, then, the phenomena of precession, nutation and aberration which produce these slight variations in the right ascensions and declinations of the stars.

The right ascensions and declinations of nearly 700 stars are given from pages 26 through 55 in the *Star Almanac for Land Surveyors*. For all except ten close circumpolar stars the values are given for the beginning of the month, with a special interpolation table, "Interpolation Table for Stars", near the end of the almanac*. For the ten circumpolar stars, the R.A. and Dec. are given for every ten days throughout the year, and interpolation may easily be done by inspection. More detailed information is given in the Introduction to the *Star Almanac*.

4.10 Star catalogues

For most ordinary stellar observations made by surveyors, the *Star Almanac* is quite adequate with its tabular information on 695 stars (pp. 26-55). However, it is as well for surveyors to know that other, more comprehensive catalogues exist. Geodetic astronomers, for example, use the *Apparent Places of Fundamental Stars* (*APFS*) based upon the FK4 system and published annually by Astronomisches Rechen-Institut, Heidelberg.† There are 1535 stars (down

---

* See extract from *S.A.* on page 105.
† See extract from *APFS* on page 58.

Astronomy for surveyors

to about magnitude 6) in this catalogue and their coordinates are given to three places of decimals in the seconds of R.A. and to two places in the seconds of declination, every ten days throughout the year.

There are other, larger catalogues available for general astronomical work, such as the Boss catalogue and that of the Smithsonian Astrophysical Observatory (258 997 stars).* These give the coordinates of the stars at some reference point (epoch) in the time-scale, such as the beginning of the year 1950 (stated as 1950.0), and the person using them is left to compute the corrections for the changes in the star's position that have taken place between then and the time (epoch) of actual observation. The position of a star given in such catalogues is the Mean Place, so the Smithsonian catalogue, for example, gives the Mean Place of each star for 1950.0. Many of the stars in these larger catalogues are too faint to be observed with the theodolite which is not capable of "seeing" much beyond the sixth magnitude.

4.11 Mean Place and Apparent Place

If a surveyor happens for some reason to observe a star which is in neither the *Star Almanac* nor the *APFS,* he will need its Apparent Place for the time of his observation in order to compute a result. This involves computing the change in Mean Place between the catalogue epoch (1950.0 say) and the *beginning of the year nearest* the time of the observation. A further step is then necessary to compute the Apparent Place at the time and date of observation, from this nearest Mean Place. In computing from one Mean Place to another, proper motion and precession are allowed for during the intervening (integral) number of years. In computing from Mean Place to Apparent Place the effects of proper motion, precession, nutation and annual aberration are taken into account over the portion of a year between the nearest Mean Place and the time of observation.

In the author's experience, it sometimes happens that a student, using fourth to fifth magnitude stars in programmed observations, will unknowingly get a nearby wrong star (i.e. one not in the *Star Almanac*) in the field of view and use it, perhaps as one of a group of four for a position line fix (see Chapter 14). Rather than abandon the entire observation, it is better to try to identify the star in, say, the Smithsonian catalogue and find its Apparent Place as outlined above, thus avoiding the need to repeat the whole observation on another night. Of course, if the star happens to be in the *APFS* there is no problem in finding its coordinates. In order that the star may be identified, however, it is necessary to observe an approximate azimuth (to within a minute or two of arc), as well as altitude and time, and this means that the azimuth of the R.O. (see **8.04**)

---

* See extract from *SAO* Catalogue on page 59.

The sun and the stars

used to set the theodolite circle must be known with similar accuracy. The R.O. can usually be oriented during daylight by a simple sun observation (see **8.08** and **10.21**). Approximately correct values of the latitude and longitude of the observing station must be known.

Having computed the R.A. and declination of the unknown star as in **9.04**, they must be reduced approximately to the catalogue epoch for identification purposes. The formulae at (4.1) in section **4.12** following may be transposed to yield the desired values:

$$\alpha_0 = \alpha_t - \tau \left(\frac{d\alpha}{dT}\right)_0 \; ; \quad \delta_0 = \delta_t - \tau \left(\frac{d\delta}{dT}\right)_0$$

where α_t and δ_t are in this case the approximate Apparent Places calculated from the observed time, altitude and azimuth, and the approximate latitude and longitude known to a minute or two of arc. The values of the other quantities in the above formulae are the same as those in **4.12**.

The magnitudes of observed stars should always be estimated and entered in the field-book in case they are needed for identification purposes later. Since a star observed with a theodolite is hardly likely to be fainter than magnitude 6, and since by far the majority of stars in a catalogue like the Smithsonian are, in fact, fainter than 6, the few brighter "theodolite stars" on a page will provide a first clue to identification, once the approximate R.A. and declination have been computed, by glancing down the magnitude column.

A good star atlas, e.g. A. Bečvář's *Atlas Coeli Skalnaté Pleso* 1950.0, can also be a useful aid in identifying stars.*

4.12 Example of reduction to Apparent Place

An observed star of about magnitude 4 was identified as No. 131 019 in the Smithsonian Astrophysical Observatory's catalogue. The approximate time and the date of observation were U.T. 9^h on 1974 December 8. Find the Apparent Place then.

Mean Place 1950.0 to Mean Place 1975.0

The R.A. and declination of the star for 1950.0 from the catalogue are $4^h \; 09^m \; 25^s.339$ and $-6° \; 57' \; 59''.74$† respectively. The centennial proper motions in R.A. and declination are $+0^s.03$ and $+8''.3$ respectively. A rigorous reduction will not be attempted here as the observation was not done for first-order work. Therefore terms involving squares will be omitted from the Mean Place formulae which follow overleaf.

---

* For small but useful charts see extracts from *S.A.* on pages 60 and 61.
† See page 59.

$$\alpha_t = \alpha_0 + \tau \left(\frac{d\alpha}{dT}\right)_0 \quad ; \quad \delta_t = \delta_0 + \tau \left(\frac{d\delta}{dT}\right)_0 \tag{4.1}$$

The suffix zero indicates the values at the catalogue epoch (1950.0 in this case); the suffix t indicates the values at the desired epoch (1975.0 in this case). α = R.A.; δ = declination; $\tau = \frac{1}{100}(t - t_0)$. T is in centuries from 1900.0.

$$\frac{d\alpha}{dT} = m + \tfrac{1}{15}.n.\sin\alpha.\tan\delta + \mu$$

$$\frac{d\delta}{dT} = n.\cos\alpha + \mu'$$

$$m = 307^s.2337 + 0^s.186\,30.T + 0^s.000\,008.T^2$$

$$n = 2004''.685 - 0''.8533.T - 0''.000\,37.T^2.$$

μ and μ' are the centennial proper motions, and they are therefore 100 times the annual values given in the Smithsonian catalgoue; in this case they are $+0^s.03$ and $+8''.3$ respectively. In this example:

$\tau = \frac{1}{100}(1975.0 - 1950.0) = +0.25$

$T = \frac{1}{100}(1950.0 - 1900.0) = +0.50$

$m = 307^s.2337 + 0^s.186\,30.(0.5) + 0^s.000\,008.(0.5)^2 = +307^s.326\,85$

$n = 2004''.685 - 0''.8533.(0.5) - 0''.000\,37.(0.5)^2 = +2004''.258\,26$

$\alpha_0 = 62°.355\,579; \delta_0 = -6°.966\,594$

$$\left(\frac{d\alpha}{dT}\right)_0 = +292^s.893\,62$$

$$\left(\frac{d\delta}{dT}\right)_0 = +938''.241\,68.$$

The Mean Place for *1975.0* is given by α_t and δ_t:

$\alpha_t = 4^h\,09^m\,25^s.339 + (0.25 \times 292^s.893\,62) = 4^h\,10^m\,38^s.562$

$\delta_t = -6°\,57'\,59''.74 + (0.25 \times 938''.241\,68) = -6°\,54'\,05''.18.$

(The values obtained from a rigorous reduction are $4^h\,10^m\,38^s.580$ and $-6°\,54'\,06''.38$, the differences being due to the second-order terms not included in the above reduction. These additional terms may be calculated from the full formulae given in the *Fourth Fundamental Catalogue* (1963) by the publishers of the *APFS*.)

The sun and the stars

Mean Place 1975.0 to Apparent Place of date

The Apparent Place of date is for 1974 December 8 at approximate U.T. 9h.

For this part of the calculation we may use either Besselian Day Numbers or Independent Day Numbers. These Day Numbers are to be found in the *Astronomical Ephemeris*, or the *American Ephemeris* for the year concerned (1974 in this case), and we shall use the Besselian Numbers.

The formulae for the Apparent Place reduction are:

$$\alpha_{Ap} = \alpha_t + \tau.\mu_\alpha + A.a + B.b + C.c + D.d + E$$

$$\delta_{Ap} = \delta_t + \tau.\mu_\delta + A.a' + B.b' + C.c' + D.d',$$

where

α_{Ap} = the Apparent R.A. and δ_{Ap} = the Apparent declination.

α_t and δ_t are the Mean Place R.A. and declination already found.

τ is the fraction of the year away from the Mean Place epoch; it may be positive or negative, depending on whether that epoch is earlier or later than the date of the observation. It is found by interpolation from the Besselian Day Number table.

μ_α and μ_δ are the *annual* proper motions in R. A. and declination, respectively.

A, B, C, D and E are the Besselian Day Numbers, found by linear interpolation from the daily values given in the *Astronomical Ephemeris*. (Note that these Numbers are given for Ephemeris Time, or E.T., where E.T. = U.T. + ΔT. ΔT may be found from

$$\Delta T = 32^s.18 + \text{I.A.T.} - \text{U.T.1},$$

where I.A.T. = International Atomic Time (see **5.02** and **5.09**).

a, b, c, d and a', b', c', d' are values computed from the following formulae:

$a = \frac{1}{15}(m/n + \sin\alpha.\tan\delta);$ $\qquad a' = \cos\alpha;$

$b = \frac{1}{15}(\cos\alpha.\tan\delta);$ $\qquad b' = -\sin\alpha;$

$c = \frac{1}{15}(\cos\alpha.\sec\delta);$ $\qquad c' = \tan\epsilon.\cos\delta - \sin\alpha.\sin\delta;$

$d = \frac{1}{15}(\sin\alpha.\sec\delta);$ $\qquad d' = \cos\alpha.\sin\delta,$

ϵ = the obliquity of the ecliptic.

The values of m and n may be computed from the formulae already given, but along with ϵ they can be got more quickly from the table of Precessional Constants near the beginning of the *Astronomical Ephemeris*. In the present case, from this table, $m = 3^s.073\ 74; n = 1^s.336\ 03\ (20''.0404); \epsilon = 23°.442\ 534$. The values of α and δ used for computing, a, b, etc. are, strictly speaking, the Apparent Place values, but in this non-rigorous reduction, the Mean Place values (for 1975.0) may be used. In our example:

Astronomy for surveyors

$$a = +0.1462; \quad a' = +0.4593;$$
$$b = -0.0037; \quad b' = -0.8883;$$
$$c = +0.0308; \quad c' = +0.5372;$$
$$d = +0.0597; \quad d' = -0.0552.$$

From the *Astronomical Ephemeris*, by interpolation,

$$A = +4''.907; \quad B = +3''.536; \quad C = +4''.584; \quad D = +19''.877;$$
$$E = +0''.0023; \quad \tau = -0.0647.$$

The apparent place is then:

α_t	$4^h\ 10^m\ 38^s.562$	δ_t	$-6°\ 54'\ 05''.18$
$\tau\mu_\alpha$	-0.000	$\tau\mu_\delta$	-0.005
$A.a$	$+0.717$	$A.a'$	$+2.254$
$B.b$	-0.013	$B.b'$	-3.141
$C.c$	$+0.141$	$C.c'$	$+2.463$
$D.d$	$+1.187$	$D.d'$	-1.097
E	$+0.002$		
α_{Ap}	$4^h\ 10^m\ 40^s.596$	δ_{Ap}	$-6°\ 54'\ 04''.706$

The star concerned is actually No. 103 in the *Star Almanac for Land Surveyors* (o^1 Eridani) and No. 154 in the *Apparent Places of Fundamental Stars*, which give the Apparent Place as:

	R.A.	Declination
Star Almanac	$4^h\ 10^m\ 40^s.6$	S. $6°\ 54'\ 06''$
APFS	$4\ \ 10\ \ \ 40.607$	$-6\ \ 54\ \ 05.83$

For ordinary work the discrepancies are quite tolerable.

Larger star catalogues and *The Astronomical Ephemeris* (published annually in England by H.M.S.O., and in the United States of America by the U.S. Naval Observatory, Nautical Almanac Office, under the title *The American Ephemeris*) are usually available in the bigger libraries or through interloan services.

4.13 Sample extracts from star catalogues

Following are facsimile pages from *The Star Almanac for Land Surveyors* (*S.A.*) (these being slightly reduced in size to fit this book's format), *Apparent*

The sun and the stars

Places of Fundamental Stars (*APFS*) and *Positions and Proper Motions of 258 997 Stars for the Epoch and Equinox of 1950.0* (*SAO*). The methods of interpolation in the first two are fairly obvious. In the case of the *S.A.*, a special table to aid interpolation is included (see page 105 herein). If maximum accuracy is required in star positions from the *APFS*, e.g. in first-order geodetic work, interpolation using second differences is usually necessary.

The author is indebted to the Controller of Her Majesty's Stationery Office, Norwich, England, for kind permission to reproduce material from *The Star Almanac for Land Surveyors*; to Verlag G. Braun GmbH, Karlsruhe, Federal Republic of Germany, representing the Astronomisches Rechen-Institut, for kind permission to reproduce a page from *Apparent Places of Fundamental Stars*; and to the Smithsonian Astrophysical Observatory, Washington, D.C., U.S.A., for kind permission to reproduce a page from *Positions and Proper Motions of 258 997 Stars for the Epoch and Equinox of 1950.0.*

The order in which the facsimiles appear is: pages 56 and 57 shows pages 32 and 33 of the *S.A.* for 1975; page 58 shows page 67 from the *APFS* for 1974; page 59 shows a page from the *SAO* Catalogue; and pages 60 and 61 show pages 76 and 77 (star charts) from the *S.A.* for 1984.

RIGHT ASCENSION OF STARS, 1975

No.	Mag.	R.A.	Jan.	Feb.	Mar.	Apr.	May	June	July	Aug.	Sept.	Oct.	Nov.	Dec.	Jan.
		h m	s	s	s	s	s	s	s	s	s	s	s	s	s
151	3·8	5 37	31·7	31·6	31·2	30·7	30·3	30·2	30·4	31·0	31·9	32·7	33·5	34·2	34·5
152	2·7	5 38	46·8	46·6	46·0	45·3	44·7	44·5	44·7	45·2	46·1	47·1	48·0	48·6	48·9
153	2·0	5 39	32·1	32·0	31·6	31·1	30·7	30·6	30·8	31·4	32·2	33·1	33·9	34·6	34·9
154	3·8	5 43	27·5	27·3	26·8	26·2	25·7	25·5	25·7	26·3	27·1	28·0	28·8	29·5	29·8
155	3·7	5 45	51·5	51·4	51·0	50·4	50·0	49·8	50·0	50·6	51·4	52·3	53·1	53·7	54·1
156	2·2	5 46	36·5	36·4	36·0	35·4	35·0	34·8	35·1	35·6	36·4	37·3	38·1	38·8	39·1
157	3·9	5 46	44·1	43·7	42·8	41·8	40·9	40·4	40·5	41·1	42·1	43·3	44·4	45·2	45·4
158	4·4	5 49	25·0	24·5	23·5	22·4	21·3	20·7	20·7	21·4	22·5	23·8	25·0	25·8	26·1
159	4·2	5 49	48·4	48·4	47·9	47·2	46·7	46·6	46·9	47·7	48·7	49·9	50·9	51·8	52·3
160	3·2	5 50	07·0	06·8	06·2	05·4	04·8	04·5	04·6	05·2	06·1	07·0	08·0	08·6	08·9
161	3·9	5 50	17·0	16·9	16·4	15·8	15·3	15·1	15·3	15·9	16·7	17·6	18·4	19·1	19·4
162 d	0—1	5 53	51·5	51·4	51·1	50·5	50·1	50·0	50·3	50·8	51·7	52·5	53·4	54·1	54·5
163	3·8	5 55	18·2	18·1	17·7	17·1	16·6	16·5	16·7	17·2	18·0	18·9	19·7	20·4	20·7
164	4·4	5 56	41·3	41·1	40·6	39·8	39·2	38·9	39·0	39·5	40·4	41·4	42·3	43·0	43·3
165	3·9	5 57	32·0	31·9	31·3	30·3	29·6	29·3	29·7	30·6	31·9	33·4	34·8	36·0	36·6
166	2·1	5 57	45·0	45·0	44·4	43·7	43·1	42·9	43·2	44·0	45·1	46·4	47·5	48·5	49·1
167	2·7	5 58	04·0	04·0	03·5	02·9	02·3	02·2	02·5	03·2	04·2	05·3	06·4	07·3	07·8
168	4·0	5 58	25·3	25·0	24·3	23·5	22·7	22·4	22·4	23·0	23·9	24·9	25·9	26·7	27·0
169	4·3	6 02	38·6	38·7	38·3	37·7	37·3	37·1	37·4	38·0	38·9	39·8	40·8	41·6	42·1
170	4·4	6 06	11·1	11·2	10·8	10·3	09·8	09·7	09·9	10·5	11·3	12·2	13·1	13·9	14·3
171	3–4	6 13	24·7	24·8	24·4	23·9	23·4	23·2	23·5	24·0	24·9	25·8	26·8	27·6	28·1
172	4·4	6 13	49·9	49·9	49·6	49·0	48·5	48·3	48·5	49·1	50·0	51·1	52·0	52·9	53·5
173	4·4	6 17	29·4	29·4	28·7	27·7	26·8	26·3	26·6	27·5	28·9	30·6	32·2	33·5	34·4
174	3·1	6 19	23·5	23·4	22·9	22·3	21·7	21·3	21·4	21·9	22·6	23·6	24·5	25·2	25·7
175	4·0	6 21	14·3	14·2	13·7	13·0	12·4	12·0	12·1	12·5	13·3	14·2	15·2	16·0	16·4
176	3·2	6 21	29·5	29·6	29·2	28·7	28·2	28·0	28·2	28·8	29·6	30·6	31·5	32·3	32·9
177	2·0	6 21	38·2	38·1	37·8	37·2	36·7	36·4	36·5	37·0	37·7	38·6	39·5	40·2	40·6
178	4·5	6 22	29·0	29·0	28·7	28·2	27·7	27·5	27·7	28·2	29·0	29·8	30·7	31·4	31·9
179 d	—0·9	6 23	26·5	26·3	25·5	24·4	23·4	22·8	22·7	23·1	24·1	25·3	26·5	27·4	27·8
180	4·1	6 27	31·3	31·4	31·1	30·5	30·1	29·9	30·1	30·6	31·4	32·3	33·3	34·1	34·6
181	1·9	6 36	18·6	18·7	18·4	17·9	17·4	17·2	17·4	17·9	18·6	19·5	20·4	21·3	21·8
182	3·2	6 36	62·2	62·1	61·5	60·7	59·9	59·4	59·3	59·7	60·5	61·6	62·6	63·5	64·0
183	3·2	6 42	26·3	26·5	26·2	25·6	25·1	24·9	25·0	25·5	26·3	27·3	28·3	29·2	29·8
184	3·4	6 43	55·6	55·8	55·5	55·0	54·5	54·3	54·4	54·9	55·6	56·5	57·4	58·2	58·8
185 d	—1·6	6 44	04·9	05·0	04·6	04·1	03·5	03·3	03·3	03·7	04·4	05·3	06·1	06·9	07·4
186	3·3	6 47	59·3	59·0	58·0	56·6	55·2	54·2	53·9	54·2	55·2	56·6	58·2	59·4	60·0
187	3·8	6 48	56·7	56·8	56·3	55·7	55·0	54·6	54·6	54·9	55·7	56·6	57·5	58·4	58·9
188	2·8	6 49	21·6	21·5	20·8	19·8	18·9	18·2	18·0	18·4	19·2	20·3	21·5	22·5	23·1
189	3·6	6 51	11·4	11·6	11·3	10·7	10·1	09·9	10·0	10·5	11·3	12·3	13·4	14·4	15·1
190	4·2	6 53	04·0	04·1	03·8	03·2	02·7	02·4	02·5	02·9	03·5	04·4	05·2	06·0	06·6
191	1·6	6 57	40·9	41·0	40·6	40·0	39·3	39·0	38·9	39·3	39·9	40·8	41·8	42·6	43·1
192	3·7	7 00	45·7	45·8	45·4	44·8	44·2	43·8	43·8	44·1	44·8	45·6	46·6	47·4	48·0
193	3·1	7 01	61·1	61·2	60·9	60·3	59·7	59·4	59·3	59·7	60·3	61·2	62·1	62·9	63·5
194	3·9	7 02	40·2	40·4	40·1	39·6	39·1	38·9	38·9	39·4	40·1	41·0	41·9	42·8	43·5
195	4·1	7 02	39·9	40·0	39·7	39·2	38·7	38·4	38·4	38·7	39·4	40·2	41·1	41·9	42·4
196	2·0	7 07	24·8	24·9	24·5	24·0	23·4	23·0	22·9	23·2	23·9	24·7	25·7	26·5	27·1
197	3·9	7 08	62·1	61·7	60·3	58·4	56·3	54·7	53·9	54·1	55·2	57·1	59·2	60·9	61·7
198	4·1	7 10	37·6	37·8	37·5	37·0	36·6	36·3	36·3	36·7	37·3	38·1	39·0	39·8	40·4
199	4·5	7 11	53·4	53·4	52·9	52·1	51·2	50·6	50·3	50·6	51·3	52·3	53·5	54·5	55·1
200	3·8	7 13	50·1	50·3	49·9	49·4	48·8	48·4	48·3	48·6	49·2	50·1	51·0	51·9	52·4

The figures given refer to the beginning of the month, and should be interpolated to the actual date by means of the table on page 73.

The sun and the stars

DECLINATION OF STARS, 1975

No.		Name	Dec.	Jan.	F.	M.	Apr.	M.	J.	July	A.	S.	Oct.	N.	D.	Jan.
			° ′	″	″	″	″	″	″	″	″	″	″	″	″	″
151	σ	Orionis	S 2 36	53	56	59	59	58	54	50	46	43	42	44	48	53
152	α	Columbae	S 34 04	79	87	91	91	88	81	72	64	59	58	62	70	80
153	ζ^1	Orionis	S 1 57	21	25	28	28	27	23	19	15	12	11	13	18	23
154	γ	Leporis	S 22 27	26	33	37	37	35	29	22	15	10	09	13	20	28
155	ζ	Leporis	S 14 49	55	61	64	64	62	57	51	45	41	40	43	49	56
156	κ	Orionis	S 9 40	44	49	52	53	51	47	41	36	32	31	34	40	46
157	β	Pictoris	S 51 04	38	47	52	52	48	40	31	21	15	14	19	28	39
158	γ	Pictoris	S 56 10	28	38	43	44	40	32	22	12	06	05	10	19	30
159	ν	Aurigae	N 39 08	32	35	36	36	34	31	28	26	25	25	25	27	30
160	β	Columbae	S 35 46	44	52	57	57	54	47	38	30	24	23	27	36	45
161	δ	Leporis	S 20 52	55	62	66	66	64	58	51	45	40	39	43	50	58
162	*Betelgeuse* (α Ori)		N 7 24	09	06	05	04	05	07	09	12	14	15	13	10	06
163	η	Leporis	S 14 10	22	28	32	32	30	26	20	14	09	09	12	18	25
164	γ	Columbae	S 35 16	73	82	86	87	84	77	69	61	55	54	58	66	76
165	δ	Aurigae	N 54 16	63	68	71	72	69	64	58	53	50	50	51	55	60
166	β	Aurigae	N 44 56	45	49	52	52	49	46	42	38	36	36	37	39	43
167	θ	Aurigae	N 37 12	41	44	46	45	44	41	38	36	35	35	35	36	38
168	η	Columbae	S 42 48	64	73	78	79	76	69	60	51	45	44	48	56	67
169	ι	Geminorum	N 23 15	54	54	54	54	54	53	53	53	53	53	52	51	51
170	ν	Orionis	N 14 46	17	16	15	15	15	16	17	18	19	20	18	16	14
171	η	Geminorum	N 22 30	51	51	51	51	51	50	50	50	50	50	49	48	47
172	κ	Aurigae	N 29 30	27	28	29	29	28	27	25	24	23	23	22	22	23
173	2	Lyncis	N 59 01	15	22	26	27	24	19	13	07	03	01	02	06	11
174	ζ	Canis Majoris	S 30 02	71	79	84	86	83	78	70	62	57	55	58	66	75
175	δ	Columbae	S 33 25	29	37	43	44	42	36	28	20	14	12	16	24	33
176	μ	Geminorum	N 22 31	35	35	35	35	35	34	34	34	34	33	32	31	30
177	β	Canis Majoris	S 17 56	39	46	50	51	49	45	39	32	28	27	29	36	44
178	ε	Monocerotis	N 4 36	19	16	14	13	14	16	19	22	24	24	22	18	14
179	*Canopus* (α Carinae)		S 52 40	59	69	76	78	75	68	59	50	43	40	44	53	64
180	ν	Geminorum	N 20 13	40	39	39	39	39	39	39	40	40	39	38	36	35
181	γ	Geminorum	N 16 25	13	12	11	11	11	12	13	13	14	13	12	09	07
182	ν	Puppis	S 43 10	29	39	46	48	46	40	31	22	16	13	16	25	35
183	ε	Geminorum	N 25 09	20	20	21	21	21	20	20	19	18	17	15	14	13
184	ξ	Geminorum	N 12 55	19	17	17	16	17	18	19	20	21	21	19	16	13
185	*Sirius* (α CMa)		S 16 40	59	66	70	72	70	66	61	55	50	49	52	59	66
186	α	Pictoris	S 61 54	56	67	74	78	76	70	61	51	43	40	43	51	62
187	κ	Canis Majoris	S 32 28	49	58	64	66	65	60	52	44	38	36	39	46	56
188	τ	Puppis	S 50 34	69	79	86	89	87	82	73	63	56	53	56	64	75
189	θ	Geminorum	N 33 59	27	29	31	32	32	30	28	25	23	21	20	19	20
190	θ	Canis Majoris	S 12 00	29	35	39	40	39	36	31	25	21	20	23	29	36
191	ε	Canis Majoris	S 28 56	19	28	34	36	35	30	23	16	10	08	10	17	27
192	σ	Canis Majoris	S 27 53	59	67	73	76	74	70	63	55	50	48	50	57	66
193	o^2	Canis Majoris	S 23 47	51	59	64	66	65	61	55	48	42	40	43	50	58
194	ζ	Geminorum	N 20 36	22	22	22	22	23	23	23	22	22	21	19	16	14
195	γ	Canis Majoris	S 15 35	49	56	60	62	61	57	52	46	42	40	43	49	57
196	δ	Canis Majoris	S 26 21	15	23	29	31	30	26	19	12	07	04	07	14	23
197*	$γ^2$	Volantis	S 70 27	34	45	54	58	57	52	43	34	25	21	23	31	42
198	δ	Monocerotis	S 0 27	08	12	15	16	15	13	09	06	04	03	06	11	16
199	I	Puppis	S 46 42	65	76	83	87	86	81	73	64	56	53	55	63	74
200	ω	Canis Majoris	S 26 43	48	56	62	65	64	60	53	46	40	38	40	47	56

* No., mag., dist. and p.a. of companion star: **197**, 5·8, 14″, 299°

Astronomy for surveyors

APPARENT PLACES OF STARS, 1974
AT UPPER TRANSIT AT GREENWICH

No.	1115		1116		154		1117	
Name	43 Tauri		44 Tauri		o¹ Eridani		μ Persei	
Mag. Spect.	5.67	G5	5.55	F0	4.14	F2	4.28	G0
U.T.	R.A.	Dec.	R.A.	Dec.	R.A.	Dec.	R.A.	Dec.
	$4^h 07^m$	$+19° 32'$	$4^h 09^m$	$+26° 24'$	$4^h 10^m$	$-6° 53'$	$4^h 12^m$	$+48° 20'$
I -4.1	41.275 $_{+19}\atop{-23}$	35.70 $_{-11}\atop{-15}$	17.138 $_{+22}\atop{-23}$	58.39 $_{+28}\atop{+20}$	37.792 $_{+5}\atop{-34}$	75.30 $_{-153}\atop{-141}$	62.130 $_{+16}\atop{-46}$	51.36 $_{+152}\atop{+134}$
I 5.9	41.252 $_{-61}$	35.55 $_{-19}$	17.115 $_{-64}$	58.59 $_{+12}$	37.758 $_{-69}$	76.71 $_{-124}$	62.084 $_{-100}$	52.70 $_{+111}$
I 15.9	41.191 $_{-99}$	35.36 $_{-23}$	17.051 $_{-104}$	58.71 $_{+2}$	37.689 $_{-104}$	77.95 $_{-107}$	61.984 $_{-156}$	53.81 $_{+84}$
I 25.8	41.092 $_{-131}$	35.13 $_{-29}$	16.947 $_{-138}$	58.73 $_{-11}$	37.585 $_{-133}$	79.02 $_{-86}$	61.828 $_{-200}$	54.65 $_{+52}$
II 4.8	40.961 $_{-153}$	34.84 $_{-34}$	16.809 $_{-163}$	58.62 $_{-23}$	37.452 $_{-154}$	79.88 $_{-63}$	61.628 $_{-233}$	55.17 $_{+19}$
II 14.8	40.808 $_{-172}$	34.50 $_{-39}$	16.646 $_{-181}$	58.39 $_{-35}$	37.298 $_{-170}$	80.51 $_{-41}$	61.395 $_{-257}$	55.36 $_{-17}$
II 24.7	40.636 $_{-175}$	34.11 $_{-43}$	16.465 $_{-185}$	58.04 $_{-47}$	37.128 $_{-175}$	80.92 $_{-16}$	61.138 $_{-262}$	55.19 $_{-51}$
III 6.7	40.461 $_{-169}$	33.68 $_{-45}$	16.280 $_{-180}$	57.57 $_{-57}$	36.953 $_{-169}$	81.08 $_{+6}$	60.876 $_{-253}$	54.68 $_{-83}$
III 16.7	40.292 $_{-156}$	33.23 $_{-46}$	16.100 $_{-164}$	57.00 $_{-64}$	36.784 $_{-158}$	81.02 $_{+32}$	60.623 $_{-233}$	53.85 $_{-112}$
III 26.7	40.136 $_{-127}$	32.77 $_{-43}$	15.936 $_{-136}$	56.36 $_{-67}$	36.626 $_{-132}$	80.70 $_{+56}$	60.390 $_{-193}$	52.73 $_{-135}$
IV 5.6	40.009 $_{-94}$	32.34 $_{-38}$	15.800 $_{-100}$	55.69 $_{-68}$	36.494 $_{-102}$	80.14 $_{+78}$	60.197 $_{-147}$	51.38 $_{-152}$
IV 15.6	39.915 $_{-54}$	31.96 $_{-30}$	15.700 $_{-57}$	55.01 $_{-64}$	36.392 $_{-66}$	79.36 $_{+102}$	60.050 $_{-90}$	49.86 $_{-164}$
IV 25.6	39.861 $_{-6}$	31.66 $_{-18}$	15.643 $_{-8}$	54.37 $_{-56}$	36.326 $_{-22}$	78.34 $_{+124}$	59.960 $_{-25}$	48.22 $_{-168}$
V 5.6	39.855 $_{+41}$	31.48 $_{-5}$	15.635 $_{+41}$	53.81 $_{-46}$	36.304 $_{+21}$	77.10 $_{+143}$	59.935 $_{+39}$	46.54 $_{-164}$
V 15.5	39.896 $_{+89}$	31.43 $_{+10}$	15.676 $_{+92}$	53.35 $_{-31}$	36.325 $_{+67}$	75.67 $_{+162}$	59.974 $_{+107}$	44.90 $_{-157}$
V 25.5	39.985 $_{+137}$	31.53 $_{+26}$	15.768 $_{+142}$	53.04 $_{-16}$	36.392 $_{+111}$	74.05 $_{+175}$	60.081 $_{+173}$	43.33 $_{-141}$
VI 4.5	40.122 $_{+178}$	31.79 $_{+41}$	15.910 $_{+186}$	52.88 $_{+2}$	36.503 $_{+152}$	72.30 $_{+186}$	60.254 $_{+230}$	41.92 $_{-123}$
VI 14.4	40.300 $_{+218}$	32.20 $_{+57}$	16.096 $_{+227}$	52.90 $_{+18}$	36.655 $_{+189}$	70.44 $_{+192}$	60.484 $_{+285}$	40.69 $_{-101}$
VI 24.4	40.518 $_{+252}$	32.77 $_{+71}$	16.323 $_{+262}$	53.08 $_{+36}$	36.844 $_{+223}$	68.52 $_{+191}$	60.769 $_{+332}$	39.68 $_{-75}$
VII 4.4	40.770 $_{+275}$	33.48 $_{+82}$	16.585 $_{+288}$	53.44 $_{+50}$	37.067 $_{+247}$	66.61 $_{+187}$	61.101 $_{+368}$	38.93 $_{-49}$
VII 14.4	41.045 $_{+298}$	34.30 $_{+90}$	16.873 $_{+311}$	53.94 $_{+64}$	37.314 $_{+269}$	64.74 $_{+177}$	61.469 $_{+399}$	38.44 $_{-22}$
VII 24.3	41.343 $_{+309}$	35.20 $_{+95}$	17.184 $_{+324}$	54.58 $_{+75}$	37.583 $_{+283}$	62.97 $_{+158}$	61.868 $_{+418}$	38.22 $_{+6}$
VIII 3.3	41.652 $_{+315}$	36.15 $_{+96}$	17.508 $_{+330}$	55.33 $_{+81}$	37.866 $_{+289}$	61.39 $_{+138}$	62.286 $_{+429}$	38.28 $_{+30}$
VIII 13.3	41.967 $_{+318}$	37.11 $_{+95}$	17.838 $_{+334}$	56.14 $_{+87}$	38.155 $_{+294}$	60.01 $_{+112}$	62.715 $_{+435}$	38.58 $_{+55}$
VIII 23.3	42.285 $_{+312}$	38.06 $_{+90}$	18.172 $_{+327}$	57.01 $_{+89}$	38.449 $_{+289}$	58.89 $_{+80}$	63.150 $_{+430}$	39.13 $_{+78}$
IX 2.2	42.597 $_{+303}$	38.96 $_{+81}$	18.499 $_{+319}$	57.90 $_{+87}$	38.738 $_{+281}$	58.09 $_{+48}$	63.580 $_{+421}$	39.91 $_{+97}$
IX 12.2	42.900 $_{+292}$	39.77 $_{+72}$	18.818 $_{+307}$	58.77 $_{+85}$	39.019 $_{+270}$	57.61 $_{+11}$	64.001 $_{+407}$	40.88 $_{+117}$
IX 22.2	43.192 $_{+273}$	40.49 $_{+60}$	19.125 $_{+289}$	59.62 $_{+79}$	39.289 $_{+253}$	57.50 $_{-23}$	64.408 $_{+384}$	42.05 $_{+132}$
X 2.1	43.465 $_{+256}$	41.09 $_{+48}$	19.414 $_{+271}$	60.41 $_{+73}$	39.542 $_{+234}$	57.73 $_{-56}$	64.792 $_{+361}$	43.37 $_{+145}$
X 12.1	43.721 $_{+234}$	41.57 $_{+37}$	19.685 $_{+248}$	61.14 $_{+68}$	39.776 $_{+214}$	58.29 $_{-88}$	65.153 $_{+331}$	44.82 $_{+157}$
X 22.1	43.955 $_{+208}$	41.94 $_{+25}$	19.933 $_{+221}$	61.82 $_{+61}$	39.990 $_{+187}$	59.17 $_{-115}$	65.484 $_{+294}$	46.39 $_{+165}$
XI 1.1	44.163 $_{+182}$	42.19 $_{+15}$	20.154 $_{+193}$	62.43 $_{+54}$	40.177 $_{+161}$	60.32 $_{-135}$	65.778 $_{+256}$	48.04 $_{+171}$
XI 11.0	44.345 $_{+151}$	42.34 $_{+8}$	20.347 $_{+161}$	62.97 $_{+49}$	40.338 $_{+131}$	61.67 $_{-151}$	66.034 $_{+211}$	49.75 $_{+175}$
XI 21.0	44.496 $_{+116}$	42.42 $_{+0}$	20.508 $_{+124}$	63.46 $_{+44}$	40.469 $_{+97}$	63.18 $_{-159}$	66.245 $_{+159}$	51.50 $_{+173}$
XI 30.9	44.612 $_{+81}$	42.42 $_{-4}$	20.632 $_{+88}$	63.90 $_{+38}$	40.566 $_{+63}$	64.77 $_{-162}$	66.404 $_{+108}$	53.23 $_{+169}$
XII 10.9	44.693 $_{+40}$	42.38 $_{-9}$	20.720 $_{+43}$	64.28 $_{+32}$	40.629 $_{+25}$	66.39 $_{-159}$	66.512 $_{+47}$	54.92 $_{+160}$
XII 20.9	44.733 $_{-1}$	42.29 $_{-13}$	20.763 $_{+1}$	64.60 $_{+25}$	40.654 $_{-12}$	67.98 $_{-148}$	66.559 $_{-12}$	56.52 $_{+144}$
XII 30.9	44.732 $_{-40}$	42.16 $_{-17}$	20.764 $_{-41}$	64.85 $_{+17}$	40.642 $_{-50}$	69.46 $_{-136}$	66.547 $_{-70}$	57.96 $_{+126}$
XII 40.9	44.692 $_{-81}$	41.99 $_{-21}$	20.723 $_{-85}$	65.02 $_{+7}$	40.592 $_{-87}$	70.82 $_{-119}$	66.477 $_{-129}$	59.22 $_{+101}$
Mean Place	38.882	30.20	14.614	51.73	35.650	75.67	58.876	41.54
sec δ, tan δ	+1.061	+0.355	+1.117	+0.497	+1.007	−0.121	+1.505	+1.124
dα(ψ), dδ(ψ)	+0.069	+0.19	+0.073	+0.18	+0.058	+0.18	+0.088	+0.18
dα(ε), dδ(ε)	−0.011	+0.88	−0.015	+0.89	+0.004	+0.89	−0.034	+0.89
Dble. Trans.	November 23		November 23		November 24		November 24	

58

131000 NUMBER	MAGNITUDES m_{pg}	m_v	EPOCH 1950 α_{1950} h m s	μ s	σ_μ ″ 001	δ_{1950} ° ′ ″	μ' ″	$\sigma_{\mu'}$ ″ 001	σ_{1950} ″	ORIGINAL EPOCH α_2 s	σ ″ 01	ep.	δ_2 ″	σ' ″ 01	ep.	SP.	SOURCE CAT.	STAR NUMBER	- 0° 4H DM NUMBER
1		5.6T	4 7 56.008	-0.0006	03	-7 3 11.61	-0.012	03	0.19	56.040	09	94.3	10.94	09	93.6	G5	GC	5027	B* -7 758
2		8.4A	7 57.293	-0.0016	11	-4 3 52.48	-0.065	11	0.23	57.319	16	33.8	51.42	16	33.8	F8	17	1049	A -4 751
3		7.5T	8 15.656	0.0040	13	-5 0 6.23	-0.128	14	0.68	15.460	17	01.2	0.22	17	02.9	G0	GC	5033K	B* -5 841
4		8.5A	8 22.763	0.0011	13	-4 26 3.50	-0.049	13	0.27	22.746	16	33.8	2.70	16	33.8	G5	17	1052	A -4 753
5		5.9T	8 23.099	0.0021	06	-8 56 55.54	0.010	05	0.24	23.014	09	09.4	55.95	09	08.5	K0	GC	5035	-9 837
6		9.2T	8 24.189	0.0012	16	-2 37 5.50	-0.037	17	0.73	24.137	18	07.1	3.92	18	07.5	A2	GC	5036	-2 826
7		8.9A	8 25.126	-0.0021	13	-7 19 50.37	0.004	13	0.27	25.159	16	33.8	50.44	16	33.8	K0	16	1038	A -7 761
8		7.8A	8 36.838	-0.0034	15	-5 23 54.89	-0.048	13	0.28	36.893	16	33.8	54.12	16	33.8	K0	17	1053	A -5 843
9		9.0A	8 40.664	0.0023	13	-3 30 4.06	0.016	13	0.27	40.627	16	33.8	4.31	16	33.8	G2 F	17	1054	A -3 702
10		8.5A	8 49.969	0.0013	12	-3 5 35.17	-0.010	12	0.25	49.949	16	33.8	35.01	16	33.8	K2	17	1055	A -3 703
11		8.5A	8 57.658	-0.0009	12	-3 23 2.75	0.001	12	0.25	57.673	16	33.8	2.77	16	33.8	K2	17	1056	A -3 704
12		6.7T	8 58.394	-0.0004	11	-0 33 3.29	-0.013	10	0.61	58.416	17	94.9	2.56	18	93.5	K0	GC	5049	-0 653
13		9.0A	9 0.903	-0.0017	12	-3 52 42.82	-0.048	12	0.25	0.931	16	33.8	42.04	16	33.8	G2 F	17	1057	A -4 757
14		9.1A	9 3.620	-0.0007	12	-6 5 51.59	-0.031	12	0.25	3.630	16	33.8	51.09	16	33.8	F0	16	1039	A -6 840
15		8.8A	9 6.362	0.0003	11	-7 5 46.15	-0.029	11	0.23	6.357	16	33.8	45.68	16	33.8	MB	16	1040	A 7 763
16		9.0A	9 8.399	-0.0049	18	-9 57 1.41	0.002	20	0.35	8.478	18	33.8	1.45	16	33.8	K5 J	11	994	A -10 856
17		8.8A	9 9.631	-0.0006	09	-0 1 41.12	-0.038	11	0.21	9.638	17	37.7	40.65	17	37.7	K0	21	903	A -0 654
18		9.0A	9 16.763	-0.0001	11	-3 47 37.98	-0.014	11	0.23	16.765	16	33.8	37.75	16	33.8	F0 F	17	1058	A -4 758
19		4.1H	9 25.339	0.0003	01	-6 57 59.74	0.083	01	0.04	25.329	02	13.3	2.83	02	12.7	F2	F4	154	-7 764
20		6.6T	9 31.501	0.0040	10	-8 57 54.91	0.012	11	0.52	31.309	18	02.0	55.49	17	01.7	G5	GC	5058	-9 843
21		9.1A	9 33.385	-0.0005	13	-6 45 40.98	0.015	13	0.27	33.393	16	33.8	41.23	16	33.8	K0 R	16	1042	A -7 765
22		8.9A	9 33.783	-0.0014	12	-8 17 0.18	0.003	12	0.25	33.805	16	33.8	0.23	16	33.8	A0	16	1044	A -8 807
23		8.7A	9 37.950	-0.0026	12	-9 24 1.51	0.017	12	0.25	37.992	16	33.8	1.79	16	33.8	G0	16	1045K	A -9 844
24		9.0A	9 46.515	-0.0048	13	-5 42 13.62	0.004	13	0.27	46.593	16	33.8	13.69	16	33.8	K0 R	17	1059	A -5 848
25		6.9T	9 51.330	-0.0011	15	-4 32 16.30	-0.005	16	0.76	51.384	17	01.6	16.05	18	03.1	K5	GC	5065	-4 763
26		8.7A	9 51.823	-0.0024	12	-4 4 18.02	-0.019	12	0.25	51.862	16	33.8	17.71	16	33.8	G0	17	1061	A -4 762
27		8.9A	9 56.179	-0.0003	12	-4 25 21.61	0.006	12	0.25	56.184	16	33.8	21.70	16	33.8	K5 F	17	1062	A -4 765
28		7.6Q	10 1.824	0.0025	12	-2 38 25.70	0.123	11	0.61	1.697	15	99.3	31.99	17	98.8	K0	GC	5069	-2 832
29		9.3A	10 4.209	-0.0019	12	-6 38 19.48	-0.029	12	0.25	4.240	16	33.8	19.02	16	33.8	F8 R	16	1047	A -6 842
30		9.0A	10 13.430	-0.0013	11	-4 15 36.11	0.001	11	0.23	13.452	16	33.8	36.12	16	33.8	A2	17	1064	A -4 767
31		9.2Q	10 13.820	0.0010	08	-6 57 42.19	-0.026	07	0.42	13.768	17	99.4	40.74	15	94.2	K2	GC	5073K	B* -7 768
32		8.9A	10 17.609	-0.0024	12	-4 8 9.83	0.006	12	0.25	17.648	16	33.8	9.92	16	33.8	F5 F	17	1065	A -4 768
33		7.1T	10 29.113	-0.0007	12	-6 30 49.09	-0.131	11	0.57	29.146	15	01.5	42.64	15	00.7	K0	GC	5078	-6 847
34		9.0A	10 30.112	-0.0015	13	-2 2 54.36	-0.018	13	0.27	30.137	16	33.8	54.07	16	33.8	K0	17	1066	A -2 834
35		9.0A	10 30.578	-0.0006	09	-0 11 59.56	-0.016	11	0.21	30.585	17	37.7	59.36	17	37.7	F5 R	21	906	A -0 656
36		7.8A	10 33.274	-0.0002	13	-4 26 19.10	0.009	13	0.27	33.278	16	33.8	19.24	16	33.8	G5	17	1067	A -4 770
37		7.4A	10 42.816	0.0013	09	-4 0 40.71	0.010	09	0.22	42.795	16	33.8	40.87	16	33.8	MA	17	1068	A -4 771
38		9.0A	10 46.422	-0.0016	17	-4 2 21.56	-0.011	17	0.31	46.448	16	33.8	21.38	16	33.8	M0 F	17	1069	A -4 773
39		8.5A	10 47.185	-0.0019	12	-4 23 40.53	0.089	12	0.25	47.217	16	33.8	41.96	16	33.8	G5	17	1070	A -4 774
40		8.8A	10 55.001	-0.0008	12	-9 35 48.44	-0.039	12	0.25	55.014	16	33.8	47.81	16	33.8	F5	16	1050	A -9 849
41		6.3T	11 5.668	0.0002	07	-1 16 32.18	0.001	08	0.41	5.658	14	99.3	32.23	15	01.0	B5	GC	5097	B* -1 600
42		9.0A	11 8.432	-0.0012	09	-0 19 3.50	-0.028	11	0.21	8.447	17	37.7	3.16	17	37.7	K0 R	21	909	A -0 658
43		9.0A	11 24.324	-0.0011	11	-9 28 42.50	-0.011	11	0.23	24.342	16	33.8	42.32	16	33.8	G0 R	16	1051	A -9 852
44		9.0A	11 36.940	-0.0010	09	-0 28 13.66	-0.006	11	0.21	36.953	17	37.7	13.59	17	37.7	K0 R	21	911	A -0 660
45		8.5A	11 45.825	-0.0013	09	-1 44 53.80	-0.009	11	0.22	45.847	16	33.8	53.66	16	33.8	K0	17	1071	A -1 603
46		8.7A	11 47.778	0.0016	12	-2 37 34.86	-0.186	11	0.24	47.753	16	33.8	31.85	16	33.8	G5	17	1072	A -2 841
47		8.9A	11 49.384	-0.0017	09	-0 28 53.74	-0.006	11	0.21	49.405	17	37.7	53.67	17	37.7	A0	21	913	A -0 662
48		8.4A	11 58.317	0.0001	12	-7 5 57.86	0.001	12	0.25	58.315	16	33.8	57.88	16	33.8	K0	16	1052	A -7 773
49		9.4A	12 0.430	0.0007	11	-8 58 40.45	-0.004	11	0.23	0.418	16	33.8	40.39	16	33.8	K2 R	16	1053	A -9 854
50		7.8A	12 5.518	-0.0000	13	-2 15 52.96	0.013	13	0.27	5.518	16	33.8	53.17	16	33.8	K5	17	1073	A -2 844
51		9.0A	4 12 22.728	0.0005	12	-7 18 43.09	0.008	12	0.25	22.720	16	33.8	43.22	16	33.8	A3	16	1054	A -7 776
52		8.4A	12 23.810	0.0039	13	-5 31 17.78	0.029	13	0.27	23.746	16	33.8	18.25	16	33.8	F8	17	1074	A -5 857
53		8.4A	12 26.139	0.0002	09	-1 7 18.46	-0.039	11	0.21	26.137	17	37.7	17.98	17	37.7	G5	21	915	A -1 604
54		9.0A	12 27.827	-0.0029	09	-0 39 39.32	-0.016	11	0.21	27.862	17	37.7	39.12	17	37.7	K0 R	21	916	A -0 663
55		8.8A	12 38.256	0.0009	09	-1 20 41.55	-0.023	11	0.21	38.245	17	37.7	41.27	17	37.7	K2	21	917	A -1 605
56		9.0A	12 38.940	-0.0016	12	-9 47 2.41	-0.021	12	0.25	38.966	16	33.8	2.07	16	33.8	K7	16	1055	A -10 874
57		9.0A	12 40.167	0.0061	13	-4 32 29.15	-0.097	13	0.27	40.068	16	33.8	27.57	16	33.8	K0 F	17	1075	A -4 782
58		8.9A	12 40.961	-0.0026	12	-9 46 33.61	-0.004	12	0.25	41.003	16	33.8	33.55	16	33.8	A2	16	1056	A -10 875
59		8.4A	12 43.068	-0.0006	13	-3 17 39.67	-0.057	13	0.27	43.078	16	33.8	38.75	16	33.8	F8	17	1076	A -3 725
60		9.0A	12 46.265	0.0051	13	-4 28 42.28	0.091	13	0.27	46.182	16	33.8	43.75	16	33.8	G5 F	17	1077	A -4 784
61		8.6A	12 50.389	-0.0006	12	-9 36 56.30	-0.076	12	0.25	50.398	16	33.8	55.07	16	33.8	K5	16	1058	A -9 856
62		8.9A	12 52.201	-0.0030	11	-7 21 29.23	-0.002	09	0.22	52.249	16	33.8	29.20	16	33.8	A0	16	1057	A -7 779
63		4.5T	12 58.161	-0.1501	02	-7 43 45.89	-3.420	02	0.10	+ 5.679	04	99.9	-53.82	04	99.7	G5	GC	5138K	B* -7 780
64		9.2A	13 2.719	-0.0006	12	-8 19 58.40	-0.055	12	0.25	2.728	16	33.8	57.51	16	33.8	G0 R	16	1059	A -8 814
65		9.7Q	13 3.655	-0.1469	10	-7 44 8.92	-3.440	12	0.54	10.425	17	03.9	-26.16	17	02.7	A	GC	5140K	B* -7 781
66		7.9A	13 4.131	0.0000	09	-0 7 13.14	0.011	11	0.21	4.131	17	37.7	13.27	17	37.7	F0	21	918	A -0 666
67		9.0A	13 5.873	0.0016	12	-4 51 23.70	0.026	12	0.25	5.847	16	33.8	24.11	16	33.8	K0 F	17	1078	A -5 859
68		9.0A	13 7.134	-0.0010	12	-4 57 23.01	0.022	12	0.25	7.151	16	33.8	23.36	16	33.8	M4P F	17	1079	A -5 860
69		8.8A	13 11.316	0.0005	09	-1 42 4.48	-0.030	09	0.22	11.309	16	33.8	4.00	16	33.8	K	17	1080	A -1 607
70		8.5A	13 14.568	-0.0012	11	-3 39 7.97	-0.008	11	0.23	14.588	16	33.8	7.83	16	33.8	A2	17	1081	A -3 727
71		7.6A	13 18.429	0.0023	12	-8 45 37.07	-0.003	12	0.25	18.391	16	33.8	37.02	16	33.8	K0	16	1062	A -8 815
72		9.0A	13 24.822	-0.0004	09	-0 34 36.80	-0.033	11	0.21	24.826	17	37.7	36.39	17	37.7	F5	21	921	A -0 667
73		9.0A	13 25.042	-0.0006	12	-4 34 30.74	-0.025	09	0.23	25.052	16	33.8	30.33	16	33.8	K2 F	17	1082	A -4 786
74		8.6A	13 41.673	-0.0005	11	-3 15 19.87	-0.012	11	0.23	41.682	16	33.8	19.68	16	33.8	F0	17	1083	A -3 730
75		9.0A	13 44.176	0.0002	13	-5 7 47.64	0.007	13	0.27	44.173	16	33.8	47.76	16	33.8	F2 R	17	1084	A -5 863
76		9.0A	13 47.640	0.0002	13	-5 28 42.16	0.010	13	0.27	47.637	16	33.8	42.33	16	33.8	F8 R	17	1085	A -5 864
77		8.8A	13 47.654	-0.0019	13	-8 16 7.01	-0.025	13	0.27	47.684	16	33.8	6.61	16	33.8	G5 R	16	1064	A -8 819
78		8.8A	13 49.283	-0.0003	11	-8 19 25.51	-0.009	11	0.23	49.288	16	33.8	25.37	16	33.8	K0	16	1065	A -8 820
79		9.0A	13 49.310	-0.0061	13	-5 42 21.50	-0.028	13	0.27	49.408	16	33.8	21.05	16	33.8	F8 R	17	1086	A -5 865
80		8.3A	13 57.343	-0.0043	09	-0 41 36.59	-0.190	11	0.21	57.396	17	37.7	34.25	17	37.7	G0	21	925	A -0 668
81		9.0A	14 0.939	-0.0011	09	-4 31 50.73	0.006	09	0.22	0.958	16	33.8	50.82	16	33.8	G2 F	17	1087	A -4 789
82		8.2A	14 4.041	-0.0008	12	-2 29 58.28	-0.023	12	0.25	4.054	16	33.8	57.91	16	33.8	A0	17	1088	A -2 848
83		8.4A	14 6.770	0.0012	12	-6 51 1.45	-0.020	12	0.25	6.751	16	33.8	1.13	16	33.8	G5	16	1066	A -7 785
84		9.4A	14 10.036	-0.0019	13	-2 24 45.55	0.002	13	0.27	10.067	16	33.8	45.58	16	33.8	G3 F	17	1089	A -2 850
85		8.6A	14 12.670	-0.0016	09	-1 15 58.87	-0.022	11	0.21	12.690	17	37.7	58.60	17	37.7	K0	21	926	A -1 609
86		8.9A	14 14.011	-0.0013	09	-0 51 12.47	-0.022	11	0.21	14.027	17	37.7	12.20	17	37.7	A2	21	927	A -1 608
87		8.9A	14 15.510	-0.0008	15	-3 14 42.41	-0.005	15	0.29	15.520	16	33.8	42.33	16	33.8	F5	17	1090	A -3 732
88		8.6A	14 16.637	-0.0008	12	-9 34 35.93	-0.017	12	0.25	16.650	16	33.8	35.66	16	33.8	K0	16	1067	A -9 859
89		9.2A	14 20.036	0.0014	09	-2 24 0.88	-0.002	09	0.22	20.014	16	33.8	0.85	16	33.8	F3 F	17	1091	A -2 852
90		8.9A	14 46.505	0.0011	09	-2 23 44.70	-0.019	09	0.22	46.487	16	33.8	44.39	16	33.8	A2	17	1092	A -2 853
91		9.2A	14 47.682	-0.0033	11	-7 30 58.21	-0.013	11	0.23	47.736	16	33.8	58.00	16	33.8	K0 R	16	1068	A -7 787
92		6.1T	14 52.388	-0.0017	06	-6 35 37.99	0.003	05	0.28	52.466	10	03.2	38.14	10	00.6	K0	GC	5183	-6 862
93		9.0A	14 53.705	-0.0007	13	-1 39 43.08	0.003	13	0.27	53.717	16	33.8	43.11	16	33.8	G0	17	1093	A -1 611
94		8.9A	15 3.608	-0.0008	09	-1 1 4.93	-0.034	11	0.21	3.616	17	37.7	4.51	17	37.7	F5	21	930	A -1 612
95		8.8A	15 9.126	-0.0003	11	-1 58 20.06	-0.037	11	0.23	9.131	16	33.8	19.46	16	33.8	F5	17	1094	A -2 855
96		9.0A	15 13.245	-0.0011	14	-0 17 29.03	-0.025	14	0.24	13.259	17	37.7	28.72	17	37.7	F8 R	21	931	A* -0 675
97		8.3A	15 16.781	-0.0039	13	-5 16 7.57	0.041	13	0.27	16.844	16	33.8	8.24	16	33.8	G0	17	1095	A -5 871
98		9.0A	15 32.872	-0.0016	09	-4 38 26.68	-0.025	09	0.22	32.898	16	33.8	26.27	16	33.8	G0 F	17	1096	A -4 793
99		9.0A	15 34.819	0.0012	09	-0 15 12.50	0.010	11	0.21	34.804	17	37.7	12.62	17	37.7	F8 R	21	932	A -0 677
00		8.9A	15 42.126	0.0005	12	-5 2 8.53	-0.018	12	0.25	42.118	16	33.8	8.24	16	33.8	G8 F	17	1097	A -5 874

Astronomy for surveyors

STAR CHARTS

NORTHERN STARS

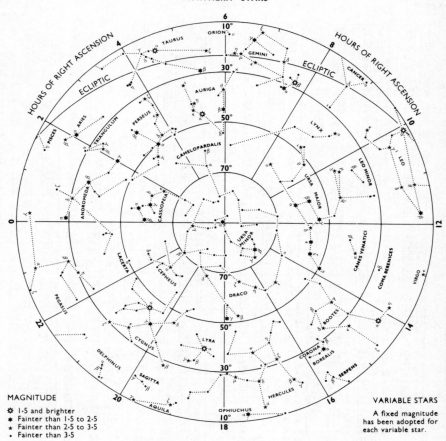

MAGNITUDE
- ☼ 1·5 and brighter
- ✹ Fainter than 1·5 to 2·5
- ★ Fainter than 2·5 to 3·5
- · Fainter than 3·5

VARIABLE STARS
A fixed magnitude has been adopted for each variable star.

EQUATORIAL STARS (R.A. 12ʰ to 24ʰ)

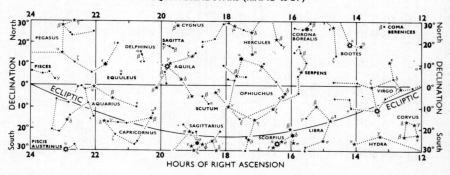

60

The sun and the stars
STAR CHARTS
SOUTHERN STARS

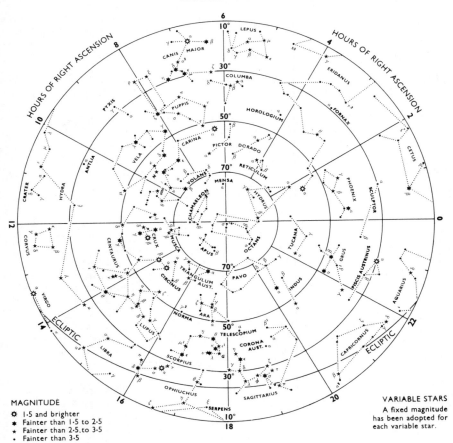

MAGNITUDE
✱ 1·5 and brighter
✶ Fainter than 1·5 to 2·5
★ Fainter than 2·5 to 3·5
· Fainter than 3·5

VARIABLE STARS
A fixed magnitude has been adopted for each variable star.

EQUATORIAL STARS (R.A. 0ʰ to 12ʰ)

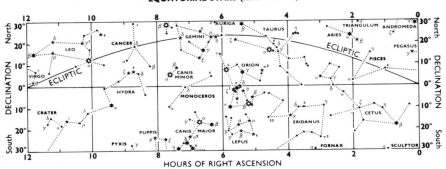

HOURS OF RIGHT ASCENSION

61

References

Astronomisches Rechen-Institut, *Apparent Places of Fundamental Stars*, containing the 1535 stars in the *Fourth Fundamental Catalogue (FK 4)*. Published annually. For sale by Verlag G. Braun, Karl-Friedrich-Strasse 14–18, Karlsruhe, Germany.

Bečvář, A., *Atlas Coeli Skalnaté Pleso* 1950.0, Praha, 1948.

Bečvář, A., *Atlas Eclipticalis 1950.0*, Praha, 1964.

Bečvář, A., *Atlas Australis 1950.0*, Praha, 1964.

(All three above atlases are also published by Sky Publishing Corporation, Harvard, U.S.A.)

Bennett, G. G., Yearly Sets of Fourier Coefficients for the Evaluation of Astronomical Data. *Survey Review*, **24** (189), July 1978.

Bennett, G. G., A Solar Ephemeris for use with Programmable Calculators. *Australian Surveyor*, **30**(3), Sept. 1980.

Boss, B., *General Catalogue of 33,342 Stars for the Epoch 1950*. Carnegie Institution of Washington, U.S.A., 1936, 5 vols. Publication no. 468.

Fricke, W., and Kopff, A., *Fourth Fundamental Catalogue (FK 4)*. Astronomisches Rechen-Institut, Heidelberg 10, 1963.

H.M.S.O., *The Astronomical Ephemeris*. Published annually, London. Published also in the U.S.A. as *The American Ephemeris*, for sale by the Superintendent of Documents, U.S. Govt Printing Office, Washington, D.C. 20402.

H.M.S.O., *Explanatory Supplement to The Astronomical Ephemeris and The American Ephemeris*. London. Also available in the U.S.A. from the Superintendent of Documents, U.S. Govt Printing Office, Washington, D.C. 20402.

H.M.S.O., *The Star Almanac for Land Surveyors*. London. Overseas orders to: The Government Bookshop, P. O. Box 276, London SW8 5DT. Obtainable in U.S.A. from: Pendragon House Inc., 2595 East Bayshore Road, Palo Alto, California 94303; and in Canada from: Pendragon House Ltd., 2525 Dunwin Drive, Mississauga, Ontario L5L 1T2.

Meeus, J., *Tables of Moon and Sun*. Kesselburgh Sterrenwacht (Private Observatory), Kessel-Lo, Belgium, 1962.

Newcomb, S., Tables of the motion of the earth on its axis around the sun. *Astronomical Papers prepared for the use of the American Ephemeris and Nautical Almanac (A.P.A.E.)*, VI(1), Washington, 1895.

Smithsonian Astrophysical Observatory, *Positions and Proper Motions of 258,997 Stars for the Epoch and Equinox of 1950.0*. Smithsonian Publication no. 4652, 1966. For sale by the Superintendent of Documents, U.S. Govt Printing Office, Washington, D.C. 20402 (in sets of 4 volumes only).

United States Department of the Interior, Bureau of Land Management, *The Ephemeris*. For sale by the Superintendent of Documents, U.S. Government Printing Office, Washington, D.C. 20402.

EXERCISES

1. Determine the meridian altitude of the sun at a place in Latitude $30°$, (a) at the equinoxes, (b) during the summer solstice.

Ans. $60°$ and $83° \ 26\frac{1}{2}'$.

The sun and the stars

2. Find the latitude of the place where the greatest altitude of the sun in midsummer is $60°$.

Ans. $53° \, 26\tfrac{1}{2}'$.

3. At a place in Latitude $80°$ N., on a certain day the sun at midday just appears above the horizon. Find the sun's declination. Find also the altitude of the sun at midday when its declination is N. $20°$.

Ans. S. $10°$ and $30°$.

4. On 5 February 1980 an observed star was identified as one in the *SAO* Catalogue whose 1950.0 coordinates were: R.A. $1^h \, 06^m \, 17.251^s$, Dec. $-55° \, 30' \, 45.72''$. The annual values of μ and μ' were $+0.002^s$ and $+0.03''$ respectively. Tabular values extracted from the *Astronomical Ephemeris* for 1980 were:

Besselian Day Numbers (interpolated from the tables for the date and time of observation):

A	B	C	D	E	τ
-1.028	$+7.941$	-13.456	$+14.295$	-0.0011	$+0.0970$

(τ in these tables is the fraction of the year from 1980.0.) The obliquity of the ecliptic, ϵ, was given as $23° \, 26' \, 30.78''$. Reduce the 1950.0 coordinates to the Mean Place for 1980.0 and then to the Apparent Place for 1980 February 5.

Ans. Mean Place 1980.0:
R.A. $1^h \, 07^m \, 32.9^s$.
Dec. $-55° \, 21' \, 08.6''$.
Apparent Place 1980 Feb 5:
R.A. $1^h \, 07^m \, 31.0^s$.
Dec. $-55° \, 21' \, 30''$.

CHAPTER 5
Time

BASIC KINDS OF TIME

5.01 Sidereal time

We have already noted that the earth, as it rotates upon its axis, also moves around the sun in an approximately circular orbit some 299×10^6 kilometres in diameter. The *nearest* star is about 4.1×10^{13} kilometres from the sun. The angle subtended by the radius of the orbit at this star is $\text{arc}^{-1}\{(149.5 \times 10^6)/(4.1 \times 10^{13})\}$, or 0.75 seconds of arc (the star's parallax). From the point of view of the star, the earth oscillates through twice this very small angle every six months, so even from the nearest star the earth appears to be almost stationary. Both star and solar system have proper motions, and these are quite large in terms of kilometres per second, and they are not parallel, but in spite of this the apparent motion of the earth viewed from the nearest star is still extremely small. Most of the stars are a great deal further away than the nearest one, hence their parallaxes are very much smaller and certainly not measurable with survey equipment.

We may therefore picture the earth as being fixed in space, rotating steadily on its axis, with the stars also fixed in space on the celestial sphere an infinite distance away. The overall impression on an observer on the earth will be that the stars rise and set with a perfect regularity which is a measure of the earth's velocity of rotation. The interval of time between successive upper transits of any particular star across the observer's meridian constitutes a *sidereal day* which can be subdivided into sidereal hours, minutes and seconds.

For a long time the monitoring of star transits by large, fixed observatories was the best method of accurate time-keeping. It was not a very convenient time-scale from the civil point of view, since it bears a changing relationship to the rising and setting of the sun, but the astronomers were easily enough able to use it to regulate civil time which is related to the somewhat irregular transits of the sun across the meridian.

The First Point of Aries, or the equinox, moves with the stars, and when it is on the meridian of a place at upper transit the Local Apparent Sidereal Time (L.A.S.T.) is 0^h. However, the equinox does not maintain its position precisely, but is affected to a small degree by nutation. If a mean position of the equinox is taken, we get 0^h Local Mean Sidereal Time (L.M.S.T.) when this is on the

Time

meridian of a place at upper transit. Apparent Sidereal Time minus Mean Sidereal Time gives us the Equation of the Equinoxes which can be found, if required, in Table II of *Apparent Places of Fundamental Stars*. However, for ordinary work the surveyor need concern himself only with Apparent Sidereal Time, which is related to the quantity R on pages 2-25 of the *Star Almanac for Land Surveyors*.

5.02 Atomic time

In recent times physicists have developed atomic clocks which are so accurate that they easily reveal irregularities in the earth's rotational velocity, i.e. in sidereal time as determined by observatories. International Atomic Time (I.A.T.) is thus now the most uniform kind of time available to the world. Its rate is quite independent of that of any form of astronomically determined time, but civil time is based upon it. The Système International (SI) unit of time is the atomic second which was defined in 1967 by the 13th General Conference of Weights and Measures as "the duration of 9 192 631 770 periods of the radiation corresponding to the transition between the two hyperfine levels of the ground state of the caesium atom 133".

Coordinated Universal Time (U.T.C.), upon which civil time is based, is kept at the same rate as I.A.T. but is displaced an integral number of seconds from it; (I.A.T. − U.T.C.) = +19 seconds exactly in January 1980, and gaining at about 1^s per year.

Since civil time must be related to the sun's apparent motion which is somewhat irregular, it follows that U.T.C. must be adjusted slightly at fixed intervals. Alterations to (I.A.T. − U.T.C.) are therefore made by introducing a "leap second" at approximate times announced well in advance, but usually at the last U.T.C. second on either 31 December or 30 June (possibly on 31 March or 30 September). U.T.C. is broadcast by radio time-signal stations and further reference to it will be made later.

5.03 Solar time

It is the day as determined by the sun that controls our habits and rules our lives. The *solar day*, or period of time between successive transits of the sun across the meridian is, however, variable in length, and it is impossible to regulate a clock so that it shall always indicate exactly 12 o'clock just when the sun is on the meridian at upper transit. The reason for this may be seen from a consideration of Fig. 5.1, which shows in an exaggerated way the movement of the earth in its orbital revolution round the sun. Suppose that, when the earth is in the position marked 1, the sun is directly overhead to an observer at A, and that, if it could be seen, the star F would appear in the same direction. As the earth revolves on its axis it also travels forward in its orbit, so that at

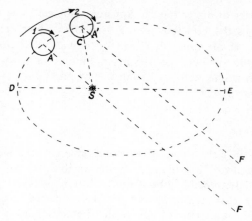

Fig. 5.1

the end of a sidereal day (one complete revolution later) it is in the position marked 2. The observer has been carried round to the point A' so that the same star F appears vertically overhead, $A'F$ being parallel to AF since the star is at practically an infinite distance. The interval between these two positions marks a sidereal day. But to bring the sun overhead to the same observer, he must wait till he is carried round the extra distance $A'C$. The solar day will then be longer than the sidereal day by the length of time the earth requires to turn through this extra distance. Thus, whilst the sidereal day is the time taken by the earth to make a complete revolution on its axis, the solar day is the time taken to make a little more than a revolution. Considering this matter further, it becomes obvious that if the earth makes x revolutions on its axis in following its orbit from position 1 in the diagram, through position 2 and right round back to position 1 (i.e. in one year) it will make $(x-1)$ revolutions with respect to the sun in the same time, since it has circled round the sun once during this period, in the same direction as the direction of its own rotation.

Now the earth does not move in a circular but in an elliptic orbit round the sun, which is at one of the foci, so that sometimes it is nearer to the sun than at others. When it is nearer to the sun, it is a deduction from the law of gravitation that it must travel faster in its path than when it is farther away. The result is that the extra distance, $A'C$, through which the earth has to turn in the interval of time that has to be added on to the sidereal day to give the solar day, is not always the same, and the solar day is therefore not of constant length.* If a celestial body were to move in right ascension, its period of revolution about the pole would still be constant, although not the same as that of

---

* The time between successive upper transits of the sun at any place is called the apparent solar day, and the time kept by the true sun can be called apparent time. See end of **12.15**(b).

Time

the stars, provided the movement was a uniform one. The difficulty with the sun as a timekeeper is that its motion in right ascension is variable.

5.04 Mean solar time (mean time)

The right ascension of the real sun changes by 360° in the course of a year, but the rate of change is not always the same. We might conceive of an imaginary body travelling with the sun, so that its right ascension changes by the same total amount in the course of the whole year, but having its motion in right ascension perfectly uniform. Such an imaginary sun would form an ideal timekeeper; we could regulate our clocks to mark noon when it should be on the meridian, and it would have the great practical advantage that the time so indicated would never be very far different from that of the actual sun. Such an imaginary sun is, in fact, used; it is called the *mean sun*, and the time indicated by it is called *mean solar time*. The mean sun is pictured as moving along the equator with uniform speed, so that its motion is the average of that of the actual sun in right ascension. A *mean solar day* is the interval between two successive transits of the imaginary mean sun across the meridian.

5.05 Equation of Time

The difference between mean solar and solar time at any particular instant is known as the *Equation of Time*. It will thus be seen to be the difference between the hour angles of the true sun and the imaginary mean sun. A graph showing its variation throughout the year is shown in Fig. 5.2. From the graph

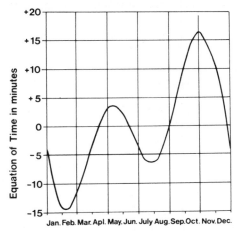

Fig. 5.2   Curve of Equation of Time

The sign is in the sense (Solar − Mean Solar)

Astronomy for surveyors

it will be seen that the maximum negative value of the Equation of Time is $14^m\ 19^s$ on 12 February and the maximum positive value is $16^m\ 24^s$ on 4 November, with two secondary, or smaller extremes between. Thus the difference between Solar and Mean Solar Time is never more than about $16\frac{1}{2}$ minutes. It is zero on four occasions during the year, in mid-April, mid-June, early September and late December.

In field astronomy for surveyors the Equation of Time itself is seldom used nowadays; instead, it is incorporated in the quantity E tabulated in *The Star Almanac for Land Surveyors* on pages 2-25. If, for any reason, it *is* required, it may be obtained by subtracting 12 hours from E; this gives its sign in the sense (Solar Time − Mean Solar Time). Values of the Equation of Time are also tabulated in *The Astronomical Ephemeris* (published annually by H.M.S.O., London). The quantity E is the difference between the Greenwich Hour Angle of the Sun and Universal Time.

The difference between Mean Solar and Solar Time is due to two causes:

(1) The earth does not move in a circle round the sun, but in an ellipse, as already explained and shown in Fig. 5.1. The result is that the motion of the earth in its orbit is not uniform, but varies with its distance from the sun, being fastest when it is nearest.
(2) The imaginary mean sun is assumed to move at uniform speed along the celestial equator, whereas the actual sun moves at variable speed along the ecliptic. It will be seen from a study of Fig. 4.3 that even if the actual sun moved at constant speed, such uniform motion along the ecliptic would not represent uniform motion in right ascension, and therefore would not correspond to uniform motion of the mean sun along the equator.

The second of these causes produces somewhat the greater effect, and the Equation of Time is due to the combined effect of the two.

5.06 Systems of time measurement

From what has already been stated, we see that there are three main systems of measuring time which concern the surveyor:

(1) Star or Sidereal time, as determined by the revolution of the earth on its axis, with respect to the fixed stars, or by the apparent revolution of the stars about the earth.
(2) Atomic time, as defined in **5.02** above, and not dependent on any form of astronomical time. This determines the *rate* of Coordinated Universal Time (U.T.C.) as broadcast by time-signal stations, although there is an integral number of seconds difference between the two. Calendar dates are determined by U.T.C. and there is a changing but determinate relationship between sidereal time and U.T.C.

Time

(3) Sun time, divided into two kinds;
   (a) *Solar*, as measured by the revolution of the earth on its axis with respect to the visible sun (sundial, or apparent time); the days are of variable length, but longer than the sidereal day.
   (b) *Mean Solar*, as measured by the revolution of the earth on its axis with respect to the imaginary mean sun; the days are of equal length, about $3^m\ 56^s$ longer than the sidereal day. This is the kind of time kept by our clocks. Mean solar time results merely from the evening out, over a year, of the irregularities present in solar time, and the two are related to each other by the Equation of Time, or by the quantity E in the *Star Almanac for Land Surveyors* (on pages 2-25).

One of the common types of observation in field astronomy is the determination of the error of a watch or chronometer on the correct local time by observation of the stars or of the sun. The "correct local time" referred to is usually the time kept by the mean sun (mean solar time). Observation of the stars will give us sidereal time, and observation of the visible sun will give us solar time, neither of which is the kind we really want. Therefore the main time problems which arise in field astronomy are:

(A) The conversion of a sidereal time into the corresponding mean solar time at the same instant; and
(B) The conversion of a solar time into the corresponding mean solar time at the same instant.

We have already seen that the Equation of Time, or the quantity called E which incorporates the Equation of Time, is involved in (B) above. The *Star Almanac* tabulates anther quantity called R on pages 2-25 which is required in the conversion (A) above. We shall refer to this later.

5.07 Local sidereal time

The local sidereal time (L.S.T.) for the purposes of this book is L.A.S.T. as explained in **5.01** above. At any place it is reckoned by counting as 0 hours (0^h) the instant when the First Point of Aries (♈) last crossed the meridian of the place at upper transit. It is clear from this that the L.S.T. is the local hour angle of the First Point of Aries. In Chapter 2 we saw that hour angles are customarily measured from the meridian *westwards*; in the same chapter we saw that right ascensions are reckoned *eastwards* from the First point of Aries. Therefore, if say, 3 sidereal hours have elapsed since the First Point of Aries was on the meridian, its hour angle is 3 hours, and the L.S.T. is 3 hours; furthermore, at this instant we have to go back from the First Point of Aries 3 sidereal hours to the east to return to the meridian, so we could say that the right ascension of the meridian was then 3 hours. Thus, L.S.T. = local H.A. of ♈ = R.A. of the meridian.

Astronomy for surveyors

A star whose R.A. was 3 hours would be on the meridian at the particular instant we have been considering, so we see that the L.S.T. is at once given by the R.A. of any star on the meridian at that moment.

As the earth rotates uniformly on its axis from west to east, it follows that the further east a place is situated the sooner will the First Point of Aries cross the meridian, and therefore the *later* will be its L.S.T. than that of places further west. All places on the same meridian of longitude have their 0^h L.S.T. at the same instant, and, as the earth turns, one meridian after another is brought opposite to the First Point of Aries. Thus the interval of time between 0^h L.S.T. at two different places will depend upon their difference of longitude.

As the earth turns through 360° relative to the fixed stars in 24 sidereal hours, it follows that a difference of 15° of longitude corresponds to a difference of 1 sidereal hour in time, 15' of arc to a difference of 1 minute of time, and 15" of arc to 1 second of time.

Thus if we know the longitude and the local sidereal time at one place A, we can readily compute the local sidereal time at any other place B whose longitude is given. We have only to convert the difference of longitude into time, at the rate of 15° per hour, and add this to the time at A if B is to the east, or subtract it if B is to the west, from A.

EXAMPLE. If the longitude of A is 36° 03' 37" E. and the local sidereal time is 11^h 31^m 17^s, find the local sidereal time at B in longitude 3° 27' 13" E. at the same instant.

The difference of longitude = 32° 36' 24".

To convert this into time, we simply divide by 15, giving us, as the difference in time between the two places, 2^h 10^m 25.6^s; or, more conveniently, we use the "Conversion of Time to Arc" table on page 69 of the *Star Almanac*:

$$32° 30' \quad = 2^h \ 10^m$$
$$6' \ 24" = \quad\quad 25.6^s$$
$$\overline{32° \ 36' \ 24" = 2^h \ 10^m \ 25.6^s}$$

As B is to the west from A, this has to be subtracted from 11^h 31^m 17^s, giving us as the L.S.T. at B, 9^h 20^m 51.4^s.

Should one longitude be east from Greenwich and the other west, we must add them instead of subtracting, in order to get the angle between the meridians.

EXAMPLE. If the sidereal time at A, Long. 35° E., is 12^h 30^m, find the sidereal time at the same instant at B, Long. 27° W.

Ans. 8^h 22^m.

Greenwich Sidereal Time (G.S.T.) is just the local sidereal time of the meridian of Greenwich.

Time

5.08 Local mean time

The *local mean time* (L.M.T.) at any place is reckoned by counting as 12 hours (noon) the instant when the mean sun last crossed the meridian of the place at upper transit. We could put this another way by saying that the L.M.T. is reckoned by counting as 0^h (midnight) the instant when the mean sun last crossed the meridian of the place at *lower* transit. Since the mean sun's hour angle is 0^h when it is on the meridian at upper transit, it is clear that

$$\text{L.M.T.} = \text{local H.A. of mean sun} + 12^h.$$

We have seen that our ordinary watch and clock time is the same kind of time as that kept by the mean sun. Both the civil and astronomical days begin at midnight and end the following midnight. In civil practice we often divide the 24 hours into two periods of 12 hours each with noon half-way between; times between midnight and noon are ante-meridian (a.m.) and between noon and midnight they are post-meridian (p.m.). Astronomers, however, do not adopt this division, but reckon the day from 0^h to 24^h. It should be noted that 0^h on, say, 16 September, is the same instant as 24^h on 15 September.

As the earth rotates uniformly on its axis from west to east, it follows that the further east a place is situated the sooner will the sun cross the meridian, and, therefore, the *later* will be the local time than at places further west. All places on the same meridian of longitude have their noons at the same instant, and, as the earth turns, one meridian after another is brought opposite to the mean sun. Thus, the interval of time between the local mean noons at two different places will depend upon their difference of longitude.

As the earth turns evenly through $360°$ with respect to the mean sun in 24 mean solar hours, it follows that a difference of $15°$ of longitude corresponds to a difference of 1 hour of mean solar time, $15'$ of arc corresponds to a difference of 1 minute of time, and $15''$ of arc to a difference of 1 second of time.

Note. The above argument has followed exactly the same line as that under "Local Sidereal Time", and as the earlier part of this chapter has shown that sidereal and solar times are different, the student may feel confused when told that $15°$ of longitude difference is equal to 1 hour of sidereal time and also to 1 hour of mean solar time. However, a little thought will settle the matter. The earth turns uniformly through $360°$ in 24 sidereal hours with respect to the fixed stars, infinitely distant; it also turns uniformly through $360°$ in 24 mean solar hours with respect to the mean sun, but because it is considered as moving uniformly along a circular orbit round the mean sun in the same directional sense as its own rotation, we have seen that it takes a little longer to make the complete revolution with respect to the mean sun. A longitude difference of, say, $90°$ is still one-quarter of a revolution, however, whether it be with respect to the fixed stars or to the mean sun, even though it takes slightly longer to turn through this angle with respect to the mean sun. We may therefore apply a longitude difference to the L.S.T. at one place to get the L.S.T. at the same instant at another place; or the same longitude difference to the L.M.T. at the first place to get the L.M.T. at the second place. (It will be appreciated, of course, that we cannot apply a longitude difference to the L.S.T. at one place in order to get the corresponding L.M.T. at another!)

Thus if we know the longitude and the Local Mean Time at one place A, we can readily compute the L.M.T. at any other place B whose longitude is given. We have only to convert the difference of longitude into time, at the rate of 15° per hour, and add this to the time at A if B is to the east, or subtract it if B is to the west, from A.

EXAMPLE. If the longitude of A is 63° 22′ 40″ E., and the L.M.T. is 5 September, $2^h\ 38^m\ 18^s$ a.m., find the time at B in longitude 16° 46′ 10″ E.

The difference of longitude = 46° 36′ 30″.

Converting this into time, using the table on p. 69 of the *Star Almanac*:

$$
\begin{array}{rl}
46°\ 30' & = 3^h\ 06^m \\
06'\ 30'' & =\ \ \ \ \ \ \ \ \ \ 26.0^s \\
\hline
46°\ 36'\ 30'' & = 3^h\ 06^m\ 26.0^s
\end{array}
$$

As B is to the west from A, this has to be subtracted from the L.M.T. at A:

$$
\begin{array}{rl}
\text{L.M.T. at } A = & 5 \text{ September, } 2^h\ 38^m\ 18^s \\
\text{Longitude difference } = & 3^h\ 06^m\ 26^s \\
\hline
\therefore \text{Corresponding L.M.T. at } B = & 4 \text{ September, } 23^h\ 31^m\ 52^s
\end{array}
$$

Should one longitude be east from Greenwich and the other west, we must add them instead of subtracting, in order to get the angle between the meridians.

EXAMPLE. A ship sails eastwards from New York (Long. 74° W.) on 2 January at 8 a.m. L.M.T. and arrives in Melbourne (Long. 145° E.) at 6 p.m. L.M.T. on 12 February. Find the time occupied by the voyage.

Ans. $40^d\ 19^h\ 24^m$.

5.09 Greenwich Mean Time (Universal Time)

The Local Mean Time on the meridian of Greenwich (Longitude 0°) was known for many years as Greenwich Mean Time (G.M.T.). However, the Royal Observatory was moved from Greenwich to Herstmonceux Castle in Sussex many years ago, and while the Greenwich meridian (0°) is still almost in its old original position,

Time

it has been effectively redefined in the light of modern work. Although G.M.T. still seems a good enough name, and many will regret its passing as a major break with tradition, the more favoured name today is Universal Time (U.T.), and the time to be used in reducing our observations is U.T.1.

It is not a difficult problem for a large observatory to find its L.A.S.T. by precise observation. If a reliable value of λ, the longitude of the observatory, is known, this L.A.S.T. may be converted to G.A.S.T. (Greenwich Apparent Sidereal Time) by the formula:

$$G.A.S.T. = L.A.S.T. - \lambda.$$

Then G.A.S.T. may be converted to U.T. by use of a table (Table II in *Apparent Places of Fundamental Stars*). This U.T. is known as U.T.0 for the particular observatory.

It has been known for quite a long time that the body of the earth moves slightly in relation to the axis of rotation and that this polar motion causes small periodic changes in the latitudes and longitudes of places. The amount of movement is monitored and this allows a small correction to be made to each observatory's U.T.0 to yield what is called its U.T.1. These values of U.T.1 provided by more than 50 cooperating observatories around the world are used to provide a weighted mean which is being continuously updated. It is this mean value of U.T.1 which the surveyor needs for reducing his observations.

It would be inconvenient, however, if the surveyor had to wait for published values of U.T.1 before he could arrive at his results for ordinary (i.e. non-geodetic) work. It is not difficult for observatories to predict U.T.1 to the nearest 0.1^s, and in order to help surveyors and others who may need it, the time-signal stations code their transmission of U.T.C. in a simple manner so that U.T.1 may be derived immediately from it to this order of accuracy. More will be said about this in Chapter 6 where time-signals are discussed.

From now onwards in this book, U.T. will be used for U.T.1, as is done in the *Star Almanac for Land Surveyors*, except where it needs to be specified as U.T.1.

5.10 Local Standard Time (Zone Time)

The towns in a particular country, unless they happen to be due north or south of one another, will lie on different meridians of longitude, and therefore they will all have different L.M.T.'s at any one instant. To avoid the confusion arising from such a state of affairs, most countries now adopt the system of using, throughout their territory, the time on one particular meridian passing through, or near, the country, that lies an exact number of hours or half-hours from Greenwich. This is called Local Standard Time (L.Std.T.) or the Zone Time of the country. U.T. (G.M.T.) is the Zone Time for Britain (but note that Britain's civil clocks are 1 hour ahead of G.M.T. during the summer months).

Astronomy for surveyors

For example, the standard time for New Zealand is taken as the time of the meridian 180° E., or 12^h ahead of U.T.C. This meridian does not pass through New Zealand's land mass at all, but lies a little distance to the east of it. However, the east–west extent of the country is not sufficient to make the resulting Zone Time inconvenient for any place. Since the meridian 180° E. lies wholly to the east of New Zealand, the L.M.T.'s of all places in the country are behind the Local Standard Time, i.e. are *slow* on L.Std.T. During the summer months in New Zealand the clocks are advanced one hour, making civil time 13^h fast on U.T.C.

In countries of considerable extent in longitude (e.g. U.S.A. and Australia) one single zone time would cause the times of sunrise and sunset in the widely-separated eastern and western areas to be unrealistic, and such countries are therefore subdivided into two or more zones, each with its own Standard Meridian.

TIME TRANSFORMATIONS

5.11 To change standard time to local mean time

This problem has really been discussed already, for the difference between local standard time and local mean time at any place is that due to the difference of longitude between the given place and the standard time meridian used. For places east of the Standard Meridian, local mean time is later than (ahead of) standard time, and for places to the west the local time is earlier (behind).

EXAMPLE. The local standard time at Dunedin, New Zealand (Long. 170° 31' E.), was $4^h 28^m 11^s$ p.m. What was the L.M.T. at this instant?

$$\begin{aligned}\text{Longitude of Standard Meridian, east} &= 180° \ 00' \ 00'' \\ \text{Longitude of local meridian, east} &= 170° \ 31' \ 00'' \\ \hline \text{Difference} &= 9° \ 29' \ 00''\end{aligned}$$

Convert this into time, using the table on page 69 on the *Star Almanac*:

$$\begin{aligned} 9° \ 15' &= 0^h \ 37^m \\ 14' &= 56^s \\ \hline 9° \ 29' &= 0^h \ 37^m \ 56^s \end{aligned}$$

Since the local meridian lies west of the Standard Meridian, this amount must be subtracted from the L.Std.T. to get L.M.T.

$$\begin{aligned} \text{Given L.Std.T.} &= 16^h \ 28^m \ 11^s \\ \text{Longitude difference, in time} &= 0^h \ 37^m \ 56^s \\ \hline \therefore \text{Required L.M.T.} &= 15^h \ 50^m \ 15^s \\ \text{i.e.} &= 3^h \ 50^m \ 15^s \ \text{p.m.} \end{aligned}$$

Time

To change L.M.T. to L.Std.T. is merely the reverse of the above operation.

The standard time meridian for New Zealand being 180° E., find the local mean time at a place in the city of Dunedin (Long. 170° 30′ 54″ E.) when the standard time is $8^h\ 25^m\ 10^s$ p.m.

Ans. $19^h\ 47^m\ 13.6^s$

The Eastern Standard Time meridian for the United States of America is 75° W. If the local mean time is $10^h\ 17^m\ 18^s$ at a place in the State of New York the longitude of which is 73° 58′ W., find the Eastern Standard Time.

Ans. $10^h\ 13^m\ 10^s$.

5.12 To convert a given interval of mean solar time to sidereal time and vice versa

It will be seen from the consideration of Fig. 5.1 that in the course of its complete orbital revolution round the sun the earth will make exactly one turn less with respect to the sun than it does with respect to the fixed stars. There are approximately $365\frac{1}{4}$ mean solar days in the year, and, therefore, in the same period there are $366\frac{1}{4}$ sidereal days. More exactly, according to Bessel, the year contains 365.242 22 solar days, and hence 365.242 22 solar days = 366.242 22 sidereal days.

Therefore, if m be the measure of any interval in mean time and s the corresponding measure in sidereal time,

$$\frac{m}{s} = \frac{365.242\ 22}{366.242\ 22}$$

Thus, if m be given, s can be found, or vice versa.

Tables which facilitate the reduction are given in the *Star Almanac* near the end, under the heading "Interpolation table for R".

When tables are not used, the simplest way to make computation is as follows:

To convert an interval of mean solar time to sidereal time, *add* 9.8565 seconds for each mean solar hour. Dividing by 60, this gives us 0.1643 second to be added for each minute and 0.0027 second for each second of mean time.

Thus, to convert an interval of $6^h\ 33^m\ 17^s$ of mean solar time into the equivalent interval of sidereal time, we have:

$$\begin{aligned}
6 \times 9.8565 &= 59.139 \\
33 \times 0.1643 &= 5.422 \\
17 \times 0.0027 &= 0.046 \\
\hline
64.607 \text{ seconds} &= 1^m\ 04.6^s
\end{aligned}$$

Astronomy for surveyors

The addition of this to the given solar time gives us $6^h\ 34^m\ 21.6^s$ as the equivalent sidereal interval.

For the ordinary work of surveyors only one decimal place is required and the working should not extend beyond two places.

To convert an interval of sidereal time to the equivalent interval of mean solar time, *subtract* 9.8296 seconds for each sidereal hour. Dividing by 60, we get 0.1638 seconds to be subtracted for each sidereal minute, or 0.0027 second for each second.

Thus, to find the interval of solar time equivalent to an interval of $4^h\ 33^m\ 17^s$ of sidereal time, we have:

$$4 \times 9.8296 = 39.318$$
$$33 \times 0.1638 = 5.405$$
$$17 \times 0.0027 = 0.046$$
$$\overline{44.769}\ \text{seconds}$$

Subtracting this from the given interval of sidereal time gives $4^h\ 32^m\ 32.2^s$ as the equivalent mean solar time interval.

In order to make the conversion by using the interpolation table for R, we must first understand what this quantity R is. On pages 2-25 of the *Star Almanac** we notice that a value of R is given four times a day at the Universal Times 0^h, 6^h, 12^h and 18^h, for every day of the year. At 0^h U.T. on each day, R is really the Local Sidereal Time at Greenwich (G.S.T.), or the local hour angle of the First Point of Aries at Greenwich (G.H.A. Aries). As the days go by, it will be noticed from the table that the value of R is increasing, and if we take the difference between the values of R at the 6-hourly intervals, we will see that the increase is a uniform one of between $0^m\ 59.1^s$ and $0^m\ 59.2^s$ in 6 hours, or between $3^m\ 56.5^s$ and $3^m\ 56.6^s$ per day. If we multiply the mean daily increase of $3^m\ 56.55^s$ by $365\frac{1}{4}$, the number of mean solar days in a year, we shall get 24 hours, which is the one revolution lost during the year as explained in Fig. 5.1. Thus the R table really gives us two things:

(a) A direct connection between sidereal time and mean solar time at 0^h U.T. each day, because the value of R at 0^h U.T. is the G.S.T. or the G.H.A. Aries; in fact, by adding the U.T. to the value of R opposite it in the table, the G.H.A. Aries is obtained at each of the U.T.'s 0^h, 6^h, 12h, 18^h, so that:

$$\text{G.H.A. Aries} = \text{G.S.T.} = \text{U.T.} + R$$

(since R is defined as the difference between G.H.A. Aries and U.T.). This formula holds for *any* U.T., but for a value lying between two adjacent

---

* See extract from *S.A.* on pages 98-101.

Time

U.T.'s in the table we must interpolate for R using the table on pages 70 to 72 of the *Almanac*.*

(b) The uniform gain of sidereal time on mean solar time. We may get this gain for a 24-hour or a 6-hour interval from the main table as has been done above, but for intervals of less than 6 hours we must use the interpolation table for R.

Because we are given the gain of sidereal time on mean solar time in this interpolation table we may use it to convert intervals of mean solar time to sidereal times and vice versa.

Taking the first example at the beginning of this section, let us convert the interval 6^h 33^m 17^s of mean solar time into the equivalent interval of sidereal time, using the interpolation table for R, columns headed "solar" containing bold type figures.

The maximum interval in the table is 6 hours, so we must first find the gain for 6 hours. The table is a "critical" one, in that it gives the values between which the increase in R is exactly 0.1 second. Our first interval of 6 hours is between the 5^h 59^m 27.5^s and 5^h 60^m 04.0^s in the table, and the gain is given as 59.1 seconds. The balance of the given interval, 0^h 33^m 17^s lies between 0^h 33^m 10.5^s and 0^h 33^m 47.0^s and the gain is given as 5.5 seconds.

$$
\begin{aligned}
\text{Gain in } 6^h &= 59.1^s \\
\text{Gain in } 0^h 33^m\ 17^s &= 5.5^s \\
\text{Gain in } 6^h\ 33^m\ 17^s &= 1^m\ 04.6^s \\
\text{Given mean solar interval} &= 6^h\ 33^m\ 17^s \\
\therefore \text{Equivalent sidereal interval} &= 6^h\ 34^m\ 21.6^s
\end{aligned}
$$

The reverse process of converting an interval of sidereal time to mean solar time is done in the same way, but of course there is not a gain but a decrement, which is found in the interpolation table for R under the column headings "sidereal", and subtracted. Taking the second example above, where it was required to convert a sidereal interval of 4^h 33^m 17^s to the equivalent mean time (M.T.) interval, we enter the interpolation table for R, under "sidereal" with 4^h 33^m 17^s:

$$
\begin{aligned}
\text{Given sidereal interval} &= 4^h\ 33^m\ \ 17^s \\
\text{Decrement from table} &= \underline{\ \ \ \ \ \ \ \ \ -44.8} \\
\text{Equivalent M.T. interval} &= 4^h\ 33^m\ \ 32.2^s
\end{aligned}
$$

* See extract from S.A. on pages 102–104.

Astronomy for surveyors

Taking a further example, let us convert a sidereal interval of $10^h\ 38^m\ 47^s$ to a mean time interval:

Decrement for 6^h from table	=	59.0^s
Decrement for $4^h\ 38^m\ 47^s$ from table	=	45.7
Total decrement (for $10^h\ 38^m\ 47^s$)	=	104.7
	=	$-01^m\ 44.7^s$
Given sidereal interval		$10^h\ 38^m\ 47.0^s$
Equivalent M.T. interval	=	$10^h\ 37^m\ 02.3^s$

5.13 Given the local standard time (zone time) at any instant, to determine the local sidereal time

We have already seen that

$$\text{U.T.} + R = \text{G.H.A. Aries} = \text{G.S.T.}$$

The *Star Almanac* is thus designed for the meridian of Greenwich, so for time-conversion problems for the meridian of any other place, we must first apply its longitude to get the time at Greenwich at the same instant; then we can enter the *Star Almanac* to make the required conversion for the Greenwich meridian; finally we must apply the longitude of the place to get the new kind of time at the place concerned. The steps involved in finding the L.S.T. corresponding to a given L.Std.T. are therefore:

(1) Apply the longitude of the Standard (Zone) Meridian to the given L.Std.T. to get the U.T. of the given instant.
(2) Find the value of R corresponding to this U.T. and add it to the U.T. to get G.S.T.
(3) Apply the actual longitude of the meridian of the place to the G.S.T. to get the corresponding L.S.T.

EXAMPLE. Given that the Local Standard Time at Sydney (Long. 151° 12′ 15″ E.) on 16 October 1975 was $8^h\ 43^m\ 52^s$ p.m., find the Local Sidereal Time at the same instant. The Standard Time of Sydney (Australian Eastern Standard Time) is 10^h ahead of U.T. (see Fig. 5.3).

First, change the longitude from arc to time, using the table on page 69 of the *Star Almanac*.

90°	=	$6^h\ 00^m\ 00^s$
61°	=	$4^h\ 04^m\ 00^s$
12′ 15″	=	49^s
151° 12′ 15″	=	$10^h\ 04^m\ 49^s$

Time

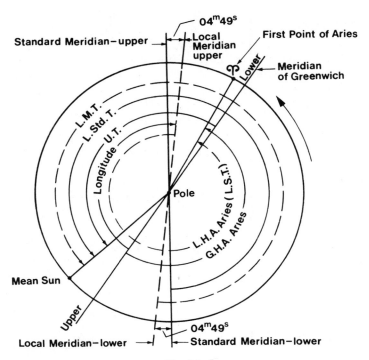

Fig. 5.3

1. Given L.Std.T., 16 October 1975	20ʰ	43ᵐ	52ˢ	1.
2. Longitude of Standard Meridian, E.	−10	00	00	2.
3. Corresponding U.T., 16 October	10	43	52	3.
4. R at U.T. 6^h, 16 October*	1	36	33.0	4.
5. Gain in R in $4^h\ 43^m\ 52^s$ (*Star Almanac*, p. 72)†			46.6	5.
6. U.T. $+R$ = G.S.T. = G.H.A. Aries	12	21	11.6	6.
7. Longitude of Sydney, E.	+10	04	49	7.
8. L.S.T. required = Local H.A. Aries	22	26	00.6	8.

Explanation:
Lines 1, 2 and 3: To the given *Standard* Time we must apply the longitude of the *Standard* Meridian to find the corresponding U.T. In this case, since Greenwich is west of the place, longitude must be subtracted to get U.T.

* See extract from *S.A.* on page 100.
† See extract from *S.A.* on page 104.

Astronomy for surveyors

Line 4: This value is extracted directly from the *Star Almanac* for 16 October 1975 (p. 20); it is taken out for 6^h U.T., since this is the nearest tabulated U.T. below the U.T. obtained in line 3.*

Line 5: This amount is found directly from the Interpolation Table for R on page 72 of the *Star Almanac*.† $4^h\ 43^m\ 52^s$ is the balance of the U.T. beyond 6^h. The sum of the values in lines 4 and 5 ($1^h\ 37^m\ 19.6^s$) really gives the value of R at U.T. $10^h\ 43^m\ 52^s$ on 16 October.

Line 6: The sum of the quantities in lines 3, 4 and 5.

Line 7: This is the longitude of the actual meridian of the place, *not* the longitude of the Standard Meridian; it is added to the G.S.T. because the place is east of Greenwich.

It should be noted from lines 6, 7 and 8 above that we have applied a longitude difference to the G.H.A. Aries to get the Local H.A. of Aries at the same instant.

Note. If, instead of the Local Standard Time, we are given the Local Mean Time and have to find the corresponding Local Sidereal Time, the only difference lies in the first two lines of the working, e.g.:

1. Given L.M.T., 16 October, 1975 $20^h\ 48^m\ 41^s$ 1.
2. Longitude of Sydney, E. $-10\ \ \ 04\ \ \ 49$ 2.
3. Corresponding U.T., 16 October $10\ \ \ 43\ \ \ 52$ 3.

Having thus found the U.T. by applying the actual longitude of the meridian of the place to the given L.M.T., we proceed as in lines 4, 5, 6, 7 and 8 of the original working.

A further example of determining the L.S.T. corresponding to a given L.Std.T. is given below.

EXAMPLE. Find the L.S.T. at Mt. Hamilton (Long. 121° 38′ 44″ W.) on 26 July 1975, the L.Std.T. being $9^h\ 17^m\ 32^s$ p.m. The Standard Meridian is 120° W. (See Fig. 5.4.)

Longitude to time:

$$
\begin{array}{rl}
90° & = 6^h\ 00^m\ 00^s \\
31°\ 30' & = 2\ \ \ 06\ \ \ 00 \\
08'\ 44'' & = 35 \\
\hline
121°\ 38'\ 44'' & = \overline{8\ \ \ 06\ \ \ 35}
\end{array}
$$

---

* See extract from *S.A.* on page 100.
† See extract from *S.A.* on page 104.

Time

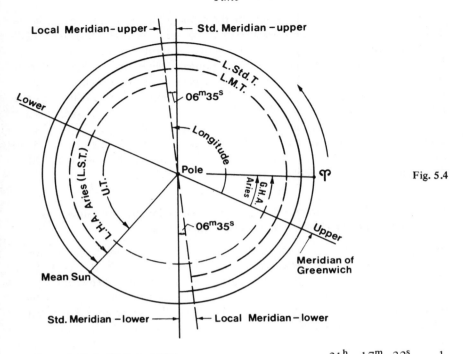

Fig. 5.4

1.	Given L.Std., 26 July 1975	21^h	17^m	32^s	1.
2.	Longitude of Standard Meridian	+8	00	00	2.
3.	Corresponding U.T., 27 July	5	17	32	3.
4.	R at U.T. 0^h, 27 July*	20	16	13.1	4.
5.	Gain in R in $5^h\ 17^m\ 32^s$ (*Star Almanac*, p. 72†)			52.2	5.
6.	U.T. + R = G.S.T. = G.H.A. Aries	25	34	37.3	6.
7.	Longitude of Mt. Hamilton, W.	−8	06	35.0	7.
8.	L.S.T. = L.H.A. Aries	17	28	02.3	8.

The above working follows exactly the same pattern as in the previous example, except that since the place is west of Greenwich the longitude has to be added to local time to get U.T. (lines 1, 2 and 3); later when the G.S.T. has been found, the longitude has to be subtracted to get L.S.T. (lines 6, 7 and 8). It will be noticed that when the values in lines 3, 4 and 5 are added, the sum is greater than 24^h. The actual G.S.T. as shown on a sidereal clock there would be $1^h\ 34^m\ 37.3^s$, since the clock will register only up to 24^h. However, rather than subtract 24^h and then have to add it again in order to be able to subtract the longitude, it is left as $25^h\ 34^m\ 37.3^s$.

* See extract from *S.A.* on page 99.
† See extract from *S.A.* on page 104.

Astronomy for surveyors

There is occasionally confusion in the student's mind about changing the *date* when addition of quantities makes a time-value exceed 4^h, or when 24^h has to be added to a time to allow subtraction of another quantity. This is only the case when we are dealing with Local Standard Time, since it is this kind of time which governs the date; sidereal time has nothing to do with the date. In the above problem, in lines 1, 2 and 3, the addition of the longitude to the L.Std.T. gives a value in excess of 24^h, therefore the U.T. at the same instant must be on the next day at Greenwich; a change of date from 26 to 27 July is necessary, and we must enter the *Star Almanac* on the 27, not the 26. However, because the G.S.T. in line 6 exceeds 24^h, we are not involved in a further date change, since it is *sidereal* time we are now dealing with.

5.14 Given the local sidereal time at a place whose longitude is known, to determine the corresponding local standard time

By application of the longitude to the given L.S.T. we can find the corresponding G.S.T. From the *Star Almanac* we can find a value of U.T. $+R$ ($=$ G.S.T.) at U.T. 0^h, 6^h, 12^h, or 18^h which will lie less than 6 hours below the G.S.T. so obtained. By subtracting the (U.T. $+ R$) value from the first G.S.T. we shall get a *sidereal* interval after U.T. 0^h, 6^h, 12^h, or 18^h. We therefore convert this sidereal interval to a mean solar time interval using the interpolation table for R and add it to the U.T. concerned (0^h, 6^h, 12^h, or 18^h) to get the U.T. corresponding to the L.S.T. originally given. Application of the longitude of the Standard Meridian to this U.T. will now give us the required L.Std.T.

In this type of problem we shall know the date on which the given L.S.T. occurred, and we must finish with a L.Std.T. on the *same* date. This means that we must be careful to enter the *Star Almanac* on the correct date when we are finding our (U.T. $+R$) value at 0^h, 6^h, 12^h or 18^h. It pays to do some simple mental arithmetic before finishing the working from this stage.

The steps involved in finding the L.Std.T. corresponding to a given L.S.T. are, therefore:

(1) Apply the longitude to the given L.S.T. to get the corresponding G.S.T.
(2) In the *Star Almanac* find a tabulated value of U.T. on the date concerned which, when added to R in the same line, will yield a value less than 6 hours below the G.S.T. found in (1) above.
(3) Subtract the tabulated (U.T. $+ R$) obtained in (2) above from the G.S.T. obtained in (1). Convert the resulting sidereal interval into a mean solar time interval using the Interpolation Table for R, and add it to the tabulated U.T. to give the U.T. corresponding to the given L.S.T.
(4) Apply the longitude of the Standard Meridian of the place to this U.T. to give the L.Std.T. corresponding to the given L.S.T.

Step 2 and the first operation of step 3 above (subtracting the tabulated

Time

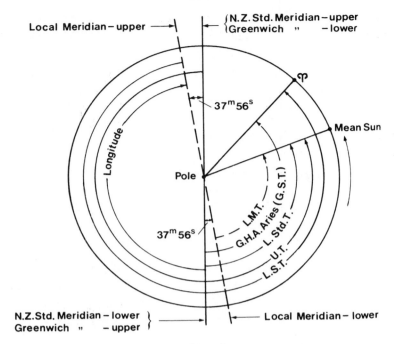

Fig. 5.5

U.T. + *R* from the G.S.T.) should be done mentally. If the sum of G.S.T. − (U.T. + *R*), the U.T. and the longitude of the Standard Meridian when East exceeds 24^h, we must look for a value of U.T. + *R* in step 2 on the date *prior* to that given in the problem. On the other hand, if the sum of G.S.T. − (U.T. + *R*) and the U.T. is less than the longitude of the Standard Meridian when West, we must look for a value of U.T. + *R* in step 2 on the date *after* that given.*

EXAMPLE. Given that the L.S.T. at a place in Longitude E. $11^h\ 22^m\ 04^s$ on 19 April 1975 was $20^h\ 36^m\ 11^s$, find the corresponding L.Std.T. The standard meridian of the place is 12^h E. (See Fig. 5.5.)

1. Given L.S.T., 19 April 1975	20^h	36^m	11^s	1.
2. Longitude, E.	−11	22	04	2.
3. Corresponding G.S.T.	9	14	07	3.
4. Add 24^h (because R > G.S.T.)	24	00	00	4.

* In most practical cases it is likely that the L.Std.T. corresponding to the given L.S.T. will be known to within a few minutes; hence the corresponding date and U.T. for entering the *Star Almanac* can be found without ambiguity.

Astronomy for surveyors

5. G.S.T.	33	14	07	5.
6. U.T. + R at U.T. 18^h, 18 April (*S.A.*, p. 9)	31	44	54.9	6.
7. Sidereal interval after U.T. 18^h, 18 April	1	29	12.1	7.
8. Reduction to mean solar time-interval (*S.A.* p. 70)			−14.6	8.
9. M.T. interval after U.T. 18^h, 18 April	1	28	57.5	9.
10. Add to U.T. 18^h, 18 April	18	00	00	10.
11. U.T. corresponding to given L.S.T., 18 April	19	28	57.5	11.
12. Longitude of Standard Meridian, E.	+12	00	00	12.
13. L.Std.T. corresponding to given L.S.T., 19 April	7	28	57.5	13.

Explanation:

Lines 1, 2 and 3 show the application of the longitude to L.S.T. to obtain G.S.T.

Line 4: When we look in the *Star Almanac* for 19 April 1975, we find that the value of R is about 13^h 47^m. Since this is greater than the G.S.T. in line 3, and later we shall be subtracting U.T. + R from this G.S.T., we must add 24^h to make the subtraction possible.

Lines 6 and 7: Looking in the *Star Almanac* for 19 April, we see that (U.T. + R) for U.T. 18^h gives 31^h 48^m 51.5^s which is less than 6^h below the G.S.T. in line 5. At this stage, however, we mentally subtract 31^h 48^m 52^s from 33^h 14^m 07^s to get about 1^h 25^m; adding this to 18^h (U.T.) and 12^h (the east longitude of the Standard Meridian of the place), we get approximately 31^h 25^m, a figure which exceeds 24^h. Thus we must take U.T. + R at 18^h U.T. *on 18 April*. This gives 31^h 44^m 54.9^s which, of course, is the Greenwich Sidereal Time at 18^h U.T. on 18 April, and on subtracting it from the G.S.T. in line 4 we get the sidereal interval after 18^h U.T. on 18 April.

Lines 7, 8 and 9: show the conversion of the sidereal interval into a mean solar time interval (see **5.12**). Refer to table on page 102.

Lines 10 and 11: Add the mean solar interval to 18^h U.T. to give the U.T. of the original L.S.T.

Lines 12 and 13: show the application of the Longitude of the Standard Meridian to the U.T. to give the L.Std.T. required *on 19 April*. (Had we used the (U.T. + R) first taken out for 19 April, we would have finished the problem incorrectly with a L.Std.T. on 20 April.)

EXAMPLE. Given that the L.S.T. at a place in longitude E. 7^h 08^m 22^s on 17 July 1975 was 10^h 22^m 46^s, find the corresponding L.Std.T. The Standard Meridian of the place is 7^h E. (See Fig. 5.6.)

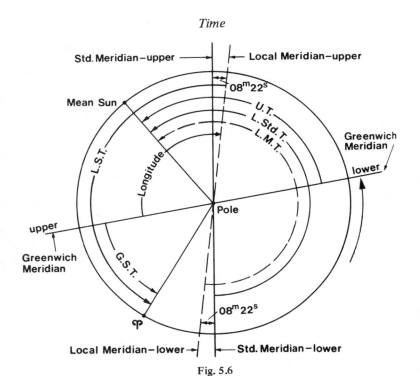

Fig. 5.6

		h	m	s	
1.	Given L.S.T., 17 July 1975	10	22	46	1.
2.	Longitude, E.	−7	08	22	2.
3.	Corresponding G.S.T.	3	14	24	3.
4.	Add 24^h (because $R > $G.S.T.)	24	00	00	4.
5.	G.S.T.	27	14	24	5.
6.	U.T. $+ R$ at U.T. 6^h, 17 July (see p. 99)	25	37	46.7	6.
7.	Sidereal interval after U.T. 6^h, 17 July	1	36	37.3	7.
8.	Reduction to mean solar time-interval (see p. 102)			−15.8	8.
9.	M.T. interval after U.T. 6^h, 17 July	1	36	21.5	9.
10.	Add to U.T. 6h, 17 July	6	00	00	10.
11.	U.T. corresponding to given L.S.T., 17 July	7	36	21.5	11.
12.	Longitude of Standard Meridian, E.	+7	00	00	12.
13.	L.Std.T. corresponding to given L.S.T.	14	36	21.5	13.

Explanation: As for the last example, except that G.S.T. $-$ (U.T. $+ R$) gives approximately $1^h\ 37^m$ which, when added to 6^h (U.T.) and 7^h (east longitude)

Astronomy for surveyors

gives $14^h\ 37^m$; this does not exceed 24^h so we take out U.T. $+R$ for 17 July, the date given in the problem.

EXAMPLE. Given that the L.S.T. at a place in Long. W. $8^h\ 06^m\ 12^s$ on 9 September 1975 was $17^h\ 14^m\ 54^s$, find the corresponding L.Std.T. The Standard Meridian of the place is 8^h West. (See Fig. 5.7.)

1. Given L.S.T., 9 September 1975	17^h	14^m	54^s	1.
2. Longitude, W.	+8	06	12	2.
3. Corresponding G.S.T.	25	21	06	3.
4. U.T. $+ R$ at 0^h U.T., 10 September	23	13	38.0	4.
5. Sidereal interval after 0^h U.T., 10 September	2	07	28.0	5.
6. Reduction to M.T. interval			−20.9	6.
7. M.T. interval after U.T. 0h, 10 September	2	07	07.1	7.
8. U.T. corresponding to given L.S.T., 10 September	2	07	07.1	8.
9. Longitude of Standard Meridian, W.	−8	00	00	9.
10. L.Std.T. corresponding to given L.S.T., 9 September	18	07	07.1	10.

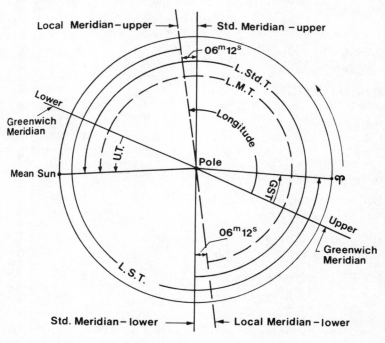

Fig. 5.7

Time

Explanation: As for previous examples, except that G.S.T. − (U.T. + R) = approximately $2^h\ 07^m$, and this plus 0^h (U.T.) gives a value less than 8^h (the west longitude of the Standard Meridian), so we had to find a value of U.T. + R for 10 September, the day after that given in the problem. Several unnecessary lines are also omitted from the working as given in the previous problems.

5.15 To find the Local Standard Time of the upper transit of the First Point of Aries at any place whose longitude is known

This is really a special case of the previous type of problem, since the upper transit of the First Point of Aries over the local meridian marks 0^h L.S.T.

EXAMPLE. Find the L.Std.T. of the upper transit of the First Point of Aries across the meridian of a place in Long. $11^h\ 22^m\ 04^s$ E. on 17 July 1975. The longitude of the Standard Meridian is 12^h E. (See Fig. 5.8.)

1. Given L.S.T. 17 July 1975	0^h	00^m	00^s	1.
2. Longitude, E.	−11	22	04	2.
3. Corresponding G.S.T.	12	37	56	3.
4. Add 24^h, because $R >$ G.S.T.	24	00	00	4.
5. G.S.T. of local transit	36	37	56	5.
6. U.T. + R at 12^h U.T., 16 July	31	34	49.2	6.
7. Sidereal interval after U.T. 12^h, 16 July	5	03	06.8	7.
8. Reduction to M.T. interval			−49.7	8.
9. M.T. interval after U.T. 12^h, 16 July	5	02	17.1	9.
10. Add to U.T. 12^h, 16 July	12	00	00	10.
11. ∴ U.T. of local transit, 16 July	17	02	17.1	11.
12. Longitude of Standard Meridian, E.	12	00	00	12.
13. L.Std.T. of local transit of ♈, 17 July	5	02	17.1	13.

(*Note.* Since {G.S.T. − (U.T. + R)} + 12^h (U.T.) + 12^h (E. Longitude) exceed 24 hours, we must take out U.T. + R for 12^h U.T. on 16 July.)

5.16 Use of the foregoing transformations

It has already been established that the hour angle of a star plus its Right Ascension is equal to the Local Sidereal Time (see **2.09**), i.e.

$$H.A. + R.A. = L.S.T.$$

If, therefore, we know accurately the L.Std.T. when we observe a star, e.g. from

Astronomy for surveyors

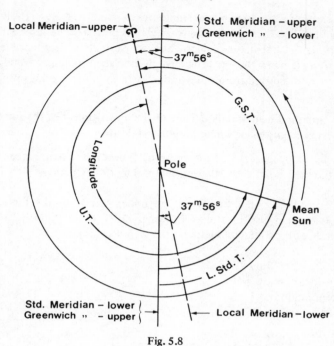

Fig. 5.8

a chronometer, we can find the L.S.T. corresponding to this by the method in **5.13**. From the *Star Almanac* we can find the **R.A.** of the star, so that the **H.A.** may then be determined from

$$\text{H.A.} = \text{L.S.T.} - \text{R.A.}$$

This H.A. may be converted from time into arc to give us one of the angles necessary for solving our spherical triangle.

On the other hand, if the solution of our spherical triangle gives us the hour angle of the star, then by adding this to the R.A. we can find the L.S.T., which in turn leads us to the L.Std.T. via the transformation in **5.14**. In this way we are able to check a watch or chronometer.

For some observations we must know the local time when certain stars are on the meridian. Now when a star is on the meridian its H.A. is zero, so that the equation H.A. + R.A. = L.S.T. reduces to R.A. = L.S.T. Thus we use the transformation in **5.14**, starting with an L.S.T. equal to the R.A. for the star concerned, and finishing with the L.Std.T. required.

Time

5.17 Given the Local Standard Time at a place whose longitude is known, to find the True Sun's Local Hour Angle at the same instant

So far we have been considering the relationships between Standard Time and Sidereal Time, and transformations which would be necessary when we observe stars. Now we turn to the time transformations we must make when we use the sun for observations.

The hour angle of the mean sun with respect to the local meridian, or the Standard Meridian, of a place, plus 12 hours, gives us L.M.T., or L.Std.T. The Equation of Time relates the hour angles of the Mean and True Suns, and it is incorporated in the quantity E in the *Star Almanac*. We would therefore expect that, given a L.M.T. or a L.Std.T. and E, we could find the True Sun's Local Hour Angle, and this is so, because

$$\text{L.H.A. True Sun} = \text{L.M.T.} + E$$

and

$$\text{G.H.A. True Sun} = \text{G.M.T.} + E = \text{U.T.} + E.$$

The second equation is merely a special case of the first. We know that

$$\text{L.M.T.} = \text{L.H.A. Mean Sun} + 12^h$$

and

$$E = \text{Eq. T} + 12^h$$
$$\therefore \text{L.M.T.} + E = \text{L.H.A. Mean Sun} + 12^h + \text{Eq. T} + 12^h$$
$$= (\text{L.H.A. Mean Sun} + \text{Eq. T}) + 24^h$$
$$= \text{L.H.A. True Sun}.$$

(The 24^h may be ignored since it only adds 360° to the hour angle.)

In making these transformations we have usually to find a value of E which lies between the tabulated values in the *Star Almanac*. Unlike R, E does not change uniformly; remembering the curve of the Equation of Time, we realize also that at certain times of the year it is increasing and at others it is decreasing. If we look at the Table of E for, say, 23 November 1975,* we see that, in a column just to the right of and lying between the values of E, are the differences in units of tenths of a second for the 6-hour period (see extract from table on next page).

To perform the interpolation, we use another table in the *Star Almanac* on page 73 called "Interpolation Table for Sun".† The tabular 6-hourly differences head the columns, while time differences are given from 0 to 6^h for

---

* See extract from *S.A.* on page 101.
† See extract from *S.A.* on page 105.

Astronomy for surveyors

U.T.		R	Dec.	E	
d	h			h m s	
23	0			12 13 51.0	
					40
Sun	6			13 47.0	
					41
	12			13 42.9	
					41
	18			13 38.8	
					41
24	0			12 13 34.7	

every 10 minutes down the left-hand side. To find the value of E at, say, U.T. 7^h 16^m 53^s on 23 November we first take the value of E at U.T. 6^h (12 13 47.0). We then enter the Interpolation Table for Sun with a 40 difference along the top and a time interval of 1^h 16^m 53^s down the side. There is no column headed 41, so we take the nearest one to it, viz. 42; we also take the nearest time interval to 1^h 16^m 53^s, namely 1^h 20^m. With these two "arguments" we get a value of 9 from the table. Since the tabular differences are in units of tenths of a second, this 9 represents 0.9 second. A glance at the E table shows that E is decreasing with increase in U.T. Thus we must subtract 0.9 from 12^h 13^m 47.0^s to get the value of E at U.T. 7^h 16^m 53^s (12^h 13^m 46.1^s).

Note. A more precise value of the change in E may, if required, be obtained by interpolation between the values of the tabular differences and the 10-minute time intervals. For instance, in the above example, 0.88 sec. would be closer than 0.9 sec., but for ordinary work it is quite unnecessary to go beyond the nearest tenth of a second.

Since the values of E given in the *Star Almanac* are for the meridian of Greenwich, we must, in these transformations as in those involving R, go from local time to U.T., make the transformation for the Greenwich meridian, and then return to the local meridian.

EXAMPLE. Given that the L.Std.T. at a place in Long. 9^h 47^m 25^s E. was 5^h 16^m 53^s p.m. on 23 November 1975, find the True Sun's Local Hour Angle at this instant. The Standard Meridian is 10^h E. (See Fig. 5.9.)

1. Given L.Std.T., 23 November 1975	17^h 16^m 53^s	1.
2. Longitude of Standard Meridian, E.	10 00 00	2.
3. Corresponding U.T. 23 November	7 16 53	3.
4. E at U.T. 6^h, 23 November (*S.A.*, p. 23)*	12 13 47.0	4.
5. Change in E in 1^h 16^m 53^s (*S.A.*, p. 73)†	−00.9	5.

* See extract from *S.A.* on page 101.
† See extract from *S.A.* on page 105.

Time

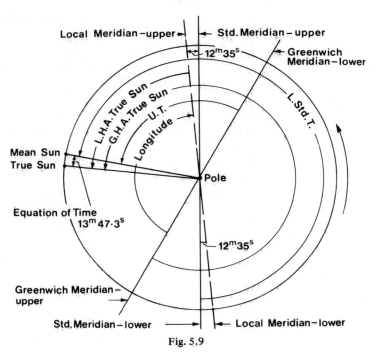

Fig. 5.9

6. E at U.T. 7^h 53^m 53^s, 23 November	12	13	46.1	6.
7. U.T. (line 3)	7	16	53	7.
8. G.H.A. True Sun (= U.T. + E = line 7 + line 6)	19	30	39.1	8.
9. Longitude of place, E.	9	47	25	9.
10. L.H.A. True Sun	29	18	04.1	10.
i.e.	5	18	04.1	

Explanation:

Lines 1, 2 and 3 give the application of the longitude of the Standard Meridian to the given L.Std.T. to obtain the corresponding U.T.

Lines 4, 5 and 6 are concerned with the determination of E at this U.T., from the *Star Almanac* tables (used as illustration earlier).

Line 8 gives the sum of the U.T. and E, which is equal to the Greenwich Hour Angle of the True Sun, and in lines 9 and 10 the longitude of the place is applied to this G.H.A. to yield the L.H.A.

Astronomy for surveyors

5.18 Given the True Sun's Local Hour Angle at a place whose longitude is known, to find the Local Standard Time at the same instant

This is the reverse of the previous operation. By applying the longitude to the given L.H.A. Sun we shall get the G.H.A. Sun which is equal to U.T. + E. If, now, we can find E we shall be able to deduce the U.T. and hence, by application of the longitude of the Standard Meridian, the L.Std.T.

Since E is always about 12^h in value, if the G.H.A. is less than E we must add 24^h to it. Next we look in the table for E on the date concerned for a value of E which, when added to the U.T. opposite to it in the table, yields a sum under 6 hours less than the G.H.A. already found. This sum is then subtracted from the G.H.A. to give us a time-interval; the tabular difference for E is found from the table, and with these two values (the time-interval and the tabular difference) the interpolation table for the sun is entered to obtain a correction for E. The E which we previously found from the table is now corrected by this amount, and the new value is subtracted from the G.H.A. to yield the U.T. Application of the longitude of the Standard Meridian to this U.T. will give the required L.Std.T.

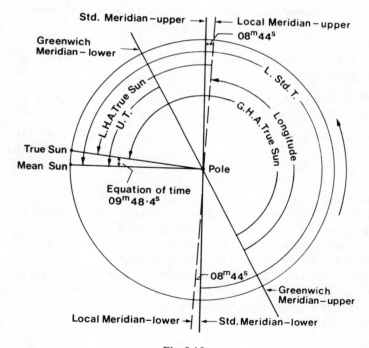

Fig. 5.10

Time

EXAMPLE. At a place in longitude 9^h 51^m 16^s W. the sun's Local Hour Angle at a certain moment on 12 March 1975 was 5^h 42^m 07^s. Calculate the L.Std.T. of this moment. The standard meridian is 10^h W. (See Fig. 5.10.)

1. Given L.H.A. Sun, 12 March 1975	5^h	42^m	07^s	1.
2. Longitude, W.	+9	51	16	2.
3. G.H.A. Sun, 13 March	15	33	23	3.
4. U.T. $+E$ at U.T. 0^h, 13 March*	11	50	12.0	4.
5. G.H.A. Sun $-$ (U.T. $+E$)	3	43	11.0	5.
6. E at U.T. 0^h, 13 March	11	50	12.0	6.
7. Change in E in 3^h 43^m 11.0^s†			+02.4	7.
8. E at moment concerned (line 6 + line 7)	11	50	14.4	8.
9. G.H.A. (line 3) $-E$ (line 8) = U.T. of moment, 13 March	3	43	08.6	9.
10. Longitude of Standard Meridian, W.	-10	00	00	10.
11. L.Std.T. of moment concerned, 12 March	17	43	08.6	11.

Explanation:

Lines 1, 2 and 3 show application of west longitude to given L.H.A. Sun to give corresponding G.H.A. Sun. A little mental arithmetic is necessary here to determine the date for entering the *Star Almanac* to find the (U.T. $+E$) in the next line. By adding 12^h to the given L.H.A. Sun we get the approximate mean solar time of the instant concerned, in this case, about 17^h 42^m on 12 March. When we add the longitude (approximately 9^h 51^m) to this, we get an approximate U.T. on 13 March. Thus we see that, for the next step, we must enter the *Star Almanac* on 13 March.

Line 4: The tabulated U.T. for which U.T. $+E$ on 13 March is less than 6 hours below the G.H.A. Sun in line 3 is 0^h in this case.

Lines 5 and 6 are self-explanatory.

Line 7: This change is found from the Interpolation Table for Sun by entering it with tabular difference 40 and time difference 3^h 43^m 11.0^s as arguments.

Lines 8, 9, 10 and 11 are self-explanatory.

EXAMPLE. At a place in Long. 11^h 28^m 15^s E. the sun's L.H.A. at a certain moment on 31 October 1975 was 20^h 58^m 12^s. Calculate the Local Standard Time of this moment. The Standard Meridian is 12^h E.

1. Given L.H.A. Sun, 31 October 1975	20^h	58^m	12^s	1.
2. Longitude, E.	-11	28	15	2.

* See extract from *S.A.* on page 98.
† See extract from *S.A.* on page 105.

Astronomy for surveyors

3. Corresponding G.H.A. Sun, 30 October	9	29	57	3.
4. Add 24^h, since $E >$ G.H.A. Sun	24	00	00	4.
5. G.H.A. Sun, 30 October	33	29	57	5.
6. U.T. $+ E$ at U.T. 18^h, 30 October	30	16	17.6	6.
7. G.H.A. Sun $-$ (U.T. $+ E$)	3	13	39.4	7.
8. E at U.T. 18^h, 30 October	12	16	17.6	8.
9. Change in E in $3^h\ 13^m\ 39.4^s$			+00.5	9.
10. E at moment concerned (line 8 + line 9)	12	16	18.1	10.
11. G.H.A. (line 5) $- E$ (line 10) = U.T. of moment, 30 October	21	13	38.9	11.
12. Longitude of Standard Meridian, E.	+12	00	00	12.
13. L.Std.T. of moment concerned, 31 October	9	13	38.9	13.

Explanation:
This operation follows the same pattern as in the previous example, except that the given L.H.A. Sun $+ 12^h$ gives the approximate mean solar time as 9 a.m. on 31 October 1975. When the east longitude is subtracted from this we get an approximate U.T. of $21^h\ 30^m$ on 30 October, so that the U.T. $+ E$ in line 6 must be taken on 30 October if we are to finish with a L.Std.T. on 31 October.

The student should draw the circular diagram himself as an exercise.

5.19 To find the Local Standard Time of the upper transit of the sun across the meridian on any particular date

The sun will transit the meridian when the Local Mean Time $= 24^h - E$, because

$$24^h - E = 24^h - (12^h + \text{Eq. T.})$$
$$= 12^h - \text{Eq. T.}$$
$$= 12^h - (\text{L.H.A. True Sun} - \text{L.H.A. Mean Sun}).$$

But at transit, L.H.A. True Sun is zero.

$$\therefore 24^h - E = 12 + \text{L.H.A. Mean Sun (at transit of True Sun)}$$
$$= \text{L.M.T. (at transit of True Sun)}.$$

A quick, but approximate, way of finding the L.M.T. of transit (in any longitude) is to enter the *Star Almanac* on the date concerned, with U.T. $= 12^h$, for the value of E. This is then subtracted from 24^h to get the approximate L.M.T. of transit, and the L.M.T. is changed to L.Std.T. by applying the longitude difference between the local meridian and the Standard Meridian.

Time

EXAMPLE. Find the L.Std.T. of transit of the sun on 11 January 1975 at a place in Long. $8^h\ 16^m\ 10^s$ E. The longitude of the Standard Meridian is 8^h E. (Fig. 5.11).

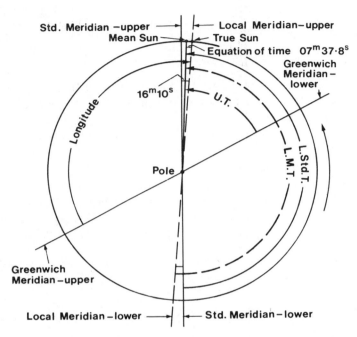

Fig. 5.11

E at U.T. 12^h, 11 January 1975 (*S.A.* p.2) approx.	11^h	52^m	14^s	
$24^h - E$ = approximate L.M.T. of transit	12	07	46	
Standard Meridian slow on L.M.T.		−16	10	
∴ Approximate L.Std.T. of transit	11	51	36	

For greater accuracy we must take the longitude into consideration and utilize the fact that G.H.A. Sun = U.T. + E

1. At upper transit, L.H.A. Sun	24^h	00^m	00^s	1.
2. Longitude E.	−8	16	10	2.
3. ∴ at upper transit, G.H.A. Sun (= U.T. + E)	15	43	50	3.
4. U.T. +E at U.T. 0^h on 11 January 1975	11	52	26.0	4.

Astronomy for surveyors

5. Difference	3^h	51^m	24.0^s	5.
6. Change in E in this difference			−03.8	6.
7. ∴ E at transit	11	52	22.2	7.
8. Line 3 − line 7, i.e. (U.T. +E) − E, = U.T.	3	51	27.8	8.
9. Longitude of Standard Meridian, E.	8	00	00	9.
10. ∴ L.Std.T. of transit	11	51	27.8	10.

Explanation:

Line 6: −03.8^s is obtained from the Interpolation Table for the sun on page 73 of the *Star Almanac*, entering this with the time difference of $3^h\ 51^m\ 24^s$ and the 6-hourly change of 6.1^s.

Line 7: $11^h\ 52^m\ 22.2^s$ = E at U.T. 0^h on 11 January minus 03.8^s.

5.20 Use of transformations in 5.17–5.19

If an accurate L.Std.T. of an observation of the sun has been obtained, we can determine its L.H.A. and so obtain one of the elements necessary in the solution of the astronomical triangle. On the other hand, if the L.H.A. has been calculated from the astronomical triangle, we can find the L.Std.T. and so check a watch or chronometer.

If we wish to observe the sun as it crosses the local meridian we must know the local time of transit as determined in **5.19**.

5.21 Coefficients for computing R, E, sun's declination and semi-diameter

As from 1977 inclusive, the *Star Almanac* tabulates for each year, on page 75, the monthly coefficients for calculating the quantities R, E, Declination of the sun, and Semi-diameter of the sun directly by small programmable computer. This obviates any need for interpolation from the main tables on pages 2–25 of the *S.A.* when frequent use of astronomical observations is being made. An explanation and examples are given in the *S.A.* on page 74.

For those with good computing facilities (e.g. HP-41CV or a microcomputer) it is now easily possible to calculate the solar ephemeris for at least the period to the end of this century. See "A Solar Ephemeris for use with Programmable Calculators" by G. G. Bennett, *The Australian Surveyor*, 1980, **30**(3), p. 147. A suitable program in Level II BASIC, using the algorithm given by Bennett, and double precision, is listed in Appendix II. With input of year, month, day and UT, it computes the sun's R.A., declination, G.H.A. and semi-diameter in about half a minute. A machine language routine would be much quicker. The program could easily be incorporated as a sub-routine in main programs for reducing solar observations. However, if an HP-41CV is available, the ephemeris can be carried in the field in the computer's memory for on-the-spot reductions without the need for tables.

Time

5.22 Reference summary

For convenience of reference, the paragraphs in which various time transformations are described are listed below.

Given	To Find	Section
L.Std.T.	L.M.T.	5.11
Mean Solar Interval	Sidereal Interval and vice versa	5.12
L.Std.T.	L.S.T.	5.13
L.S.T.	L.Std.T.	5.14
Date	L.Std.T. of upper transit of Aries	5.15
L.Std.T.	Sun's L.H.A.	5.17
Sun's L.H.A.	L.Std.T.	5.18
Date	L.Std.T. of sun's upper transit	5.19

5.23 Sample extracts from the Star Almanac

Following are further facsimile pages (slightly reduced) from the *Star Almanac for Land Surveyors* for 1975 and 1984. These cover many of the worked examples included in this chapter and should help the reader to see how the various quantities were obtained.

The author is indebted to the Controller of Her Majesty's Stationery Office, Norwich, England, for kind permission to reproduce this material.

Astronomy for surveyors

SUN—MARCH, 1975

U.T.		R		Dec.		E		U.T.		R		Dec.		E
d	h	h m s		° ′		h m s		d	h	h m s		° ′		h m s
1	0	10 32 42·9	S	7 54·3		11 47 25·4		9	0	11 04 15·4	S	4 49·6		11 49 10·3
Sat.	6	33 42·1		7 48·6 ⁵⁷		47 28·3 ²⁹	Sun.	6	05 14·5		4 43·7 ⁵⁸		49 14·0 ³⁷
	12	34 41·2		7 43·0 ⁵⁶		47 31·2 ²⁹		12	06 13·7		4 37·9 ⁵⁸		49 17·8 ³⁸
	18	35 40·4		7 37·3 ⁵⁷		47 34·0 ²⁸		18	07 12·8		4 32·0 ⁵⁹		49 21·5 ³⁷
				⁵⁷		³⁰						⁵⁹		³⁸
2	0	10 36 39·5	S	7 31·6		11 47 37·0		10	0	11 08 11·9	S	4 26·1		11 49 25·3
Sun.	6	37 38·6		7 25·9 ⁵⁷		47 39·9 ²⁹	Mon.	6	09 11·1		4 20·3 ⁵⁸		49 29·0 ³⁷
	12	38 37·8		7 20·1 ⁵⁸		47 42·9 ³⁰		12	10 10·2		4 14·4 ⁵⁹		49 32·9 ³⁹
	18	39 36·9		7 14·4 ⁵⁷		47 46·0 ³¹		18	11 09·4		4 08·5 ⁵⁹		49 36·7 ³⁸
				⁵⁷		³⁰								³⁸
3	0	10 40 36·1	S	7 08·7		11 47 49·0		11	0	11 12 08·5	S	4 02·6		11 49 40·5
Mon.	6	41 35·2		7 03·0 ⁵⁷		47 52·1 ³¹	Tues.	6	13 07·6		3 56·7 ⁵⁹		49 44·4 ³⁹
	12	42 34·3		6 57·2 ⁵⁸		47 55·2 ³¹		12	14 06·8		3 50·8 ⁵⁹		49 48·3 ³⁹
	18	43 33·5		6 51·5 ⁵⁷		47 58·3 ³¹		18	15 05·9		3 45·0 ⁵⁸		49 52·2 ³⁹
				⁵⁸		³²						⁵⁹		³⁹
4	0	10 44 32·6	S	6 45·7		11 48 01·5		12	0	11 16 05·0	S	3 39·1		11 49 56·1
Tues.	6	45 31·7		6 40·0 ⁵⁷		48 04·7 ³²	Wed.	6	17 04·2		3 33·2 ⁵⁹		50 00·1 ⁴⁰
	12	46 30·9		6 34·2 ⁵⁸		48 07·9 ³²		12	18 03·3		3 27·3 ⁵⁹		50 04·0 ³⁹
	18	47 30·0		6 28·4 ⁵⁸		48 11·2 ³³		18	19 02·4		3 21·4 ⁵⁹		50 08·0 ⁴⁰
				⁵⁷		³²								⁴⁰
5	0	10 48 29·2	S	6 22·7		11 48 14·4		13	0	11 20 01·6	S	3 15·5		11 50 12·0
Wed.	6	49 28·3		6 16·9 ⁵⁸		48 17·7 ³³	Thur.	6	21 00·7		3 09·6 ⁵⁹		50 16·0 ⁴⁰
	12	50 27·4		6 11·1 ⁵⁸		48 21·1 ³⁴		12	21 59·9		3 03·7 ⁶⁰		50 20·1 ⁴¹
	18	51 26·6		6 05·3 ⁵⁸		48 24·4 ³³		18	22 59·0		2 57·7 ⁶⁰		50 24·1 ⁴⁰
				⁵⁸		³⁴						⁵⁹		⁴¹
6	0	10 52 25·7	S	5 59·5		11 48 27·8		14	0	11 23 58·1	S	2 51·8		11 50 28·2
Thur.	6	53 24·9		5 53·7 ⁵⁸		48 31·2 ³⁴	Fri.	6	24 57·3		2 45·9 ⁵⁹		50 32·3 ⁴¹
	12	54 24·0		5 47·9 ⁵⁸		48 34·7 ³⁵		12	25 56·4		2 40·0 ⁵⁹		50 36·4 ⁴¹
	18	55 23·1		5 42·1 ⁵⁸		48 38·1 ³⁴		18	26 55·5		2 34·1 ⁵⁹		50 40·5 ⁴¹
						³⁵						⁵⁹		⁴²
7	0	10 56 22·3	S	5 36·3		11 48 41·6		15	0	11 27 54·7	S	2 28·2		11 50 44·7
Fri.	6	57 21·4		5 30·5 ⁵⁸		48 45·1 ³⁵	Sat.	6	28 53·8		2 22·2 ⁶⁰		50 48·8 ⁴¹
	12	58 20·6		5 24·6 ⁵⁹		48 48·6 ³⁵		12	29 53·0		2 16·3 ⁵⁹		50 53·0 ⁴²
	18	10 59 19·7		5 18·8 ⁵⁸		48 52·2 ³⁶		18	30 52·1		2 10·4 ⁵⁹		50 57·2 ⁴²
				⁵⁸		³⁶								⁴²
8	0	11 00 18·8	S	5 13·0		11 48 55·8		16	0	11 31 51·2	S	2 04·5		11 51 01·4
Sat.	6	01 18·0		5 07·1 ⁵⁹		48 59·4 ³⁶	Sun.	6	32 50·4		1 58·5 ⁶⁰		51 05·6 ⁴²
	12	02 17·1		5 01·3 ⁵⁸		49 03·0 ³⁶		12	33 49·5		1 52·6 ⁵⁹		51 09·8 ⁴²
	18	03 16·2		4 55·4 ⁵⁹		49 06·7 ³⁷		18	34 48·6		1 46·7 ⁵⁹		51 14·1 ⁴³
	24	11 04 15·4	S	4 49·6 ⁵⁸		11 49 10·3 ³⁶		24	11 35 47·8	S	1 40·8 ⁵⁹		11 51 18·3 ⁴²

Sun's S.D. 16ʹ·1   SUNRISE

Date	South Latitude								North Latitude						
	60°	55°	50°	45°	40°	30°	20°	10°	0°	10°	20°	30°	40°	45°	50°	55°	60°
Mar.	h	h	h	h	h	h	h	h	h	h	h	h	h	h	h	h	h
1	5·2	5·4	5·5	5·6	5·7	5·8	6·0	6·1	6·2	6·2	6·3	6·5	6·6	6·7	6·8	7·0
6	5·4	5·5	5·6	5·7	5·8	5·9	6·0	6·1	6·1	6·2	6·3	6·3	6·5	6·6	6·7	6·7
11	5·6	5·7	5·8	5·8	5·9	6·0	6·0	6·1	6·1	6·2	6·2	6·2	6·3	6·3	6·4	6·4	6·5
16	5·8	5·9	5·9	6·0	6·0	6·0	6·0	6·1	6·1	6·1	6·1	6·2	6·2	6·2	6·2	6·2	6·2
21	6·0	6·0	6·0	6·0	6·0	6·1	6·1	6·1	6·1	6·1	6·0	6·0	6·0	6·0	6·0	6·0
26	6·2	6·2	6·2	6·2	6·1	6·1	6·1	6·1	6·0	6·0	6·0	5·9	5·9	5·8	5·8	5·7
31	6·4	6·4	6·3	6·3	6·2	6·2	6·1	6·1	6·0	6·0	5·9	5·8	5·8	5·7	5·7	5·6	5·5

Moon's phases: last quarter, $4^d 20^h 20^m$; new moon, $12^d 23^h 47^m$.

Time

SUN—JULY, 1975

U.T.	R	Dec.	E	U.T.	R	Dec.	E
d h	h m s	° ′	h m s	d h	h m s	° ′	h m s
17 0	19 36 47·5	N 21 20·4	11 53 59·6	25 0	20 08 20·0	N 19 50·2	11 53 33·0
Thur. 6	37 46·7	21 18·0 ^{24}	53 58·2 ^{14}	Fri. 6	09 19·1	19 47·0 ^{32}	53 32·8 $_2$
12	38 45·8	21 15·5 ^{25}	53 56·9 ^{13}	12	10 18·3	19 43·9 ^{31}	53 32·6 $_2$
18	39 44·9	21 12·9 $^{26}_{25}$	53 55·6 $^{13}_{13}$	18	11 17·4	19 40·6 $^{33}_{32}$	53 32·4 $_1$
18 0	19 40 44·1	N 21 10·4	11 53 54·3	26 0	20 12 16·5	N 19 37·4	11 53 32·3
Fri. 6	41 43·2	21 07·8 ^{26}	53 53·1 $_{12}$	Sat. 6	13 15·7	19 34·2 ^{32}	53 32·2 $_1$
12	42 42·4	21 05·2 ^{26}	53 51·9 $_{12}$	12	14 14·8	19 30·9 ^{33}	53 32·1 $_1$
18	43 41·5	21 02·6 $^{26}_{26}$	53 50·7 $^{12}_{11}$	18	15 14·0	19 27·6 $^{33}_{33}$	53 32·1 $_0$
19 0	19 44 40·6	N 21 00·0	11 53 49·6	27 0	20 16 13·1	N 19 24·3	11 53 32·1
Sat. 6	45 39·8	20 57·3 ^{27}	53 48·5 $_{11}$	Sun. 6	17 12·2	19 20·9 ^{34}	53 32·2 $_1$
12	46 38·9	20 54·6 ^{27}	53 47·4 $_{11}$	12	18 11·4	19 17·6 ^{33}	53 32·3 $_1$
18	47 38·1	20 51·9 $^{27}_{27}$	53 46·4 $^{10}_{10}$	18	19 10·5	19 14·2 $^{34}_{34}$	53 32·4 $_1$
20 0	19 48 37·2	N 20 49·2	11 53 45·4	28 0	20 20 09·6	N 19 10·8	11 53 32·6
Sun. 6	49 36·3	20 46·5 ^{27}	53 44·4 $_{10}$	Mon. 6	21 08·8	19 07·4 ^{34}	53 32·8 $_2$
12	50 35·5	20 43·7 ^{28}	53 43·5 $_9$	12	22 07·9	19 04·0 ^{34}	53 33·0 $_2$
18	51 34·6	20 40·9 $^{28}_{28}$	53 42·6 $^9_8$	18	23 07·1	19 00·5 $^{35}_{35}$	53 33·3 $_3$
21 0	19 52 33·8	N 20 38·1	11 53 41·8	29 0	20 24 06·2	N 18 57·0	11 53 33·6
Mon. 6	53 32·9	20 35·3 ^{28}	53 41·0 $_8$	Tues. 6	25 05·3	18 53·6 ^{34}	53 33·9 $_3$
12	54 32·0	20 32·4 ^{29}	53 40·2 $_8$	12	26 04·5	18 50·0 ^{36}	53 34·3 $_4$
18	55 31·2	20 29·5 $^{29}_{29}$	53 39·5 $^7_8$	18	27 03·6	18 46·5 $^{35}_{35}$	53 34·7 $_5$
22 0	19 56 30·3	N 20 26·6	11 53 38·7	30 0	20 28 02·7	N 18 43·0	11 53 35·2
Tues. 6	57 29·5	20 23·7 ^{29}	53 38·1 $_6$	Wed. 6	29 01·9	18 39·4 ^{36}	53 35·7 $_5$
12	58 28·6	20 20·8 ^{29}	53 37·4 $_7$	12	30 01·0	18 35·8 ^{36}	53 36·2 $_6$
18	19 59 27·7	20 17·8 $^{30}_{30}$	53 36·8 $_5$	18	31 00·2	18 32·2 $^{36}_{36}$	53 36·8 $_6$
23 0	20 00 26·9	N 20 14·8	11 53 36·3	31 0	20 31 59·3	N 18 28·6	11 53 37·4
Wed. 6	01 26·0	20 11·8 ^{30}	53 35·7 $_6$	Thur. 6	32 58·4	18 24·9 ^{37}	53 38·0 $_6$
12	02 25·2	20 08·8 ^{30}	53 35·2 $_5$	12	33 57·6	18 21·3 ^{36}	53 38·7 $_7$
18	03 24·3	20 05·8 $^{30}_{31}$	53 34·8 $_4$	18	34 56·7	18 17·6 $^{37}_{37}$	53 39·4 $_8$
24 0	20 04 23·4	N 20 02·7	11 53 34·4	32 0	20 35 55·8	N 18 13·9	11 53 40·2
Thur. 6	05 22·6	19 59·6 ^{31}	53 34·0 $_4$	Fri. 6	36 55·0	18 10·1 ^{38}	53 40·9 $_7$
12	06 21·7	19 56·5 ^{31}	53 33·6 $_4$	12	37 54·1	18 06·4 ^{37}	53 41·8 $_9$
18	07 20·8	19 53·4 ^{31}	53 33·3 $_3$	18	38 53·3	18 02·6 ^{38}	53 42·6 $_8$
24	20 08 20·0	N 19 50·2 ^{32}	11 53 33·0 $_3$	24	20 39 52·4	N 17 58·9 ^{37}	11 53 43·5 $_9$

Sun's S.D. 15ʹ·8   SUNSET

Date	South Latitude							North Latitude						
	60°	55°	50°	45°	40°	30°	20°	10°	0°	10°	20°	30°	40°	45°	50°	55°	60°
July	h	h	h	h	h	h	h	h	h	h	h	h	h	h	h	h	h
1	15·1	15·7	16·1	16·5	16·7	17·2	17·5	17·8	18·1	18·4	18·7	19·1	19·6	19·8	20·2	20·7	21·4
6	15·2	15·8	16·2	16·5	16·8	17·2	17·6	17·9	18·1	18·4	18·7	19·1	19·5	19·8	20·2	20·6	21·4
11	15·3	15·8	16·3	16·6	16·9	17·2	17·6	17·9	18·1	18·4	18·7	19·1	19·5	19·8	20·1	20·6	21·2
16	15·4	16·0	16·4	16·6	16·9	17·3	17·6	17·9	18·2	18·4	18·7	19·0	19·4	19·7	20·1	20·5	21·1
21	15·6	16·1	16·4	16·7	17·0	17·4	17·6	17·9	18·2	18·4	18·7	19·0	19·4	19·6	20·0	20·4	20·9
26	15·7	16·2	16·6	16·8	17·0	17·4	17·7	17·9	18·2	18·4	18·7	19·0	19·3	19·6	19·9	20·2	20·8
31	15·9	16·4	16·7	16·9	17·1	17·4	17·7	17·9	18·2	18·4	18·6	18·9	19·2	19·5	19·7	20·1	20·6

Moon's phases: full moon, 23^{d}05^{h}28^m; last quarter, 31^{d}08^{h}48^m.

Astronomy for surveyors

SUN—OCTOBER, 1975

U.T.		R		Dec.		E		U.T.		R		Dec.		E
d	h	h m s		° ′		h m s		d	h	h m s		° ′		h m s
1	0	0 36 25.6	S	2 51.4		12 09 59.8		9	0	1 07 58.0	S	5 56.4		12 12 25.3
Wed.	6	37 24.8		2 57.2 $_{58}$		10 04.6 ^{48}		Thur.	6	08 57.2		6 02.2 $_{58}$		12 29.5 ^{42}
	12	38 23.9		3 03.0 $_{58}$		10 09.5 ^{49}			12	09 56.3		6 07.9 $_{57}$		12 33.6 ^{41}
	18	39 23.0		3 08.9 $_{58}^{59}$		10 14.3 $^{48}_{48}$			18	10 55.4		6 13.6 $_{57}$		12 37.7 $^{41}_{41}$
2	0	0 40 22.2	S	3 14.7		12 10 19.1		10	0	1 11 54.6	S	6 19.3		12 12 41.8
Thur.	6	41 21.3		3 20.5 $_{58}$		10 23.9 ^{48}		Fri.	6	12 53.7		6 25.0 $_{57}$		12 45.8 ^{40}
	12	42 20.4		3 26.3 $_{58}$		10 28.7 ^{48}			12	13 52.9		6 30.7 $_{57}$		12 49.9 ^{41}
	18	43 19.6		3 32.1 $_{58}^{48}$		10 33.5 $^{48}_{47}$			18	14 52.0		6 36.4 $_{56}$		12 53.9 $^{40}_{39}$
3	0	0 44 18.7	S	3 37.9		12 10 38.2		11	0	1 15 51.1	S	6 42.0		12 12 57.8
Fri.	6	45 17.9		3 43.7 $_{58}$		10 42.9 ^{47}		Sat.	6	16 50.3		6 47.7 $_{57}$		13 01.8 ^{40}
	12	46 17.0		3 49.6 $_{59}$		10 47.6 ^{47}			12	17 49.4		6 53.4 $_{57}$		13 05.7 ^{39}
	18	47 16.1		3 55.4 $_{58}^{58}$		10 52.3 $^{47}_{47}$			18	18 48.6		6 59.1 $_{56}$		13 09.6 $^{39}_{38}$
4	0	0 48 15.3	S	4 01.2		12 10 57.0		12	0	1 19 47.7	S	7 04.7		12 13 13.4
Sat.	6	49 14.4		4 07.0 $_{58}$		11 01.6 ^{46}		Sun.	6	20 46.8		7 10.4 $_{57}$		13 17.3 ^{39}
	12	50 13.5		4 12.8 $_{58}$		11 06.2 ^{46}			12	21 46.0		7 16.0 $_{56}$		13 21.1 ^{38}
	18	51 12.7		4 18.5 $_{58}^{57}$		11 10.8 $^{46}_{46}$			18	22 45.1		7 21.6 $_{56}$		13 24.8 ^{37}
5	0	0 52 11.8	S	4 24.3		12 11 15.4		13	0	1 23 44.3	S	7 27.3		12 13 28.6
Sun.	6	53 11.0		4 30.1 $_{58}$		11 19.9 ^{45}		Mon.	6	24 43.4		7 32.9 $_{56}$		13 32.3 ^{37}
	12	54 10.1		4 35.9 $_{58}$		11 24.4 ^{45}			12	25 42.5		7 38.5 $_{56}$		13 36.0 ^{37}
	18	55 09.2		4 41.7 $_{58}^{58}$		11 28.9 $^{45}_{45}$			18	26 41.7		7 44.1 $_{57}$		13 39.6 $^{36}_{36}$
6	0	0 56 08.4	S	4 47.5		12 11 33.4		14	0	1 27 40.8	S	7 49.8		12 13 43.2
Mon.	6	57 07.5		4 53.2 $_{57}$		11 37.9 ^{45}		Tues.	6	28 39.9		7 55.4 $_{56}$		13 46.8 ^{36}
	12	58 06.6		4 59.0 $_{58}$		11 42.3 ^{44}			12	29 39.1		8 00.9 $_{55}$		13 50.4 ^{36}
	18	0 59 05.8		5 04.8 $_{58}^{57}$		11 46.7 $^{44}_{44}$			18	30 38.2		8 06.5 $_{56}$		13 53.9 $^{35}_{35}$
7	0	1 00 04.9	S	5 10.5		12 11 51.1		15	0	1 31 37.4	S	8 12.1		12 13 57.4
Tues.	6	01 04.1		5 16.3 $_{58}$		11 55.5 ^{44}		Wed.	6	32 36.5		8 17.7 $_{55}$		14 00.8 ^{34}
	12	02 03.2		5 22.0 $_{57}$		11 59.8 ^{43}			12	33 35.6		8 23.2 $_{55}$		14 04.2 ^{34}
	18	03 02.3		5 27.8 $_{58}^{57}$		12 04.1 $^{43}_{43}$			18	34 34.8		8 28.8 $_{56}$		14 07.6 ^{34}
8	0	1 04 01.5	S	5 33.5		12 12 08.4		16	0	1 35 33.9	S	8 34.4		12 14 11.0
Wed.	6	05 00.6		5 39.3 $_{58}$		12 12.7 ^{43}		Thur.	6	36 33.0		8 39.9 $_{55}$		14 14.3 ^{33}
	12	05 59.7		5 45.0 $_{57}$		12 16.9 ^{42}			12	37 32.2		8 45.4 $_{55}$		14 17.6 ^{33}
	18	06 58.9		5 50.7 $_{57}^{57}$		12 21.1 $^{42}_{42}$			18	38 31.3		8 51.0 $_{56}$		14 20.8 ^{32}
	24	1 07 58.0		5 56.4 ^{57}		12 12 25.3 ^{42}			24	1 39 30.5	S	8 56.5 ^{55}		12 14 24.0 ^{32}

Sun's S.D. 16.0   SUNRISE

Date	South Latitude									North Latitude					
	60°	55°	50°	45°	40°	30°	20°	10°	0°	10°	20°	30°	40°	45°	50°	55°	60°
Oct.	h	h	h	h	h	h	h	h	h	h	h	h	h	h	h	h	h
1	5.4	5.5	5.5	5.5	5.6	5.7	5.7	5.7	5.8	5.8	5.8	5.9	5.9	6.0	6.0	6.0	6.1
6	5.1	5.2	5.3	5.4	5.5	5.5	5.6	5.7	5.7	5.8	5.9	5.9	6.0	6.1	6.1	6.2	6.3
11	4.9	5.0	5.2	5.2	5.3	5.5	5.5	5.7	5.7	5.8	5.9	6.0	6.1	6.2	6.2	6.3	6.5
16	4.6	4.8	5.0	5.1	5.2	5.3	5.5	5.6	5.7	5.8	5.9	6.0	6.2	6.3	6.4	6.5	6.7
21	4.4	4.6	4.8	5.0	5.1	5.3	5.4	5.6	5.7	5.8	6.0	6.1	6.3	6.4	6.5	6.7	6.9
26	4.1	4.4	4.6	4.8	5.0	5.2	5.4	5.5	5.7	5.8	6.0	6.2	6.4	6.5	6.7	6.8	7.1
31	3.9	4.2	4.5	4.7	4.8	5.1	5.3	5.5	5.7	5.8	6.0	6.2	6.5	6.6	6.8	7.0	7.3

Moon's phases: new moon, $5^d 03^h 23^m$; first quarter, $12^d 01^h 15^m$.

Time

SUN—NOVEMBER, 1975

U.T.	R	Dec.	E	U.T.	R	Dec.	E
d h	h m s	° ′	h m s	d h	h m s	° ′	h m s
17 0	3 41 43.6	S 18 46.5$_{38}$	12 15 11.6$_{27}$	25 0	4 13 16.1	S 20 34.8$_{30}$	12 13 17.6$_{44}$
Mon. 6	42 42.8	18 50.3$_{37}$	15 08.9$_{29}$	Tues. 6	14 15.2	20 37.8$_{30}$	13 13.2$_{45}$
12	43 41.9	18 54.0$_{36}$	15 06.0$_{28}$	12	15 14.4	20 40.8$_{29}$	13 08.7$_{45}$
18	44 41.0	18 57.6$_{37}$	15 03.2$_{30}$	18	16 13.5	20 43.7$_{30}$	13 04.2$_{45}$
18 0	3 45 40.2	S 19 01.3$_{36}$	12 15 00.2$_{29}$	26 0	4 17 12.6	S 20 46.7$_{29}$	12 12 59.7$_{46}$
Tues. 6	46 39.3	19 04.9$_{36}$	14 57.3$_{31}$	Wed. 6	18 11.8	20 49.6$_{28}$	12 55.1$_{47}$
12	47 38.5	19 08.5$_{36}$	14 54.2$_{30}$	12	19 10.9	20 52.5$_{29}$	12 50.4$_{47}$
18	48 37.6	19 12.1$_{36}$	14 51.2$_{32}$	18	20 10.1	20 55.3$_{29}$	12 45.7$_{47}$
19 0	3 49 36.7	S 19 15.7$_{36}$	12 14 48.0$_{32}$	27 0	4 21 09.2	S 20 58.2$_{28}$	12 12 41.0$_{48}$
Wed. 6	50 35.9	19 19.3$_{35}$	14 44.8$_{32}$	Thur. 6	22 08.3	21 01.0$_{28}$	12 36.2$_{48}$
12	51 35.0	19 22.8$_{35}$	14 41.6$_{33}$	12	23 07.5	21 03.8$_{27}$	12 31.4$_{49}$
18	52 34.2	19 26.3$_{35}$	14 38.3$_{33}$	18	24 06.6	21 06.5$_{28}$	12 26.5$_{49}$
20 0	3 53 33.3	S 19 29.8$_{35}$	12 14 35.0$_{34}$	28 0	4 25 05.7	S 21 09.3$_{27}$	12 12 21.6$_{50}$
Thur. 6	54 32.4	19 33.3$_{34}$	14 31.6$_{34}$	Fri. 6	26 04.9	21 12.0$_{27}$	12 16.6$_{50}$
12	55 31.6	19 36.7$_{34}$	14 28.2$_{35}$	12	27 04.0	21 14.7$_{26}$	12 11.6$_{51}$
18	56 30.7	19 40.1$_{34}$	14 24.7$_{36}$	18	28 03.2	21 17.3$_{27}$	12 06.5$_{51}$
21 0	3 57 29.9	S 19 43.5$_{34}$	12 14 21.1$_{36}$	29 0	4 29 02.3	S 21 20.0$_{26}$	12 12 01.4$_{51}$
Fri. 6	58 29.0	19 46.9$_{34}$	14 17.5$_{36}$	Sat. 6	30 01.4	21 22.6$_{26}$	11 56.3$_{52}$
12	3 59 28.1	19 50.2$_{33}$	14 13.9$_{36}$	12	31 00.6	21 25.2$_{25}$	11 51.1$_{52}$
18	4 00 27.3	19 53.6$_{33}$	14 10.2$_{37}$	18	31 59.7	21 27.7$_{26}$	11 45.8$_{53}$
22 0	4 01 26.4	S 19 56.9$_{33}$	12 14 06.5$_{38}$	30 0	4 32 58.9	S 21 30.3$_{25}$	12 11 40.6$_{54}$
Sat. 6	02 25.6	20 00.2$_{32}$	14 02.7$_{39}$	Sun. 6	33 58.0	21 32.8$_{24}$	11 35.2$_{53}$
12	03 24.7	20 03.4$_{32}$	13 58.8$_{39}$	12	34 57.1	21 35.2$_{25}$	11 29.9$_{55}$
18	04 23.8	20 06.7$_{32}$	13 54.9$_{39}$	18	35 56.3	21 37.7$_{24}$	11 24.4$_{54}$
23 0	4 05 23.0	S 20 09.9$_{32}$	12 13 51.0$_{40}$	31 0	4 36 55.4	S 21 40.1$_{24}$	12 11 19.0$_{55}$
Sun. 6	06 22.1	20 13.1$_{32}$	13 47.0$_{41}$	Mon. 6	37 54.5	21 42.5$_{24}$	11 13.5$_{56}$
12	07 21.3	20 16.3$_{32}$	13 42.9$_{41}$	12	38 53.7	21 44.9$_{24}$	11 07.9$_{55}$
18	08 20.4	20 19.4$_{31}$	13 38.8$_{41}$	18	39 52.8	21 47.3$_{23}$	11 02.4$_{55}$
24 0	4 09 19.5	S 20 22.5$_{31}$	12 13 34.7$_{42}$	24	4 40 52.0	S 21 49.6$_{23}$	12 10 56.7$_{57}$
Mon. 6	10 18.7	20 25.6$_{31}$	13 30.5$_{42}$				
12	11 17.8	20 28.7$_{31}$	13 26.2$_{43}$				
18	12 16.9	20 31.8$_{31}$	13 21.9$_{43}$				
24	4 13 16.1	S 20 34.8$_{30}$	12 13 17.6$_{43}$				

Sun's S.D. 16ʹ.2

SUNSET

Date	South Latitude								North Latitude						
	60°	55°	50°	45°	40°	30°	20°	10°	0°	10°	20°	30°	40°	45°	50°	55°	60°
Nov.	h	h	h	h	h	h	h	h	h	h	h	h	h	h	h	h	h
1	19.6	19.3	19.0	18.8	18.6	18.4	18.1	18.0	17.8	17.6	17.4	17.2	17.0	16.8	16.6	16.4	16.1
6	19.9	19.4	19.2	18.9	18.7	18.4	18.2	18.0	17.8	17.6	17.4	17.2	16.9	16.7	16.5	16.2	15.9
11	20.1	19.6	19.3	19.1	18.8	18.5	18.2	18.0	17.8	17.6	17.4	17.1	16.9	16.6	16.4	16.1	15.7
16	20.3	19.8	19.4	19.2	18.9	18.6	18.3	18.0	17.8	17.6	17.3	17.1	16.7	16.5	16.2	15.9	15.5
21	20.5	20.0	19.6	19.3	19.0	18.6	18.3	18.1	17.8	17.6	17.3	17.0	16.7	16.4	16.2	15.8	15.3
26	20.7	20.1	19.7	19.4	19.1	18.7	18.4	18.1	17.9	17.6	17.3	17.0	16.6	16.4	16.1	15.7	15.2
31	20.9	20.3	19.8	19.5	19.2	18.8	18.4	18.1	17.9	17.6	17.3	17.0	16.6	16.3	16.0	15.6	15.1

Moon's phases: full moon, 18^{d}22^{h}28^m; last quarter, 26^{d}06^{h}52^m.

Astronomy for surveyors

70 INTERPOLATION TABLE FOR R
MUTUAL CONVERSION OF INTERVALS OF SOLAR AND SIDEREAL TIME

solar 0^h	ΔR	sidereal 0^h	solar 0^h	ΔR	sidereal 0^h	solar 1^h	ΔR	sidereal 1^h	solar 1^h	ΔR	sidereal 1^h
m s	s	m s	m s	s	m s	m s	s	m s	m s	s	m s
00 00·0	0·0	00 00·0	30 07·9	5·0	30 12·8	00 34·1	10·0	00 44·1	30 23·8	14·9	30 38·6
00 18·2	0·1	00 18·3	30 44·4	5·1	30 49·5	01 10·6	10·1	01 20·7	31 00·3	15·0	31 15·3
00 54·7	0·2	00 54·9	31 20·9	5·2	31 26·1	01 47·2	10·2	01 57·3	31 36·8	15·1	31 51·9
01 31·3	0·3	01 31·5	31 57·5	5·3	32 02·7	02 23·7	10·3	02 33·9	32 13·4	15·2	32 28·5
02 07·8	0·4	02 08·1	32 34·0	5·4	32 39·3	03 00·2	10·4	03 10·6	32 49·9	15·3	33 05·1
02 44·3	0·5	02 44·8	33 10·5	5·5	33 16·0	03 36·7	10·5	03 47·2	33 26·4	15·4	33 41·8
03 20·8	0·6	03 21·4	33 47·0	5·6	33 52·6	04 13·3	10·6	04 23·8	34 02·9	15·5	34 18·4
03 57·4	0·7	03 58·0	34 23·6	5·7	34 29·2	04 49·8	10·7	05 00·4	34 39·5	15·6	34 55·0
04 33·9	0·8	04 34·6	35 00·1	5·8	35 05·8	05 26·3	10·8	05 37·1	35 16·0	15·7	35 31·6
05 10·4	0·9	05 11·3	35 36·6	5·9	35 42·5	06 02·8	10·9	06 13·7	35 52·5	15·8	36 08·3
05 46·9	1·0	05 47·9	36 13·1	6·0	36 19·1	06 39·4	11·0	06 50·3	36 29·0	15·9	36 44·9
06 23·5	1·1	06 24·5	36 49·7	6·1	36 55·7	07 15·9	11·1	07 26·9	37 05·6	16·0	37 21·5
07 00·0	1·2	07 01·1	37 26·2	6·2	37 32·3	07 52·4	11·2	08 03·6	37 42·1	16·1	37 58·1
07 36·5	1·3	07 37·8	38 02·7	6·3	38 09·0	08 28·9	11·3	08 40·2	38 18·6	16·2	38 34·8
08 13·0	1·4	08 14·4	38 39·2	6·4	38 45·6	09 05·4	11·4	09 16·8	38 55·1	16·3	39 11·4
08 49·6	1·5	08 51·0	39 15·8	6·5	39 22·2	09 42·0	11·5	09 53·4	39 31·7	16·4	39 48·0
09 26·1	1·6	09 27·6	39 52·3	6·6	39 58·8	10 18·5	11·6	10 30·0	40 08·2	16·5	40 24·6
10 02·6	1·7	10 04·2	40 28·8	6·7	40 35·5	10 55·0	11·7	11 06·7	40 44·7	16·6	41 01·3
10 39·1	1·8	10 40·9	41 05·3	6·8	41 12·1	11 31·5	11·8	11 43·3	41 21·2	16·7	41 37·9
11 15·6	1·9	11 17·5	41 41·9	6·9	41 48·7	12 08·1	11·9	12 19·9	41 57·8	16·8	42 14·5
11 52·2	2·0	11 54·1	42 18·4	7·0	42 25·3	12 44·6	12·0	12 56·5	42 34·3	16·9	42 51·1
12 28·7	2·1	12 30·7	42 54·9	7·1	43 02·0	13 21·1	12·1	13 33·2	43 10·8	17·0	43 27·8
13 05·2	2·2	13 07·4	43 31·4	7·2	43 38·6	13 57·6	12·2	14 09·8	43 47·3	17·1	44 04·4
13 41·7	2·3	13 44·0	44 08·0	7·3	44 15·2	14 34·2	12·3	14 46·4	44 23·9	17·2	44 41·0
14 18·3	2·4	14 20·6	44 44·5	7·4	44 51·8	15 10·7	12·4	15 23·0	45 00·4	17·3	45 17·6
14 54·8	2·5	14 57·2	45 21·0	7·5	45 28·5	15 47·2	12·5	15 59·7	45 36·9	17·4	45 54·3
15 31·3	2·6	15 33·9	45 57·5	7·6	46 05·1	16 23·7	12·6	16 36·3	46 13·4	17·5	46 30·9
16 07·8	2·7	16 10·5	46 34·1	7·7	46 41·7	17 00·3	12·7	17 12·9	46 50·0	17·6	47 07·5
16 44·4	2·8	16 47·1	47 10·6	7·8	47 18·3	17 36·8	12·8	17 49·5	47 26·5	17·7	47 44·1
17 20·9	2·9	17 23·7	47 47·1	7·9	47 55·0	18 13·3	12·9	18 26·2	48 03·0	17·8	48 20·7
17 57·4	3·0	18 00·4	48 23·6	8·0	48 31·6	18 49·8	13·0	19 02·8	48 39·5	17·9	48 57·4
18 33·9	3·1	18 37·0	49 00·1	8·1	49 08·2	19 26·4	13·1	19 39·4	49 16·0	18·0	49 34·0
19 10·5	3·2	19 13·6	49 36·7	8·2	49 44·8	20 02·9	13·2	20 16·0	49 52·6	18·1	50 10·6
19 47·0	3·3	19 50·2	50 13·2	8·3	50 21·4	20 39·4	13·3	20 52·7	50 29·1	18·2	50 47·2
20 23·5	3·4	20 26·9	50 49·7	8·4	50 58·1	21 15·9	13·4	21 29·3	51 05·6	18·3	51 23·9
21 00·0	3·5	21 03·5	51 26·2	8·5	51 34·7	21 52·5	13·5	22 05·9	51 42·1	18·4	52 00·5
21 36·6	3·6	21 40·1	52 02·8	8·6	52 11·3	22 29·0	13·6	22 42·5	52 18·7	18·5	52 37·1
22 13·1	3·7	22 16·7	52 39·3	8·7	52 47·9	23 05·5	13·7	23 19·2	52 55·2	18·6	53 13·7
22 49·6	3·8	22 53·4	53 15·8	8·8	53 24·6	23 42·0	13·8	23 55·8	53 31·7	18·7	53 50·4
23 26·1	3·9	23 30·0	53 52·3	8·9	54 01·2	24 18·6	13·9	24 32·4	54 08·2	18·8	54 27·0
24 02·7	4·0	24 06·6	54 28·9	9·0	54 37·8	24 55·1	14·0	25 09·0	54 44·8	18·9	55 03·6
24 39·2	4·1	24 43·2	55 05·4	9·1	55 14·4	25 31·6	14·1	25 45·7	55 21·3	19·0	55 40·2
25 15·7	4·2	25 19·9	55 41·9	9·2	55 51·1	26 08·1	14·2	26 22·3	55 57·8	19·1	56 16·9
25 52·2	4·3	25 56·5	56 18·4	9·3	56 27·7	26 44·7	14·3	26 58·9	56 34·3	19·2	56 53·5
26 28·8	4·4	26 33·1	56 55·0	9·4	57 04·3	27 21·2	14·4	27 35·5	57 10·9	19·3	57 30·1
27 05·3	4·5	27 09·7	57 31·5	9·5	57 40·9	27 57·7	14·5	28 12·1	57 47·4	19·4	58 06·7
27 41·8	4·6	27 46·4	58 08·0	9·6	58 17·6	28 34·2	14·6	28 48·8	58 23·9	19·5	58 43·4
28 18·3	4·7	28 23·0	58 44·5	9·7	58 54·2	29 10·7	14·7	29 25·4	59 00·4	19·6	59 20·0
28 54·9	4·8	28 59·6	59 21·1	9·8	59 30·8	29 47·3	14·8	30 02·0	59 37·0	19·7	59 56·6
29 31·4	4·9	29 36·2	59 57·6	9·9	60 07·4	30 23·8		30 38·6	60 13·5		60 33·2
30 07·9		30 12·8	60 34·1		60 44·1						

In critical cases ascend.

Add ΔR to solar time interval (left-hand argument) to obtain sidereal time interval.
Subtract ΔR from sidereal time interval (right-hand argument) to obtain solar time interval.

Time

INTERPOLATION TABLE FOR R

MUTUAL CONVERSION OF INTERVALS OF SOLAR AND SIDEREAL TIME

solar 2^h	ΔR	sidereal 2^h	solar 2^h	ΔR	sidereal 2^h	solar 3^h	ΔR	sidereal 3^h	solar 3^h	ΔR	sidereal 3^h
m s	s	m s	m s	s	m s	m s	s	m s	m s	s	m s
00 13.5	19.8	00 33.2	30 03.2	24.7	30 27.8	00 29.4	29.7	00 59.0	30 19.1	34.6	30 53.6
00 50.0	19.9	01 09.9	30 39.7	24.8	31 04.4	01 05.9	29.8	01 35.7	30 55.6	34.7	31 30.2
01 26.5	20.0	01 46.5	31 16.2	24.9	31 41.1	01 42.4	29.9	02 12.3	31 32.1	34.8	32 06.9
02 03.1	20.1	02 23.1	31 52.7	25.0	32 17.7	02 19.0	30.0	02 48.9	32 08.6	34.9	32 43.5
02 39.6	20.2	02 59.7	32 29.3	25.1	32 54.3	02 55.5	30.1	03 25.5	32 45.2	35.0	33 20.1
03 16.1	20.3	03 36.4	33 05.8	25.2	33 30.9	03 32.0	30.2	04 02.2	33 21.7	35.1	33 56.7
03 52.6	20.4	04 13.0	33 42.3	25.3	34 07.6	04 08.5	30.3	04 38.8	33 58.2	35.2	34 33.4
04 29.2	20.5	04 49.6	34 18.8	25.4	34 44.2	04 45.1	30.4	05 15.4	34 34.7	35.3	35 10.0
05 05.7	20.6	05 26.2	34 55.4	25.5	35 20.8	05 21.6	30.5	05 52.0	35 11.3	35.4	35 46.6
05 42.2	20.7	06 02.9	35 31.9	25.6	35 57.4	05 58.1	30.6	06 28.6	35 47.8	35.5	36 23.2
06 18.7	20.8	06 39.5	36 08.4	25.7	36 34.1	06 34.6	30.7	07 05.3	36 24.3	35.6	36 59.9
06 55.2	20.9	07 16.1	36 44.9	25.8	37 10.7	07 11.1	30.8	07 41.9	37 00.8	35.7	37 36.5
07 31.8	21.0	07 52.7	37 21.5	25.9	37 47.3	07 47.7	30.9	08 18.5	37 37.4	35.8	38 13.1
08 08.3	21.1	08 29.3	37 58.0	26.0	38 23.9	08 24.2	31.0	08 55.1	38 13.9	35.9	38 49.7
08 44.8	21.2	09 06.0	38 34.5	26.1	39 00.6	09 00.7	31.1	09 31.8	38 50.4	36.0	39 26.4
09 21.3	21.3	09 42.6	39 11.0	26.2	39 37.2	09 37.2	31.2	10 08.4	39 26.9	36.1	40 03.0
09 57.9	21.4	10 19.2	39 47.6	26.3	40 13.8	10 13.8	31.3	10 45.0	40 03.5	36.2	40 39.6
10 34.4	21.5	10 55.8	40 24.1	26.4	40 50.4	10 50.3	31.4	11 21.6	40 40.0	36.3	41 16.2
11 10.9	21.6	11 32.5	41 00.6	26.5	41 27.1	11 26.8	31.5	11 58.3	41 16.5	36.4	41 52.9
11 47.4	21.7	12 09.1	41 37.1	26.6	42 03.7	12 03.3	31.6	12 34.9	41 53.0	36.5	42 29.5
12 24.0	21.8	12 45.7	42 13.7	26.7	42 40.3	12 39.9	31.7	13 11.5	42 29.6	36.6	43 06.1
13 00.5	21.9	13 22.3	42 50.2	26.8	43 16.9	13 16.4	31.8	13 48.1	43 06.1	36.7	43 42.7
13 37.0	22.0	13 59.0	43 26.7	26.9	43 53.6	13 52.9	31.9	14 24.8	43 42.6	36.8	44 19.4
14 13.5	22.1	14 35.6	44 03.2	27.0	44 30.2	14 29.4	32.0	15 01.4	44 19.1	36.9	44 56.0
14 50.1	22.2	15 12.2	44 39.8	27.1	45 06.8	15 06.0	32.1	15 38.0	44 55.6	37.0	45 32.6
15 26.6	22.3	15 48.8	45 16.3	27.2	45 43.4	15 42.5	32.2	16 14.6	45 32.2	37.1	46 09.2
16 03.1	22.4	16 25.5	45 52.8	27.3	46 20.0	16 19.0	32.3	16 51.3	46 08.7	37.2	46 45.8
16 39.6	22.5	17 02.1	46 29.3	27.4	46 56.7	16 55.5	32.4	17 27.9	46 45.2	37.3	47 22.5
17 16.2	22.6	17 38.7	47 05.8	27.5	47 33.3	17 32.1	32.5	18 04.5	47 21.7	37.4	47 59.1
17 52.7	22.7	18 15.3	47 42.4	27.6	48 09.9	18 08.6	32.6	18 41.1	47 58.3	37.5	48 35.7
18 29.2	22.8	18 52.0	48 18.9	27.7	48 46.5	18 45.1	32.7	19 17.8	48 34.8	37.6	49 12.3
19 05.7	22.9	19 28.6	48 55.4	27.8	49 23.2	19 21.6	32.8	19 54.4	49 11.3	37.7	49 49.0
19 42.3	23.0	20 05.2	49 31.9	27.9	49 59.8	19 58.2	32.9	20 31.0	49 47.8	37.8	50 25.6
20 18.8	23.1	20 41.8	50 08.5	28.0	50 36.4	20 34.7	33.0	21 07.6	50 24.4	37.9	51 02.2
20 55.3	23.2	21 18.5	50 45.0	28.1	51 13.0	21 11.2	33.1	21 44.3	51 00.9	38.0	51 38.8
21 31.8	23.3	21 55.1	51 21.5	28.2	51 49.7	21 47.7	33.2	22 20.9	51 37.4	38.1	52 15.5
22 08.4	23.4	22 31.7	51 58.0	28.3	52 26.3	22 24.3	33.3	22 57.5	52 13.9	38.2	52 52.1
22 44.9	23.5	23 08.3	52 34.6	28.4	53 02.9	23 00.8	33.4	23 34.1	52 50.5	38.3	53 28.7
23 21.4	23.6	23 45.0	53 11.1	28.5	53 39.5	23 37.3	33.5	24 10.8	53 27.0	38.4	54 05.3
23 57.9	23.7	24 21.6	53 47.6	28.6	54 16.2	24 13.8	33.6	24 47.4	54 03.5	38.5	54 42.0
24 34.5	23.8	24 58.2	54 24.1	28.7	54 52.8	24 50.3	33.7	25 24.0	54 40.0	38.6	55 18.6
25 11.0	23.9	25 34.8	55 00.7	28.8	55 29.4	25 26.9	33.8	26 00.6	55 16.6	38.7	55 55.2
25 47.5	24.0	26 11.5	55 37.2	28.9	56 06.0	26 03.4	33.9	26 37.2	55 53.1	38.8	56 31.8
26 24.0	24.1	26 48.1	56 13.7	29.0	56 42.7	26 39.9	34.0	27 13.9	56 29.6	38.9	57 08.5
27 00.5	24.2	27 24.7	56 50.2	29.1	57 19.3	27 16.4	34.1	27 50.5	57 06.1	39.0	57 45.1
27 37.1	24.3	28 01.3	57 26.8	29.2	57 55.9	27 53.0	34.2	28 27.1	57 42.7	39.1	58 21.7
28 13.6	24.4	28 37.9	58 03.3	29.3	58 32.5	28 29.5	34.3	29 03.7	58 19.2	39.2	58 58.3
28 50.1	24.5	29 14.6	58 39.8	29.4	59 09.2	29 06.0	34.4	29 40.4	58 55.7	39.3	59 35.0
29 26.6	24.6	29 51.2	59 16.3	29.5	59 45.8	29 42.5	34.5	30 17.0	59 32.2	39.4	60 11.6
30 03.2		30 27.8	59 52.9	29.6	60 22.4	30 19.1		30 53.6	60 08.8		60 48.2
			60 29.4		60 59.0						

In critical cases ascend.

Add ΔR to solar time interval (left-hand argument) to obtain sidereal time interval.
Subtract ΔR from sidereal time interval (right-hand argument) to obtain solar time interval.

INTERPOLATION TABLE FOR R

MUTUAL CONVERSION OF INTERVALS OF SOLAR AND SIDEREAL TIME

solar 4^h	ΔR	sidereal 4^h	solar 4^h	ΔR	sidereal 4^h	solar 5^h	ΔR	sidereal 5^h	solar 5^h	ΔR	sidereal 5^h
m s	s	m s	m s	s	m s	m s	s	m s	m s	s	m s
00 08·8	39·5	00 48·2	30 35·0	44·5	31 19·4	00 24·7	49·4	01 14·0	30 14·3	54·3	31 08·6
00 45·3	39·6	01 24·8	31 11·5	44·6	31 56·0	01 01·2	49·5	01 50·6	30 50·9	54·4	31 45·2
01 21·8	39·7	02 01·5	31 48·0	44·7	32 32·7	01 37·7	49·6	02 27·3	31 27·4	54·5	32 21·8
01 58·3	39·8	02 38·1	32 24·5	44·8	33 09·3	02 14·2	49·7	03 03·9	32 03·9	54·6	32 58·5
02 34·9	39·9	03 14·7	33 01·1	44·9	33 45·9	02 50·7	49·8	03 40·5	32 40·4	54·7	33 35·1
03 11·4	40·0	03 51·3	33 37·6	45·0	34 22·5	03 27·3	49·9	04 17·1	33 17·0	54·8	34 11·7
03 47·9	40·1	04 28·0	34 14·1	45·1	34 59·2	04 03·8	50·0	04 53·7	33 53·5	54·9	34 48·3
04 24·4	40·2	05 04·6	34 50·6	45·2	35 35·8	04 40·3	50·1	05 30·4	34 30·0	55·0	35 25·0
05 00·9	40·3	05 41·2	35 27·2	45·3	36 12·4	05 16·8	50·2	06 07·0	35 06·5	55·1	36 01·6
05 37·5	40·4	06 17·8	36 03·7	45·4	36 49·0	05 53·4	50·3	06 43·6	35 43·1	55·2	36 38·2
06 14·0	40·5	06 54·4	36 40·2	45·5	37 25·7	06 29·9	50·4	07 20·2	36 19·6	55·3	37 14·8
06 50·5	40·6	07 31·1	37 16·7	45·6	38 02·3	07 06·4	50·5	07 56·9	36 56·1	55·4	37 51·5
07 27·0	40·7	08 07·7	37 53·3	45·7	38 38·9	07 42·9	50·6	08 33·5	37 32·6	55·5	38 28·1
08 03·6	40·8	08 44·3	38 29·8	45·8	39 15·5	08 19·5	50·7	09 10·1	38 09·2	55·6	39 04·7
08 40·1	40·9	09 20·9	39 06·3	45·9	39 52·2	08 56·0	50·8	09 46·7	38 45·7	55·7	39 41·3
09 16·6	41·0	09 57·6	39 42·8	46·0	40 28·8	09 32·5	50·9	10 23·4	39 22·2	55·8	40 18·0
09 53·1	41·1	10 34·2	40 19·4	46·1	41 05·4	10 09·0	51·0	11 00·0	39 58·7	55·9	40 54·6
10 29·7	41·2	11 10·8	40 55·9	46·2	41 42·0	10 45·6	51·1	11 36·6	40 35·3	56·0	41 31·2
11 06·2	41·3	11 47·4	41 32·4	46·3	42 18·7	11 22·1	51·2	12 13·2	41 11·8	56·1	42 07·8
11 42·7	41·4	12 24·1	42 08·9	46·4	42 55·3	11 58·6	51·3	12 49·9	41 48·3	56·2	42 44·4
12 19·2	41·5	13 00·7	42 45·4	46·5	43 31·9	12 35·1	51·4	13 26·5	42 24·8	56·3	43 21·1
12 55·8	41·6	13 37·3	43 22·0	46·6	44 08·5	13 11·7	51·5	14 03·1	43 01·3	56·4	43 57·7
13 32·3	41·7	14 13·9	43 58·5	46·7	44 45·1	13 48·2	51·6	14 39·7	43 37·9	56·5	44 34·3
14 08·8	41·8	14 50·6	44 35·0	46·8	45 21·8	14 24·7	51·7	15 16·4	44 14·4	56·6	45 10·9
14 45·3	41·9	15 27·2	45 11·5	46·9	45 58·4	15 01·2	51·8	15 53·0	44 50·9	56·7	45 47·6
15 21·9	42·0	16 03·8	45 48·1	47·0	46 35·0	15 37·8	51·9	16 29·6	45 27·4	56·8	46 24·2
15 58·4	42·1	16 40·4	46 24·6	47·1	47 11·6	16 14·3	52·0	17 06·2	46 04·0	56·9	47 00·8
16 34·9	42·2	17 17·1	47 01·1	47·2	47 48·3	16 50·8	52·1	17 42·9	46 40·5	57·0	47 37·4
17 11·4	42·3	17 53·7	47 37·6	47·3	48 24·9	17 27·3	52·2	18 19·5	47 17·0	57·1	48 14·1
17 48·0	42·4	18 30·3	48 14·2	47·4	49 01·5	18 03·9	52·3	18 56·1	47 53·5	57·2	48 50·7
18 24·5	42·5	19 06·9	48 50·7	47·5	49 38·1	18 40·4	52·4	19 32·7	48 30·1	57·3	49 27·3
19 01·0	42·6	19 43·6	49 27·2	47·6	50 14·8	19 16·9	52·5	20 09·4	49 06·6	57·4	50 03·9
19 37·5	42·7	20 20·2	50 03·7	47·7	50 51·4	19 53·4	52·6	20 46·0	49 43·1	57·5	50 40·6
20 14·1	42·8	20 56·8	50 40·3	47·8	51 28·0	20 30·0	52·7	21 22·6	50 19·6	57·6	51 17·2
20 50·6	42·9	21 33·4	51 16·8	47·9	52 04·6	21 06·5	52·8	21 59·2	50 56·2	57·7	51 53·8
21 27·1	43·0	22 10·1	51 53·3	48·0	52 41·3	21 43·0	52·9	22 35·9	51 32·7	57·8	52 30·4
22 03·6	43·1	22 46·7	52 29·8	48·1	53 17·9	22 19·5	53·0	23 12·5	52 09·2	57·9	53 07·1
22 40·2	43·2	23 23·3	53 06·4	48·2	53 54·5	22 56·0	53·1	23 49·1	52 45·7	58·0	53 43·7
23 16·7	43·3	23 59·9	53 42·9	48·3	54 31·1	23 32·6	53·2	24 25·7	53 22·3	58·1	54 20·3
23 53·2	43·4	24 36·5	54 19·4	48·4	55 07·8	24 09·1	53·3	25 02·3	53 58·8	58·2	54 56·9
24 29·7	43·5	25 13·2	54 55·9	48·5	55 44·4	24 45·6	53·4	25 39·0	54 35·3	58·3	55 33·6
25 06·2	43·6	25 49·8	55 32·5	48·6	56 21·0	25 22·1	53·5	26 15·6	55 11·8	58·4	56 10·2
25 42·8	43·7	26 26·4	56 09·0	48·7	56 57·6	25 58·7	53·6	26 52·2	55 48·4	58·5	56 46·8
26 19·3	43·8	27 03·0	56 45·5	48·8	57 34·3	26 35·2	53·7	27 28·8	56 24·9	58·6	57 23·4
26 55·8	43·9	27 39·7	57 22·0	48·9	58 10·9	27 11·7	53·8	28 05·5	57 01·4	58·7	58 00·1
27 32·3	44·0	28 16·3	57 58·6	49·0	58 47·5	27 48·2	53·9	28 42·1	57 37·9	58·8	58 36·7
28 08·9	44·1	28 52·9	58 35·1	49·1	59 24·1	28 24·8	54·0	29 18·7	58 14·5	58·9	59 13·3
28 45·4	44·2	29 29·5	59 11·6	49·2	60 00·8	29 01·3	54·1	29 55·3	58 51·0	59·0	59 49·9
29 21·9	44·3	30 06·2	59 48·1	49·3	60 37·4	29 37·8	54·2	30 32·0	59 27·5	59·1	60 26·6
29 58·4	44·4	30 42·8	60 24·7		61 14·0	30 14·3		31 08·6	60 04·0		61 03·2
30 35·0		31 19·4									

In critical cases ascend.

Add ΔR to solar time interval (left-hand argument) to obtain sidereal time interval.
Subtract ΔR from sidereal time interval (right-hand argument) to obtain solar time interval.

Time

INTERPOLATION TABLE FOR SUN

Time diff.	Tabular six-hourly difference									
	3 6 9	12 15 18	21 24 27	30 33 36	39 42 45	48 51 54	57 60 63	66 69 72
h m								
0 10	0 0 0	0 0 1	1 1 1	1 1 1	1 1 1	1 1 2	2 2 2	2 2 2
20	0 0 1	1 1 1	1 1 2	2 2 2	2 2 3	3 3 3	3 3 4	4 4 4
30	0 1 1	1 1 2	2 2 2	3 3 3	3 4 4	4 4 5	5 5 5	6 6 6
40	0 1 1	1 2 2	2 3 3	3 4 4	4 5 5	5 6 6	6 7 7	7 8 8
50	0 1 1	2 2 3	3 3 4	4 5 5	5 6 6	7 7 8	8 8 9	9 10 10
1 00	1 1 2	2 3 3	4 4 5	5 6 6	7 7 8	8 9 9	10 10 11	11 12 12
10	1 1 2	2 3 4	4 5 5	6 6 7	8 8 9	9 10 11	11 12 12	13 13 14
20	1 1 2	3 3 4	5 5 6	7 7 8	9 9 10	11 11 12	13 13 14	15 15 16
30	1 2 2	3 4 5	5 6 7	8 8 9	10 11 11	12 13 14	14 15 16	17 17 18
40	1 2 3	3 4 5	6 7 8	8 9 10	11 12 13	13 14 15	16 17 18	18 19 20
50	1 2 3	4 5 6	6 7 8	9 10 11	12 13 14	15 16 17	17 18 19	20 21 22
2 00	1 2 3	4 5 6	7 8 9	10 11 12	13 14 15	16 17 18	19 20 21	22 23 24
10	1 2 3	4 5 7	8 9 10	11 12 13	14 15 16	17 18 20	21 22 23	24 25 26
20	1 2 4	5 6 7	8 9 11	12 13 14	15 16 18	19 20 21	22 23 25	26 27 28
30	1 3 4	5 6 8	9 10 11	13 14 15	16 18 19	20 21 23	24 25 26	28 29 30
40	1 3 4	5 7 8	9 11 12	13 15 16	17 19 20	21 23 24	25 27 28	29 31 32
50	1 3 4	6 7 9	10 11 13	14 16 17	18 20 21	23 24 26	27 28 30	31 33 34
3 00	2 3 5	6 8 9	11 12 14	15 17 18	20 21 23	24 26 27	29 30 32	33 35 36
10	2 3 5	6 8 10	11 13 14	16 17 19	21 22 24	25 27 29	30 32 33	35 36 38
20	2 3 5	7 8 10	12 13 15	17 18 20	22 23 25	27 28 30	32 33 35	37 38 40
30	2 4 5	7 9 11	12 14 16	18 19 21	23 25 26	28 30 32	33 35 37	39 40 42
40	2 4 6	7 9 11	13 15 17	18 20 22	24 26 28	29 31 33	35 37 39	40 42 44
50	2 4 6	8 10 12	13 15 17	19 21 23	25 27 29	31 33 35	36 38 40	42 44 46
4 00	2 4 6	8 10 12	14 16 18	20 22 24	26 28 30	32 34 36	38 40 42	44 46 48
10	2 4 6	8 10 13	15 17 19	21 23 25	27 29 31	33 35 38	40 42 44	46 48 50
20	2 4 7	9 11 13	15 17 20	22 24 26	28 30 33	35 37 39	41 43 46	48 50 52
30	2 5 7	9 11 14	16 18 20	23 25 27	29 32 34	36 38 41	43 45 47	50 52 54
40	2 5 7	9 12 14	16 19 21	23 26 28	30 33 35	37 40 42	44 47 49	51 54 56
50	2 5 7	10 12 15	17 19 22	24 27 29	31 34 36	39 41 44	46 48 51	53 56 58
5 00	3 5 8	10 13 15	18 20 23	25 28 30	33 35 38	40 43 45	48 50 53	55 58 60
10	3 5 8	10 13 16	18 21 23	26 28 31	34 36 39	41 44 47	49 52 54	57 59 62
20	3 5 8	11 13 16	19 21 24	27 29 32	35 37 40	43 45 48	51 53 56	59 61 64
30	3 6 8	11 14 17	19 22 25	28 30 33	36 39 41	44 47 50	52 55 58	61 63 66
40	3 6 9	11 14 17	20 23 26	28 31 34	37 40 43	45 48 51	54 57 60	62 65 68
50	3 6 9	12 15 18	20 23 26	29 32 35	38 41 44	47 50 53	55 58 61	64 67 70
6 00	3 6 9	12 15 18	21 24 27	30 33 36	39 42 45	48 51 54	57 60 63	66 69 72

INTERPOLATION TABLE FOR STARS

Day of month	Monthly difference							
	1 2 3	4 5 6	7 8 9	10 11 12	14 16 18	20 22 24	26 28 30
2	0 0 0	0 0 0	0 1 1	1 1 1	1 1 1	1 1 2	2 2 2
4	0 0 0	1 1 1	1 1 1	1 1 2	2 2 2	3 3 3	3 4 4
6	0 0 1	1 1 1	1 2 2	2 2 2	3 3 4	4 4 5	5 6 6
8	0 1 1	1 1 2	2 2 2	3 3 3	4 4 5	5 6 6	7 7 8
10	0 1 1	1 2 2	2 3 3	3 4 4	5 5 6	7 7 8	9 9 10
12	0 1 1	2 2 2	3 3 4	4 4 5	6 6 7	8 9 10	10 11 12
14	0 1 1	2 2 3	3 4 4	5 5 6	7 7 8	9 10 11	12 13 14
16	1 1 2	2 3 3	4 4 5	5 6 6	7 9 10	11 12 13	14 15 16
18	1 1 2	2 3 4	4 5 5	6 7 7	8 10 11	12 13 14	16 17 18
20	1 1 2	3 3 4	5 5 6	7 7 8	9 11 12	13 15 16	17 19 20
22	1 1 2	3 4 4	5 6 7	7 8 9	10 12 13	15 16 18	19 21 22
24	1 2 2	3 4 5	6 6 7	8 9 10	11 13 14	16 18 19	21 22 24
26	1 2 3	3 4 5	6 7 8	9 10 10	12 14 16	17 19 21	23 24 26
28	1 2 3	4 5 5	6 7 8	9 10 11	13 15 17	19 21 22	24 26 28
30	1 2 3	4 5 6	7 8 9	10 11 12	14 16 18	20 22 24	26 28 30

Astronomy for surveyors

EXERCISES

If a *Star Almanac* for the current year is available, it will be a useful exercise for the student to re-work the examples given in this chapter using the up-to-date values of R and E.

1. Express in sidereal time the following intervals of mean solar time: (a) 5^h 15^m 23^s; (b) 3^h 17^m 18.4^s; (c) 0^h 52^m 33.5^s.

$$Ans. \quad (a)\ 5^h\ 16^m\ 14.8^s.$$
$$(b)\ 3^h\ 17^m\ 50.8^s.$$
$$(c)\ 0^h\ 52^m\ 42.1^s.$$

2. Express in mean solar time the following intervals of sidereal time: (a) 5^h 38^m 47.5^s; (b) 4^h 00^m 14.8^s; (c) 1^h 12^m 33.2^s.

$$Ans. \quad (a)\ 5^h\ 37^m\ 52.0^s.$$
$$(b)\ 3^h\ 59^m\ 35.4^s.$$
$$(c)\ 1^h\ 12^m\ 21.3^s.$$

3. In Longitude $148°\ 15'$ E., what is the L.M.T. corresponding to 22 September 4^h 30^m p.m. standard time, of the 150th meridian east of Greenwich? Find also the corresponding U.T.

$$Ans. \quad (1)\ 16^h\ 23^m\ or\ 4^h\ 23^m\ p.m.$$
$$(2)\ 6^h\ 30^m\ 22\ September.$$

4. The True Sun's L.H.A. at Perth, Western Australia (Long. 7^h 43^m 21.7^s E.), on 3 December 1975, is 4^h 15^m 20.3^s; find (a) the corresponding L.Std.T. (Standard Meridian is 8^h E.) and (b) the corresponding L.S.T.
Given: E at 6^h U.T., 3 December 1975 $= 12^h$ 10^m 28.0^s (decreasing 05.9^s between 6^h and 12^h U.T., 3 December 1975). R at 6^h U.T. $= 4^h$ 45^m 47.7^s.

$$Ans. \quad (a)\ 16^h\ 21^m\ 32.9^s\ or$$
$$4^h\ 21^m\ 32.9^s p.m.\ L.Std.T.$$
$$(b)\ 20^h\ 51^m\ 0.56^s\ L.S.T.$$

5. Given that R at 0^h U.T. is 14^h 40^m 40.1^s, find the U.T. of the next upper transit of the First Point of Aries at Greenwich.

$$Ans. \quad 9^h\ 17^m\ 48.3^s\ U.T.$$

6. Given that the U.T. of upper transit of the First Point of Aries at Greenwich is 11^h 19^m 41.4^s, compute the value of R at 0^h U.T. on the same day.

$$Ans. \quad 12^h\ 38^m\ 27.0^s.$$

7. The R.A. of a star being 20^h 24^m 13.7^s, compute the L.M.T. of its upper transit at Madras (Long. $80°\ 14'\ 21''$ E.) on 6 September 1975, the value of R at 12^h U.T. on that date being 22^h 59^m 50.1^s.

$$Ans. \quad 21^h\ 23^m\ 43.7^s\ L.M.T.$$

8. Given that the sidereal time at Greenwich on 20 January 1975 is 22^h 22^m 44.6^s, find the corresponding U.T. if the U.T. of upper transit of the First Point of Aries on 20 January 1975 is 16^h 02^m 21.2^s.

$$Ans. \quad 14^h\ 25^m\ 21.7^s\ U.T.$$

Time

9. Find the L.M.T. corresponding to $5^h 17^m 32^s$ L.S.T. at Moscow (Long. $37° 34' 15''$ E.), given that the value of R at 0^h U.T. on the same day was $23^h 56^m 36.0^s$.

Ans. $5^h 20^m 28.0^s$ L.M.T.

10. Find the L.Std.T. of upper transit of α Centauri at Adelaide, Australia, on 1 June 1975. R.A. = $14^h 37^m 07.4^s$; longitude of observing station = $9^h 14^m 20.3^s$ E.; Standard Meridian is $9^h 30^m$ E. The value of R at U.T. 12^h on 1 June 1975 is $16^h 37^m 24.2^s$.

Ans. $22^h 15^m 15.5^s$
or $10^h 15^m 15.5^s$ p.m. L.Std.T.

11. Find the hour angle of Algol (β Persei, No. 70) at a place in Longitude $122° 22' 15''$ W. at L.Std.T. $16^h 43^m 30^s$ on 12 April 1975. The Standard Meridian of the place is $120°$ W.; R.A. of Algol is $3^h 05^m 43.7^s$, and the value of R at 0^h U.T. on 13 April 1975 is $13^h 22^m 14.7^s$.

Ans. $2^h 50^m 39.1^s$,
or $42° 39' 47''$.

12. Find the L.S.T. at $8^h 45^m 00^s$ p.m. L.Std.T. on 6 November 1975 at a place in Longitude $11^h 22^m 04^s$ E. The Standard Meridian is 12^h E. and the value of R at U.T. 6^h on 6 November 1975 is $2^h 59^m 20.7^s$.

Ans. $23^h 06^m 51.8^s$ L.S.T.

13. Find the L.S.T. at $3^h 16^m 04.6^s$ a.m. L.Std.T. on 23 September 1975, at a place in Longitude $100° 00' 30''$ W. The Standard Meridian is $105°$ W. and the value of R at U.T. 6^h on 23 September 1975 is $0^h 05^m 52.3^s$. Find also the L.M.T. of the instant concerned.

Ans. (1) $3^h 42^m 37.0^s$ LS.T.
(2) $3^h 36^m 02.6^s$ a.m. L.M.T.

14. Find the L.Std.T. of the True Sun's upper transit at a place in Longitude $31° 06' 12''$ E. on 25 October 1975. The Standard Meridian is $30°$ E. and the value of E at U.T. 6^h on 25 October 1975 is $12^h 15^m 49.3^s$, increasing 1.8^s over the next 6^h.

Ans. $11^h 39^m 45.9^s$ L.Std.T.

15. Find the Right Ascension of the True Sun at U.T. $9^h 26^m 43^s$ on 21 August 1975. The value of E at U.T. 6^h on 21 August 1975 is $11^h 56^m 41.8^s$, increasing 3.7^s over the next 6^h; the value of R at U.T. 6^h on 21 August 1975 is $21^h 55^m 46.1^s$.

(*Hint*: G.H.A. Sun = U.T. + E
R.A. Sun = G.S.T. − G.H.A. Sun.)

Ans. $9^h 59^m 36.2^s$.

16. During the morning of 1 August 1975, at Melbourne, Victoria (Long. $9^h 39^m 54^s$ E.), a mean time chronometer was compared with a sidereal clock known to be 14.6s fast on L.S.T. It was found:

Time by sidereal clock $7^h 18^m 09.0^s$
Time by chronometer   10   41   34.3

Astronomy for surveyors

Find the error of the chronometer on Victorian Standard Time (meridian 10^h E.). The value of R at U.T. 0^h on 1 August 1975 is 20^h 35^m 55.8^s.

Ans. 0^h 20^m 20.1^s slow.

17. The sun was observed in the morning at a place in Longitude 168° 35′ 21″ E. at L. Std.T. 8^h 17^m 11.2^s a.m. (Standard Meridian 12^h E.) on 4 December 1975. Find the sun's L.H.A. The value of E at 18^h U.T. on 3 December 1975 is 12^h 10^m 16.2^s, decreasing 5.9^s over the next 6^h.

Ans. 19^h 41^m 46.5^s West H.A.
or 4^h 18^m 13.5^s East H.A.
(= 64° 33′ 23″ arc).

18. Find the L.H.A. of α Tucanae (No. 612) at a place in Longitude 117° 23′ 24″ E. at L.M.T. 12^h 32^m 16^s a.m. on 10 October 1975. The R.A. of α Tucanae is 22^h 16^m 02.3^s and the value of R at U.T. 12^h on 9 October 1975 is 1^h 09^m 56.3^s.

Ans. 3^h 26^m 56.4^s
or 51° 44′ 06″ arc.

CHAPTER 6
The timing of observations

6.01 Introduction

Having discussed the kinds of time involved in the study of field astronomy, we should now consider the methods of recording the actual moments of time in which we are interested. In this work the instants we must note (the "events") are:

(a) When a star transits one of the cross-hairs of the telescope graticule or when the edge of the sun's disc is tangential to one of the cross-hairs; or

(b) When a star transits the intersection of the main horizontal and vertical cross-hairs or when the edge of the sun's disc is tangential to both of these simultaneously.

We usually find, first, the time of the event as indicated by a suitable timepiece; we then find out if any error exists in the latter by comparing it with radio time-signals.

6.02 Timepieces

The recent rapid development of digital electronics using sophisticated, large-scale integration of components and circuits in small silicon chips has opened up a new world to the modern watchmaker. The mind boggles at the technology which can now be incorporated in a wrist-watch, and the surveyor should have little trouble in finding one with stop-watch and other functions which will help him greatly in the timing of his observations. Most of these modern watches rely upon a quartz crystal oscillator, and if they have a digital rather than an analogue display, have no moving parts to wear or become clogged. Over quite long periods—of weeks rather than hours—the best ones have excellent stability which may, in part, be due to their being temperature-controlled by constant contact with the human body.

It is possible that some members of this new breed of wrist-watches incorporate memories which will retain several successive event times for recall later, but this is perhaps more likely to be a feature of a pocket electronic clock-cum-calculator. The HP-55 calculator, unfortunately no longer in production, was capable of storing up to 10 event times, but doubtless some of the many others now available will have a similar facility.

In previous editions of this book information was given on mechanical

Astronomy for surveyors

chronometers, stop-watches and chronographs, but modern technology has made these obsolete almost overnight, so descriptions are not included here. Some use of them will doubtless persist, and anyone needing details should consult the 8th edition (1978).

While electronic watches are a comparatively new development, the small, solid-state radio receiver has been with us for at least two decades. In a sense it, too, is a time-keeper since time-signals are broadcast from local stations, and if the receiver covers the short-wave bands it may well be possible for it to be tuned in to certain radio stations which transmit time-signals continuously. The use of an extension antenna may be necessary for good results.

6.03 Simple methods of timekeeping

(a) By calling out to an assistant who is watching the timepiece which, for preference, should be a good electronic wrist-watch with a digital display which shows hours, minutes and seconds. The observer gives his assistant 5 or 10 seconds warning by calling out "coming up" or "stand by", and then at the instant when the star transits the cross-hair, he calls "up". The assistant should be able to estimate the time of "up" to half a second or better, with practice. The difference between watch time and Universal Coordinated Time (U.T.C.) as broadcast by most radio time-signal stations should be determined just before and just after the observations so that the event times may be referred to U.T.C., and any difference of rate discovered.

(b) By listening to the seconds "pips" or "tocs" on a radio receiver tuned to a time-signal station while observing. The observer sets the cross-hair slightly ahead of the star and starts counting the "pips" of the time-signal. He continues to count in step with the "pips", notes the count when the star transits the cross-hair, and stops counting when the next minute marker is heard; this is generally either a longer, or a double "pip". He will need to identify this particular minute of U.T.C. by further listening, or comparing the time-signal with a watch keeping time close to U.T.C. Suppose the event occurred halfway between 5 and 6, and the observer stopped counting at 17 on the minute signal for $20^h\ 42^m$ U.T.C. The star must have crossed the hair between 11 and 12 seconds before $20^h\ 42^m$, i.e. at $20^h\ 41^m\ 48.5^s$ U.T.C. With practice, the observer will be able to interpolate the event to perhaps 0.3^s or 0.4^s, but concentration is required to estimate and remember the fraction while still counting.

The method is a handy one for a surveyor working alone, but it can be much improved by using a portable tape recorder which is set up to record the time-signal "pips" emitted by the radio, as well as the observer's voice, while he concentrates on the event. He merely calls "up" at the appropriate time, and then identifies the next, or a subsequent minute marker by voice annotation before stopping the recorder. By playing back the tape two or three times he should be able to interpolate the "up" to 0.2^s and identify the appropriate second.

The timing of observations

(c) By using a stop-watch in conjunction with the main timepiece which might be an electronic wrist-watch or clock. Like (b) above, this is a convenient method for an observer who is working single-handed. He starts the stop-watch when the star transits the cross-hair and stops it on a known second of the watch or clock. The clock time and the stop-watch interval are both entered in the field-book; the former minus the latter will then give the correct clock time of the event. This method is best suited to a situation in which reception of radio time-signals is sometimes difficult and checks of the main timepiece against them can be made only once a day, say.

(d) By using a stop-watch in conjunction with radio time-signals. Most electronic stop-watches and some electronic wrist-watches are put into the stop-watch mode by pressing a special button (A); they then display 00^m 00.00^s. Pressing another button (B) starts the display which will run to 60^m if not stopped earlier. Pressing button B again, at some event, "freezes" the display which can be entered in a field-book; pressing B again re-starts the display at the full elapsed time from 00^m 00.00^s, after which further events may be timed using B to stop and re-start the display each time. Some digital stop-watches are, in fact, complete clocks keeping the time of day when not being used in the stop mode; they incorporate a read-out of the date and day of the week as well. The Seiko Quartz Cal. SO23 is a case in point, and is illustrated in Fig. 6.1.

Any stop-watch or other timepiece used for event timing in field astronomy

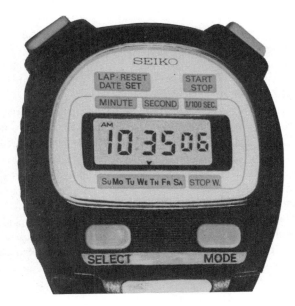

Fig. 6.1   The Seiko quartz stop-watch

must be checked for rate near the time of the observations. By starting it on a known time-signal "pip" just prior to the observations, using it in the stop-watch mode (e.g. button B above) to time the observed events, and stopping it at a later, known time-signal "pip", a direct comparison between U.T.C. and watch rates may be made, and any necessary proportional adjustment applied to the event times recorded during the running period. See **6.08**.

6.04 Personal equation

In the above simple methods, which are the ones used where the greatest accuracy is not required, it will be seen that there is considerable room for personal error. Many observers, for instance, tend habitually to anticipate by a fraction of a second the instant when a star crosses a graticule hair while others tend to delay it. This period of anticipation or delay—the "personal equation"—can usually be determined by arranging for the person concerned to do a series of time observations for longitude at a point whose position is already accurately known. Such observations, if done periodically, usually show the observer's personal equation to vary somewhat, no doubt because of varying physiological and psychological conditions. Personal equation will be involved both when the observer starts the stop-watch and when he stops it; or if there are two persons, both their personal equations will be included.

6.05 More elaborate methods of timing

These are really the province of the geodetic astronomer, but a brief description is given here by way of information.

Probably the most sophisticated modern instrument for time-keeping in geodetic astronomy is the Chronocord (see Fig. 6.2) manufactured in the United Kingdom by the Littlemore Scientific Engineering Company of Oxford. It incorporates a very stable electronic clock and a printer which may be triggered by an event timer such as an impersonal eyepiece (see section **6.06**), a manual pressel switch, or a radio time-signal. The print-out shows the event time, to a millisecond, of the Chronocord clock; by switching to the radio time-signal (from a separate receiver) the clock times of the individual "pips" are printed (usually along with some unwanted but unavoidable noise pulses); hence the offset of the clock from U.T.C. is readily obtained.

Compared with comparatively recent methods using mechanical tape or drum chronographs whose records needed hours of tedious interpretation, that using the Chronocord in conjunction with impersonal tracking eyepieces such as those on the Wild T4 and Kern DKM 3A astronomical theodolites is very simple.

The timing of observations

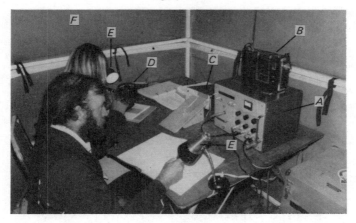

Fig. 6.2   Time-keeping in geodetic astronomy

A   Chronocord
B   Portable (transistor) short-wave radio
C   Printer module for Chronocord
D   Barometer
E   Low-voltage reading lamps
F   Tent wall

Photo by Mr Glen Rowe, printed with the permission of the Surveyor-General, N.Z. Lands and Survey Department

6.06   Impersonal telescope eyepieces

In order to reduce or eliminate the error due to the personal equation in observing, special kinds of eyepiece have been devised for use with theodolite telescopes. Brief descriptions of some of these follow.

(a) *Electro-mechanical eyepieces*

One type has a moving cross-hair operated by a micrometer knob (cf. that fitted to their DKM3-A theodolite by Kern of Aarau, Switzerland—see Fig. 6.3); the cross-hair is kept on the star as it moves across the field of view, and peripheral electric contacts built into a rotating drum touch a fixed brush as the wheel is rotated by the observer, so feeding signals to a Chronocord. A series of such signals is thus obtained as the star moves across the field of view; the face of the instrument is changed before the star reaches the centre, and a further series obtained over the same contacts as the wheel is rotated in the reverse direction to keep the cross-hair on the star. A refinement of this has the micrometer knob driven by a variable-speed electric motor, the observer keeping the cross-hair on the star by controlling the motor. Another type (the Hunter Shutter) has a graticule of fine, evenly-spaced, numbered lines, across the middle of which the star moves. A shutter, which normally blocks off the field of view through the telescope, is caused by electrical connection with a clock to swing aside every three seconds for a few hundredths of a second, so that the star is momentarily seen by the observer who calls out to a booker its position with respect to the numbered cross-hairs. As with the moving cross-hair

Astronomy for surveyors

Fig. 6.3   Kern DKM 3A astronomical theodolite in use
Note the stride level mounted on top of the instrument, and the impersonal eyepiece with its knurled knobs for moving the tracking cross-hairs.
Photo by Mr Glen Rowe, printed with the permission of the Surveyor-General, N.Z. Lands and Survey Department

eyepiece, the face is reversed and another series obtained to eliminate collimation error. (The Hunter shutter eyepiece has been manufactured by Hilger & Watts of London.) For more detailed descriptions of this type of apparatus, see *Empire Survey Review*, No. 63, **9**, 1947.

(b) *Electronic eyepieces*

The systems described in (a) above are largely mechanical and they entail personal observation by eye, so that there may still be an element of personal equation present—though probably very small. Attempts have been made more recently to replace the use of the human eye by a photo-electric cell, and some success was obtained at Oxford University (see Rushworth in *Empire Survey Review*, No. 114, **15**, 1959) with an opaque graticule carrying fine slits; the light from a star as it crossed the slits was picked up by the photo-electric cell (a photo-multiplier) whose output was fed to a recording pen to give a series of peaks which were correlated against a time-base provided by a chronometer.

Another system, called the Electronic Transit Detector, due to a Japanese experimenter Tsubokawa (Geographical Survey Institute, Chiba; Ministry of Construction, Japan, *c*. 1956), uses tiny 105° prisms ("knife-edges") in place of cross-hairs in the plane of the graticule, and two photo-electric cells, one on each side of the graticule, connected in opposition to a recorder. When a star is to one side of a prism, more light is reflected towards one side of the graticule than to the other, and more current flows through one photo-electric

The timing of observations

cell than the other, keeping the recorder pen off-centre; when the star is directly opposite the knife-edge, its light is reflected to each side, so no current flows in the recorder and the pen is on the centre-line. One of the difficulties with photo-electric cells has been the so-called "dark current" which leads to fuzziness in the output and impairs the record, but recent improvements should get over this problem. The ideal, presumably, is some suitable kind of electronic light-sensitive device recording against a time-base provided by direct reception of radio time-signals.

6.07 Radio time-signals

So far we have discussed methods of finding the clock times of events in which we are interested, but such times are of little use to us unless we can find the corresponding instants of true mean solar or sidereal time. Since astronomers, navigators and others are also concerned with this problem, many radio stations throughout the world broadcast accurate time-signals which are controlled by observatories, and by periodically comparing our chronometer times with these we can make the necessary conversions. Volume V of *The Admiralty List of Radio Signals*, published bi-annually by the Hydrographer of the British Navy and obtainable from the Agents for the sale of Admiralty Charts, lists many of these signals, giving station call-signs, transmission times and types of signal.

As the *Star Almanac for Land Surveyors* now includes annually a limited list of radio time-signals and notes on them on pages 60 and 61 (1984) issue), there is little point in including much additional information in a book such as this which is likely to get out of date in this respect before a new edition appears. Amendments to the *Star Almanac* list, and notes, are published in *Survey Review* (obtainable from C. F. Hodgson & Son Ltd., 50 Holloway Road, London N7 8JL, England).

In Australasia, stations WWV, WWVH and VNG are usually the best as far as reception is concerned. VNG, however, gives no time announcements and the surveyor must know his time to within a few minutes if he wishes to make use of this station. Between the 14th and 15th, the 29th and 30th, the 44th and 45th, and the 59th and 60th minutes of each hour there is a voice announcement from the 20th to the 50th seconds, but this merely gives the station name and transmission frequencies and serves to differentiate the quarter-hours.

Reference has been made in **5.02** and **5.09**, Chapter 5, to Coordinated Universal Time (U.T.C.) as being the basis of civil time and the kind of time broadcast by radio stations emitting time-signals. U.T.C. is given by all the stations listed in the *Star Almanac*. The civil, or Zone times of most countries are displaced by an integral number of hours or half-hours from U.T.C. It has also been stated earlier that the surveyor must use U.T.1 for reducing his observations. U.T.C. and U.T.1 are related by the quantity DUT1 as follows:

Astronomy for surveyors

$$\text{U.T.1} = \text{U.T.C.} + \text{DUT1}. \tag{6.1}$$

DUT1, to the nearest $0^s.1$, is obtained from the U.T.C. time-signals, each minute, by a special superimposed code. It is worth quoting from the *Star Almanac* notes on page 61 how this code is used:

If the seconds markers (each minute) from 1^s to n^s are emphasized, the difference (DUT1) is $+n \times 0.1$ seconds; if the seconds markers from 9^s to $(8 + m)$ seconds are emphasized, the difference (DUT1) is $-m \times 0.1$ seconds (n and m are always less than or equal to 8).* The appropriate seconds markers may be emphasized by lengthening, doubling, splitting or tone modulation of the normal seconds markers. In reducing observations it is possible either to apply corrections to all the times to reduce them to U.T.1 (e.g. by incorporating them with the corrections to the chronometer) or to apply corrections to the deduced positions. The correction in longitude in seconds of arc, measured positively to the east, is equal to *minus* fifteen times the difference DUT1 in seconds of time.

Provisional estimates of (U.T.1 − U.T.C.) in tenths of milliseconds are published in advance by various sources, including *Circular D*, Bureau International de l'Heure, and station VNG (P.M.G. Research Laboratories, Frequency Standards Division, 59 Little Collins Street, Melbourne, Victoria 3000, Australia); smoothed values in tenths of milliseconds based upon data supplied by the cooperating observatories are available fairly soon after the end of each month.

For ordinary work, U.T.1, as obtained from the coded time-signal to $0^s.1$ accuracy, is quite adequate.

6.08 Determination of clock error and rate by comparison with radio time-signals

The simplest and most obvious method of comparison is to note the time on the clock (or watch) when a signal pip marking a known time is received. With the usual continuous time-signal seconds markers, checks may be made on several successive minutes, and the fraction of a second difference between clock seconds and signal markers may be found by concentrated observation of the seconds over 10–15 seconds whilst listening to the "pips". Successive checks at intervals of several hours will enable the rate of gain or loss of the clock to be found.

---

*It may be necessary at some time in the future to alter the coding to allow for a correction of up to nine-tenths of a second.

The timing of observations

EXAMPLE. Date: 1975 January 22.   Zone Meridian: 12^h East

	Time-signal U.T.C.	Mean Time Clock Approx. L.Std.T.
(1)	$4^h\ 00^m\ 00^s$	$16^h\ 04^m\ 22.5^s$
(2)	10 00 00	22 04 24.0

The U.T.C. time-signal had the seconds markers 1 through 6 emphasized, showing that DUT1 was $+0.6^s$ to the nearest 0.1^s; hence

$$\text{U.T.1} = \text{U.T.C.} + 0.6^s$$

and we may re-write the table above as:

	Time-signal U.T.1	Clock Approx. L.Std.T.
(1)	$4^h\ 00^m\ 00.6^s$	$16^h\ 04^m\ 22.5^s$
(2)	10 00 00.6	22 04 24.0

An observation was made at $19^h\ 10^m\ 08.2^s$ clock time; it is required to convert this to U.T.1.

The clock has gained $24.0^s - 22.5^s = 01.5^s$ in 6 hours; thus by simple proportion,

$$\frac{19^h\ 10^m\ 08.2^s - 16^h\ 04^m\ 22.5^s}{6^h} \times 1.5^s = \frac{3^h\ 05^m\ 45.7^s}{6^h} \times 1.5^s$$
$$= 0.8^s.$$

The clock was fast by $(16^h\ 04^m\ 22.5^s - 4^h\ 00^m\ 00.6^s)$, or by $12^h\ 04^m\ 21.9^s$ on U.T.1 at $4^h\ 00^m\ 00.6^s$ U.T.1; therefore at the time of the observation it was $12^h\ 04^m\ 21.9^s + 0.8^s = 12^h\ 04^m\ 22.7^s$ fast on U.T.1. Hence U.T.1 at the time of the observation was $19^h\ 10^m\ 08.2^s - 12^h\ 04^m\ 22.7^s = 7^h\ 05^m\ 45.5^s$.

More refinement of the comparison may be obtained if a stop-watch is used: the rhythm of the seconds pips or tocs is picked up by moving the hand holding the watch in time with them; the watch button is then pressed at a known second (e.g. the minute marker of the time-signal, though it is often better practice to start—or stop—the watch on the third pip after the minute marker to avoid the break in rhythm caused by the omission of the 59th pip from the signals). The observer then stops the watch on an appropriate clock second; this clock time minus the stop-watch reading gives the clock time corresponding to the signal time when the watch was started. An accuracy of $\frac{1}{5}$ second can be attained with a little practice. With most time-signals, a number of successive comparisons of this kind may be made over a few minutes and a mean value taken.

Astronomy for surveyors

6.09 Accuracy of timing observations

In the above discussions we have seen that, for ordinary work using a clock and a stop-watch, we can possibly achieve a timing accuracy of the order of one-fifth of a second and perhaps slightly better (assuming that personal error is known). The *Star Almanac for Land Surveyors* gives the right ascensions of stars to the nearest tenth of a second only, and if we wish to better this we must seek the aid of more elaborate tables and equipment such as impersonal eyepieces and a Chronocord, and use comprehensive observing programmes. The only workers requiring such accuracy are geodetic surveyors engaged in high-class work where such considerations as deviations of the vertical are involved. There is obviously not much point in trying to achieve first-order accuracy in ordinary work if we are ignorant of the amount of the deviation in the area where we are working.

References

Allan, A. L., Abstracting Chronograph Times. *Empire Survey Review,* No. 115, **15**, Jan. 1960, pp. 237–9.

Anon., Time in Milliseconds. *Canadian Surveyor,* **XVIII(5)**, 1964, pp. 422–3.

Beeb, G., Personal Equation in Field Astronomy. *Australian Surveyor,* **20**, 1964–65, pp. 583–92.

Bennett, G. G., The Discrimination of Radio Time Signals in Australia: *Uniciv. Report* No. D1, June, 1963. University of New South Wales.

Buschmann, E., Accidental and Systematic Errors of Time Observations and Time Systems. *Bulletin Géodésique,* **93**, Sept. 1969, pp. 277–82.

Moreau, R. L., Photoelectric Observations in Geodetic Astronomy. *Canadian Surveyor,* **XX(4)**, 1966, pp. 282–91.

Opalski, W., Determination of Relative Personal Equations in the Observatory of the Warsaw Polytechnic. *Bulletin Géodésique,* **74**, Dec. 1964, pp. 307–14.

Robbins, A. R., Time in Geodetic Astronomy. *Survey Review,* **19**, No. 143, Jan. 1967, pp. 2–19; also **19**, No. 144, April 1967, pp. 94–5.

Robbins, A. R., *Field and Geodetic Astronomy*. Ministry of Defence, United Kingdom, 1976.

Steele, J. McA., Standard Frequency Transmissions, *Wireless World,* **68**, April 1962, pp. 160–5.

Stepec, W. A., Visible Records of Time Signals. *Canadian Surveyor,* **XXIII(5)**, 1969, pp. 512–18.

CHAPTER 7
Astronomical and instrumental corrections to observations of altitude and azimuth

7.01 General

With the theodolite we are limited to observing angles in planes which, by the construction of the instrument and attention to levelling, are as nearly as possible vertical and horizontal. It is not possible to make these planes exactly vertical and horizontal, but by doing our observations on both faces and taking means we can eliminate most of the errors arising out of these slight imperfections in the instrument. Those errors which are not eradicated by changing face must be computed and applied as corrections to the observed values. Occasionally weather conditions or curtailment of time may mean that a set of observations made on one face of the theodolite must be accepted, and in this case all the proper corrections must be calculated and applied.

Apart from such purely instrumental corrections to observations of altitude and azimuth, there are a number of others which arise from astronomical considerations.

Section 1

CORRECTIONS TO ALTITUDE OBSERVATIONS

These may be grouped as follows:
  (A) Astronomical: (i) Parallax
                      (ii) Celestial refraction
                      (iii) Semi-diameter
                      (iv) Curvature of path of celestial body
  (B) Instrumental: (v) Vertical collimation
                      (vi) Alidade bubble off-centre

7.02 Parallax corrections to altitudes

The fixed stars are so distant from us that their relative positions always appear to be the same at any instant, no matter from what point upon the earth's surface they are observed. Even with our best surveying instruments no difference can be detected, because their distance is practically infinitely great in comparison with the diameter of the earth. But with the members of our own system, the sun, the moon and the planets, we are dealing with bodies incomparably nearer to us, and their relative positions amongst the fixed stars of the sky are not precisely the same when viewed from different places at the same moment. It is, therefore, essential that their catalogued right ascensions and declinations should be referred to some definite earth point, in order that they may be readily available to all observers. The point selected is the earth's centre, because, having observed the position of a member of the solar system from any station on the earth's surface, it is an easy matter to deduce its position as it would appear from the earth's centre; and conversely if the position of the object is tabulated as it would be seen from the centre of the earth, we may readily find its position as seen from any place on the earth's surface.

The difference between the direction of a heavenly body as seen from the earth's centre and as seen from the place of observation at the same instant is known as its *parallax*.

Thus, as in Fig. 7.1, if S is, say, the sun, A the point of observation, and O the earth's centre, the parallax of the body is the angle ASO, the difference in the directions of AS and OS. If AH_1 is the direction of the horizontal at A, the altitude of S is the angle SAH_1. If OH_2 is drawn parallel to AH_1, then the difference between the angles SOH_2 and SAH_1 is equal to the difference between the angles SOB and SAB which is equal to the angle ASO. Thus, if we call q the parallax, then

$$q = \text{angle } ASO = SOH_2 - SAH_1.$$

Clearly the angle SOH_2 is always greater than the angle SAH_1. If ζ denotes the zenith distance of S as observed from A, R the earth's radius OA, and d the distance OS, then, from the triangle AOS,

$$\frac{\sin q}{\sin \zeta} = \frac{R}{d}.$$

If the body is observed on the horizon—that is to say, if $\zeta = 90°$, the corresponding value of q is called the *horizontal parallax* of S. Call this Q. Then

$$\sin Q = R/d,$$

Therefore

$$\sin q = \sin Q . \sin \zeta.$$

Since q and Q are very small, except in the case of the moon, whose parallax sometimes exceeds 1°, we may substitute the angles for their sines and write

$$q = Q.\sin \zeta = Q.\cos h, \quad \text{where } h = \text{altitude}. \tag{7.1}$$

The moon and the planets are hardly, if ever, used nowadays by the land surveyor for his astronomical observations, but the sun is often used. Its horizontal

Astronomical corrections

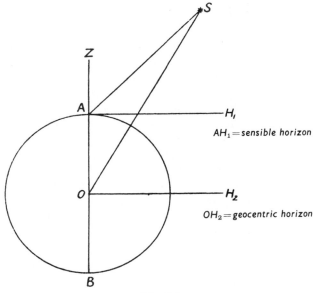

Fig. 7.1

parallax is about 9″. Parallax is greatest when the body is on the horizon and diminishes with the altitude until it becomes zero when the body is in the zenith. The values for various altitudes of the sun are tabulated below:

Sun's parallax in altitude: a critical table

Altitude	0°	19°	27°	43°	53°	60°	67°	73°	80°	86°	90°
Parallax	9″	8″	7″	6″	5″	4″	3″	2″	1″	0″	

The practice for ordinary observations is to take the sun's parallax as a constant, equal to $0.1'$, for all altitudes less than 70°. The maximum error introduced by doing this is 3″. If the parallax correction is being incorporated in a computer programme, use equation (7.1).

We see from Fig. 7.1 that the effect of the parallax upon a celestial object is to make its altitude appear less when observed from A than it would be if seen from O. Consequently, when reducing observations to the earth's centre, we must *add* the correction for parallax to the observed altitude, or

$$\text{True altitude} = \text{observed altitude} + \text{parallax}.$$

For all practical purposes parallax has no effect upon the azimuth of an object in the sky, and the correction is made to altitude only.

7.03 Celestial refraction corrections to altitudes

When a ray of light passes from one medium into a denser medium, as from air into water or from air into glass, it is bent or *refracted* towards the normal to the bounding surface. Thus, as in Fig. 7.2, if a ray of light passes from the medium A to a denser medium B, traversing the path PQR, the refracted ray QR will always make a smaller angle with the normal to the separating surface than the incident ray PQ. The direction of bending is always such that the bent or refracted ray lies in the same plane as that passing through the incident ray PQ and the normal QN. The law governing the amount of bending is that the ratio between the sines of the angles PQN and RQM is constant for these particular media, and the value of this ratio is known as the *coefficient of refraction* or the *relative refractive index* of the given pair of media.

Similarly, when a ray of light from a celestial body reaches the atmosphere surrounding the earth, it is bent slightly out of its original path. If the atmosphere were a uniform homogeneous medium with a definite upper surface, it would be comparatively easy to determine the precise amount of bending of the ray. But the density of atmospheric air diminishes with height above the earth's surface. Consequently a ray from a star S (Fig. 7.3), when it reaches the upper limit of the earth's atmosphere at A, is only very slightly bent, but the amount of bending gradually increases as it passes into the lower and denser layers of air. Its path from A to an observer on the earth's surface at O is thus a curve, and the ray ultimately reaches the observer so that it appears to him to come in the direction of OS', the tangent to the curve at O. Thus the observer sees a star apparently at S' in the celestial sphere, whereas in reality the star is at S. The effect is that the star is apparently raised above its true position,

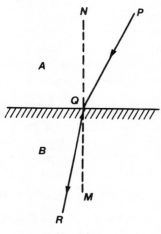

Fig. 7.2

Astronomical corrections

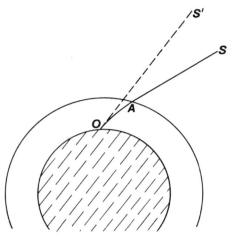

Fig. 7.3

and its apparent altitude is greater than the true altitude if it could be observed from O with no intervening atmosphere. The observed altitude of a celestial body must therefore be corrected in order to deduce its true altitude, the correction being always *subtracted* from the observed altitude. The amount of bending of the ray varies somewhat with the pressure and temperature of the air, but it is greatest for stars on the horizon, and gradually decreases to zero for a star in the zenith. For a body on the horizon the mean value of the correction is $33'$—that is to say, a star will be just visible on the horizon when it is really $33'$ below it. Thus the sun, whose diameter is about $32'$, is visible just above the horizon when it is in reality just below it.

It will be seen from the figure that since the refracted ray always lies in the plane containing the incident ray SA and the normal to the spherical bounding surface at A, then S and S' will lie in the same plane as the vertical at O. This means that refraction produces its effect entirely in altitude, and has no influence upon the apparent azimuth of a heavenly body. Thus no correction in azimuth is necessary on account of refraction.

As we do not know the exact laws which govern the pressure and temperature of the earth's atmosphere at different heights, no satisfactory computation of the amount of refraction at different altitudes can be made from theoretical conditions alone. By making different assumptions as to the character of the earth's atmosphere various formulae have been derived, but as their demonstration generally requires mathematics of a rather advanced character, we shall not attempt the problem here. In any case, as we cannot be sure of the correctness of the assumptions that have to be made in order to derive the formulae, the values of the constants used have to be obtained and checked from actual observations. There are various ways by which the amount of refraction at

Astronomy for surveyors

different altitudes may be actually measured, and for practical purposes that formula is selected which best fits the results of such measurements.

Tables from which the atmospheric refraction correction may be quickly and easily calculated are given on pages 62–64 of the *Star Almanac*. The critical table on page 62 enables the mean refraction r_0 to be found immediately by entering it with the observed altitude. Atmospheric pressure and temperature should be recorded at the time of an observation, and the tables on pages 63 and 64 then provide a correcting factor f by which r_0 is multiplied to yield the refraction correction for the pressure and temperature prevailing at the time of the observation.

EXAMPLE. A body was observed at altitude $33°\ 22'\ 16''$, the atmospheric pressure and temperature being 1009 mb and 20° C respectively. What was the true altitude?

$$\begin{aligned}
&\text{From table on page 62 of the } \textit{Star Almanac}, r_0 = \quad 88'' \\
&\text{From table on page 63 of the } \textit{Star Almanac}, f = 0.96 \\
&\text{Then refraction correction} = -(88'' \times 0.96) \quad = -84'' \\
&\text{Observed altitude} \quad\quad\quad\quad = 33°\ 22'\ 16'' \\
&\text{Refraction correction, } \psi \ = \quad\ -01\ \ 24 \\
&\quad\quad\quad\quad\quad\quad\quad\quad\quad\quad\quad\ \ \overline{33°\ 20'\ 52''}
\end{aligned}$$

It will be noticed that, in the *Star Almanac*, mean refractions are not given for observed altitudes below 10°. Generally speaking, it is unwise to observe at altitudes less than 15° (that is, where the value of the altitude is used as an element in a solution of the astronomical triangle) because of the magnitude of the correction and increasing uncertainty in its amount. As will be seen later, it is desirable where possible to make observations in pairs to stars at approximately the same altitudes but opposite azimuths in order to reduce the effect of uncertainty in the refraction corrections at altitudes even considerably higher than 15°.

There is no simple formula for computing the refraction correction, but if one is needed for a computer program the standard formula quoted by J. Saastamoinen in a paper "Contributions to the Theory of Atmospheric Refraction" (*Bulletin Géodésique*, New Series, No. 105, Sept. 1972, p. 279) may be used for zenith distances not greater than 75° (i.e. altitudes not less than 15°). The formula is (for standard conditions):

Astronomical corrections

$$\Delta\zeta_0'' = 16''.271 . \tan\zeta \left[1 + 0.0000394 . \tan^2\zeta \left(\frac{p - 0.156e}{T}\right)\right] \left(\frac{p - 0.156e}{T}\right)$$

$$-0''.0749(\tan^3\zeta + \tan\zeta) \left(\frac{p}{1000}\right) \qquad (7.2)$$

where $\Delta\zeta_0''$ = the correction in seconds of arc to the apparent zenith distance,
ζ = the apparent zenith distance,
p = the total barometric pressure in millibars,
e = the partial water vapour pressure in millibars,
T = the absolute temperature in degrees Kelvin
$(T = 273.16 + t\,°C)$.

We have noted above that it is not wise to observe altitudes below $15°$ ($\zeta = 75°$) if they are to be corrected and used in the solution of the astronomical triangle. However, should it be necessary to do this, Saastamoinen adds a further term $(+\delta'')$ to the right-hand side of (7.2) which he gives in a table on page 387, *Bulletin Géodésique*, No. 106, Dec. 1972, related to specific values of ζ and altitude of the observing station above sea-level. This term δ'' does not exceed $+1''.14$, but the table does not extend beyond $\zeta = 80°$. δ'' may be calculated but there are four rather complex terms in the formula, which is to be found on page 385 of the journal referred to immediately above.*

7.04 Semi-diameter correction to altitudes of the sun

When observing the sun with a theodolite (using a dark glass over the eyepiece to prevent damage to the eye) it appears to take up a considerable portion of the field of view, and it is not possible to estimate the position of its centre with any accuracy, particularly as it is moving. When measuring its altitude it is customary to note the instant when its apparent *lower* edge (or "limb") is tangential to the horizontal cross-hair in the centre of the field of view and to record the altitude shown on the vertical circle (with the alidade bubble levelled, or the amount of mislevelment noted); the instrument face is then quickly reversed, the sun re-sighted, and the instant noted when its apparent *upper* edge is tangential to the central portion of the horizontal cross-hair. The vertical circle is again read after attention to the alidade bubble. Then for most types of observation, if the time between the two observations is only 2 or 3 minutes, it is sufficient to take the mean of the two altitudes as the altitude of the sun's centre at the mean of the two recorded times. By doing

---

* In *Geodesy* (1975), Bomford gives, for $\zeta < 75°$, $\Delta\zeta = 16.3''\,(\tan\zeta)\,(p - 0.14e)/T$. The effect of e is generally negligible.

the observations in this way, the correction for the sun's semi-diameter and any vertical collimation error in the instrument are eliminated.

However, if it is necessary to accept for altitude a single observation to the sun on one face of the instrument, a correction to bring the observation to the sun's centre, by allowing for its semi-diameter, must be applied. The ordinary theodolite telescope reverses the image; thus when the horizontal cross-hair is tangential to the apparent lower edge of the sun it is, in fact, tangential to the sun's upper edge. In such a case, the semi-diameter must be subtracted. It must be added, of course, when tangency is to the sun's upper limb as seen through the telescope (but note that some theodolites do not invert the image).

Values of the sun's semi-diameter (in arc) are tabulated on pages 2-25 of the *Star Almanac*, at about fortnightly intervals, in the lines containing the words "Sunrise" and "Sunset". The slight variation from the mean value of 16.0' is due to the changing distance between earth and sun.

An ingenious prism attachment which fits on the theodolite telescope in front of the objective is due to Professor R. Roelofs of Delft. It is manufactured by Wild of Switzerland under the name of the Roelofs Solar Prism Attachment. This device gives four overlapping images of the sun of the form shown in Fig. 7.4.

The four bright lobes of the overlapping images form a cross against which the cross-hairs can be clearly seen; it is thus much easier to bisect the lobes of the cross with the cross-hairs than it is to obtain tangency between the limbs of the sun and the cross-hairs—particularly when simultaneous tangency of the disc with both cross-hairs must be obtained, as when azimuth and altitude are being observed at the same time.

The solar prism does away with any need for a semi-diameter correction, and it has filled a long-felt want by increasing both the ease and accuracy of sun observations for the surveyor. A more detailed description of the attachment is given on pages 67-70 of *Astronomy Applied to Land Surveying* by Roelofs (Ahrend, Amsterdam, 1950).

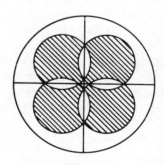

Fig. 7.4

Astronomical corrections

7.05 Corrections to altitudes for curvature of path of celestial body

The usual practice in observing altitudes is to note the vertical circle reading and the time when the star is on the horizontal cross-hair at position S_1 as in Fig. 7.5; the face of the instrument is then changed as quickly as possible and further readings of the vertical circle and time are taken when the star is at S_2. The mean of the circle readings is then taken, and after correcting for refraction is accepted for most work as the altitude at the mean of the recorded times. This places the star at $\bar{S}$, with altitude $\bar{h}$. However, the star's path is curved, and at the mean of the times it will be midway between S_1 and S_2 on the arc, i.e. at S, when it will have altitude h. Hence, if greater accuracy is required from the original observations, especially if there has been an excessive time-lag between pointings, a correction of $\Delta h = (\bar{h} - h)$ must be applied.

The second-order correction for curvature in altitude

Let an observed altitude h_i correspond to an observed time T_i. Since $h = f(T)$ and since this function is not a linear one, the mean altitude $\bar{h}$ will not correspond to the mean value $\bar{T}$ of the times.

Let h be the altitude which does correspond to $\bar{T}$. Then the interval $(h_i - h)$ will correspond to the interval $(T_i - \bar{T})$. But from a Taylor expansion,

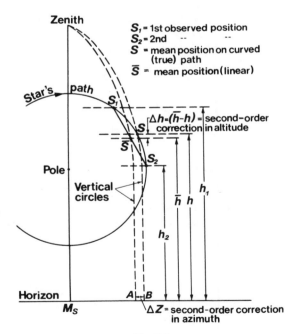

Fig. 7.5

Astronomy for surveyors

$$\Delta h = f'.\Delta T + \tfrac{1}{2}f''.(\Delta T)^2 \ldots$$

$$\therefore (h_i - h) = f'.(T_i - \overline{T}) + \tfrac{1}{2}f''.(T_i - \overline{T})^2 \ldots$$

If, now, all these differences are added together and meaned, we get, where n is the number of pointings:

$$\frac{\Sigma h_i}{n} - \frac{n.h}{n} = \frac{f'}{n}.\Sigma(T_i - \overline{T}) + \frac{1}{2n}f''.\Sigma(T_i - \overline{T})^2 \ldots,$$

neglecting higher powers, or

$$\bar{h} - h = \frac{f''}{2n}.\Sigma(T_i - \overline{T})^2, \qquad (K)$$

since differences from the mean $[\Sigma(T_i - \overline{T})]$ sum to zero.

To find the differential coefficients f' and f'', when ϕ and δ are held constant and h is varied, we start with the formula

$$\sin h = \sin\phi.\sin\delta + \cos\phi.\cos\delta.\cos t.$$

(ϕ = latitude, north+, south−; δ = declination, north+, south−; t = hour angle.)

Differentiating this:

$$\cos h.\frac{dh}{dt} = 0 - \cos\phi.\cos\delta.\sin t.$$

$$\therefore f' = \frac{dh}{dt} = -\cos\phi.\frac{\cos\delta}{\cos h}.\sin t$$

i.e.

$$f' = \cos\phi.\sin A, \qquad (7.3)$$

where A = azimuth.

Differentiating this again gives

$$f'' = \frac{d_2 h}{dt^2} = -\cos\phi.\cos\delta\left\{\sec h.\cos t + \sin t.\sec h.\tan h.\frac{dh}{dt}\right\}$$

$$= -\cos\phi.\cos\delta.\sec h.\sin t\left\{\cot t + \tan h.\frac{dh}{dt}\right\}$$

i.e.

$$f'' = f'\{\cot t + \tan h.f'\}. \qquad (7.4)$$

Substituting this in the expression (K) above, we get:

Astronomical corrections

$$\bar{h} - h = f'\{\cot t + \tan h.f'\}.\frac{\Sigma(T_i - \bar{T})^2}{2.n} \text{ in radian units,}$$

or

$$(\bar{h} - h)'' = f'\{\cot t + \tan h.f'\}.\frac{\Sigma(T_i^s - \bar{T}^s)^2 .225}{2.n.\rho} \tag{7.5}$$

in arc and time seconds units, where

$$\rho = 1/\text{arc } 1'' = 206\,265$$

and $(\bar{h} - h) = \Delta h'' =$ the second-order curvature correction in altitude, in arc-seconds.

EXAMPLE. Four pointings were made as below on a star at a place in south latitude 45° 52′ 15″. The temperature was 15 °C and the atmospheric pressure 981 mb. Find the error in accepting the mean of the observed altitudes as the correct value for the mean of the times. The sidereal clock used was 17.5^s fast on L.S.T. The R.A. and Dec. of the star were 4^h 15^m 21.3^s and S. 51° 37′ 37″, respectively.

Observed Azimuth A	Observed Altitude h	Δh	Observed L.S.T.	ΔT
231° 46′ 30″	38° 31′ 06″	+2620″	9^h 43^m 50.9^s	−321.7^s
231 34 19	38 13 26	+1560	9 46 00.4	−192.2
231 13 26	37 43 28	−238	9 49 40.7	+28.1
230 29 28	36 41 44	−3942	9 57 18.4	+485.8
Means 231 15 55.8	37 47 26		9 49 12.6	
Refraction = 75″ × 0.95	−01 11		−17.5	Chrono. fast
Corrected altitude	37° 46′ 15″		9 48 55.1	L.S.T.
ç	52 13 45		4 15 21.3	R.A. star
			5^h 33^m 33.8^s	H.A. star = t
			83° 23′ 27″	t

Δh and ΔT are differences from the mean.

We now use equation (7.5) to compute the second-order correction in altitude (the curvature correction).

From (7.3), $f' = \cos\phi . \sin A$
                           $= \cos(-45° \ 52').\sin(231° \ 16')$
                           $= -0.5432.$

From (7.4), $f'' = f'\{\cot t + \tan h.f'\}$
                           $= -0.5432\{\cot(83° \ 23') + \tan(37° \ 46').(-0.5432)\}$
                           $= (-0.5432) \times (-0.3021)$
                           $= +0.1641.$

In (7.5) the factor $\Sigma(T_i^s - \bar{T}^s)^2 .225/(2.n.\rho)$ is equal to:

$$\{(-321.7)^2 + (-192.2)^2 + (28.1)^2 + (485.8)^2\}.225/(2.4.206\,265)$$
$$= +51''.436.$$

Hence

$$(\bar{h} - h)'' = 0.1641 \times 51.436 = +8.4''.$$

Since h is the altitude corresponding to the mean of the times, it is equal to $\bar{h}$, the mean altitude, minus $08''.4$, i.e.

$$h = 37°\ 46'\ 15'' - 08'' = 37°\ 46'\ 07''.$$

Since we have accurate sidereal times of the pointings, we can check this answer by actually computing the altitude at the mean of these, using the formula:

$$\cos\zeta = \cos\omega.\cos p + \sin\omega.\sin p.\cos t,$$

where

$$\omega = (90° - \phi) = 44°\ 07'\ 45''$$
$$p = (90° - \delta) = 38\ \ 22\ \ 23$$
$$t\ \ \ \ \ \ \ \ \ \ \ \ \ \ \ = 83\ \ 23\ \ 27$$

Thus

$$\zeta = 52°\ 13'\ 53''.7 = (90° - h),$$

whence

$$h = 37°\ 46'\ 06.3''$$

which agrees closely with the answer obtained using the curvature correction.

It is not usual to have the precise local time of each pointing as we have had in the above example. We may be trying, in fact, to obtain accurate local time from an observation; measurement of the altitude of a celestial object, plus a knowledge of the latitude of the place and the object's declination, will give us the three sides of the astronomical triangle and allow us to compute the hour angle. This, in turn, will lead to the local time (e.g. H.A. + R.A. = L.S.T.), and if we have timed the altitude pointings with chronometer and stop-watch we can then find the error of the chronometer on local time. In this case the *differences* between the times of the altitude pointings will give us the ΔT's we need in order to compute the $\Sigma(T_i^s - \bar{T}^s)^2$ factor in the curvature correction formula, even though the chronometer was not showing the correct local time. See **12.12**.

It should be noted that, for an elongating star, the sign of $(\bar{h} - h)''$ is negative at altitudes above elongation (see **10.09**), and positive at those below. The

Astronomical corrections

correction vanishes at elongation and also when $Z = 90°$, i.e. when the body is on the prime vertical.

Formula (7.5) may also be used, of course, to correct the mean of a simple observation consisting of one F.L. pointing and one F.R., where there has been an excessive time-lag between them. If we call the total time-difference between the two pointings Δt, then the factor

$$\Sigma(T_i^s - \bar{T}^s)^2 .225/(2.n.\rho)$$

becomes

$$\{(\tfrac{1}{2}\Delta t)^2 + (\tfrac{1}{2}\Delta t)^2\}.225/(4.\rho) \quad (n = 2 \text{ in this case})$$

and this reduces to $\{(\Delta t)^2 .\text{arc } 1''/8\}$ if Δt is in seconds of arc, i.e. if the time-interval between faces in seconds of time is multiplied by 15.

Thus for two pointings we have

$$(\bar{h} - h)'' = f''.\{(\Delta t)^2.\text{arc } 1''/8\}. \tag{7.6}$$

Formula (7.5) may be written in another form:

$$\Delta h'' = \cos\phi.\cos Z.(\cos\phi.\cos Z.\tan h - \sin\phi).\{\Sigma(m)\}/n, \tag{7.7}$$

where Z and h are the mean values of the zenith angles and altitudes respectively, ϕ is the latitude (always positive),

$\Delta h''$ = the curvature correction in arc seconds to the mean observed altitude,
m = $2.\sin^2(\Delta T/2).\text{cosec } 1''$,
n = the number of pointings from which the mean is taken, as before,
ΔT = the difference between the mean of the observed times of the pointings and the times of each pointing (sidereal for stars and mean solar for the sun).

The value of m in seconds of arc corresponding to each ΔT may very conveniently be found from the "Table for Circum-meridian Observations" on page 68 of the *Star Almanac*. The ΔT values correspond to those given in the example earlier in this section; with their m values from the Table, we get:

ΔT			m (from Table)
321.7^s	= 05^m	21.7^s	$56''$
192.2	= 03	12.2	20
28.1	= 00	28.1	0
485.8	= 08	05.8	129
$n = 4$		Σm	205
		$(\Sigma m)/n$	$51.3''$

We note that this is, within a point or two of a second, the same value as that given earlier by the factor $\Sigma(T_i^s - \bar{T}^s)^2.225/(2.n.\rho)$.

The other factor in (7.7) is:

$$\cos\phi.\cos Z(\cos\phi.\cos Z.\tan h - \sin\phi),$$

and it computes as -0.1656, which is essentially the same value as f'' earlier, but with opposite sign. Hence

$$\Delta h'' = (-0.1656) \times 51.3''$$
$$= -8.5''$$

to be added algebraically to the mean of the observed altitudes after correction for refraction. Thus the true altitude at the mean of the observed times was $(37°\ 46'\ 15''\ -08.5'')$, or $37°\ 46'\ 06.5''$, which when rounded to the nearest second is the same as that obtained from (7.5).

Z in formula (7.7) is the zenith angle of the star, or $(A - 180°)$ in this case $= (231°\ 16' - 180°) = 51°\ 16'$.

If only two pointings to the star (or sun) are made, ΔT is half the time between the two pointings; call it $\Delta t/2$. Then

$$(\Sigma m)/n = \{(\Delta t/2) + (\Delta t/2)\}/2,$$

which is $\Delta t/2$. Hence, to obtain the curvature correction for two pointings, we find m from the *Star Almanac* table for *half* the time between the pointings and multiply it by f'' as obtained from (7.3) and (7.4).

7.06 Vertical collimation correction to altitudes

This correction is applicable only to single-face observations of altitude, since it is eliminated by taking the mean of readings on each face.

To determine the amount of the correction, or vertical circle index error, a series of left- and right-face readings is taken on a number of fixed objects at various altitudes. The means will give the correct altitudes of the objects, and the mean of the differences between the true altitudes and those yielded by the separate left- and right-face readings will give the amount of correction.

EXAMPLE

Object	*A*	*B*	*C*
L.F.	12° 42' 36"	23° 18' 54"	37° 29' 30"
R.F.	12 42 22	23 18 38	37 29 18
Mean	12 42 29	23 18 46	37 29 24
Difference between mean and single-face readings	07"	08"	06"

Astronomical corrections

Mean difference = 07″ (to be subtracted from single L.F. readings and added to single R.F. readings).

7.07 Correction to altitude due to alidade bubble being off centre

Most modern theodolites have a slow-motion alidade adjusting screw by means of which the vertical circle bubble may be centred without altering the direction of the line of sight through the telescope. If, then, immediately after sighting to a celestial object, the alidade bubble is centred by means of this screw, the altitude recorded will be free from bubble error. Some alidade bubble-tubes are graduated to enable the observer to determine the amount (if any) by which the bubble is off centre. Others have an arrangement of prisms to show the bubble split longitudinally; opposite half-ends of the bubble are brought into the viewing window, and, by turning the alidade screw, these can be brought together so that they appear as one single end of the full bubble when the latter is centred. Usually there are no divisions marked on this type of bubble tube, so that it *must* be centred before the circle is read.

Sometimes when one is using an instrument equipped with a graduated alidade bubble-tube, it is convenient to read the amount by which the bubble is off centre and make a correction later to the observed altitude. Bubble-tubes are graduated evenly outwards on both sides of the centre, and in order to be able to find the position of the centre of the bubble with respect to the centre of the tube, both ends of it must be read against the graduations. The end of the bubble towards the eye is always called the E end, and the other, towards the object, is the O end, whatever the face being used for the observation. Each separate altitude noted will thus have its accompanying E and O reading in the field-book.

The level correction is found by dividing the difference between the sums of the readings of the object end and eye end by the total number of readings, and then multiplying the result by the angular value of one division of the scale of the level tube (usually marked upon it). If the readings of the object end are greater than those of the eye end, then the zero line is tilted slightly upwards in the direction of the object, and since the altitude on the circle is measured from this line it will be too small, hence the correction is to be added. If the readings of the eye end are the greater, then the correction is to be subtracted. Thus,

$$\text{correction to altitude} = \{[\Sigma(O) - \Sigma(E)]/n\} \times v$$

where

n = number of end readings,
v = angular value of 1 bubble-tube division.

Astronomy for surveyors

Suppose that two observations were taken, one on face left and the other on face right, as follows:

	Bubble readings	
	O	E
F.L.	5	9
F.R.	7	7

then

$$\frac{\Sigma(O) - \Sigma(E)}{n} = \frac{12 - 16}{4} = -1.$$

If the angular value of 1 division was $20''$, it follows that the mean observed altitude must be reduced by $20''$ to yield the true altitude.

Even with modern instruments using split-bubble vertical circle levels, some observers prefer not to centre the bubble at each pointing, but to read and record any displacement. This can be achieved by inserting a piece of transparent drawing medium, suitably graduated in black ink, over the bubble tube. A correction, as above, is then needed.

(Note that some modern theodolites have automatic levelling of the vertical circle index; no manual bubble centring is possible and no correction is necessary.)

Section 2

CORRECTIONS TO AZIMUTH OBSERVATIONS

These may be grouped as follows:

(A) Astronomical: (i) Semi-diameter
(ii) Curvature of path of celestial body
(B) Instrumental: (iii) Horizontal collimation
(iv) Trunnion axis not level.

7.08 Semi-diameter correction to azimuths of the sun

In **7.04** we have seen that, in order to correct altitude observations to the sun's upper or lower limb, we have merely to apply the *Star Almanac* value of its semi-diameter in the correct sense. With observations for azimuth, however, the matter is not quite so simple. Thus, in Fig. 7.6, if C denotes the centre of the sun's disc, Z the zenith, ZCA the vertical trace on the celestial sphere passing through C and the zenith, ZLB the vertical trace just touching the edge of the sun's disc at L, then the error in azimuth made by sighting the limb instead of the centre is the angle CZL. But in the right-angled spherical triangle ZLC we have

Astronomical corrections

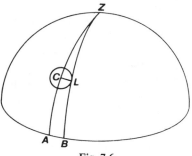

Fig. 7.6

$$\sin CL = \sin CZ . \sin CZL.$$

Now CL is an angle of about $16'$, and its circular measure does not differ appreciably from its sine. Consequently, we may write

$$CL = \sin CZ \times CZL$$

or

$$CZL = CL \times \operatorname{cosec} CZ$$
$$= CL \times \sec CA$$

or

correction in azimuth = semi-diameter × secant altitude of sun's centre (7.8)

Remembering that the telescope usually reverses the image, and that apparent tangency between the vertical cross-hair and the sun's left limb means actual tangency with the right limb, we see that when the sun's left limb is observed through the telescope, the recorded azimuth must be reduced by the amount of the correction, and vice versa.

Provided that there is not too long a time-interval between a left- and a right-face observation to the sun's left and right limbs, the mean of the recorded azimuths will give the mean azimuth of the sun's centre at the mean of the observed times. If a curvature correction (see next section, **7.09**) *is* necessary, however, it will probably also be necessary to apply a correction to the sun's azimuth because the secants of the FL and FR altitudes will differ appreciably in (7.8) above. The correction is given by

$$\Delta A'' = \tfrac{1}{2}.(\text{S.D.}).\sin 1'. \sec h . \tan h . \Delta h'', \qquad (7.9)$$

where

- $\Delta A''$ = the correction to the sun's azimuth computed from the mean altitude,
- (S.D.) = the sun's semi-diameter in *minutes* of arc,
- $\Delta h''$ = the difference in altitude of the sun's *centre* between the two pointings, in *seconds* of arc.

With $\Delta h''$ taken always as positive, the correction is positive for morning observations and negative for afternoons, when the true left limb is observed first and the true right limb observed second; the sign is reversed if the order of pointings to the limbs is reversed.

A useful table of secants is given on page 65 of the *Star Almanac*.

From a consideration of **7.04**, it will be seen that use of the Roelofs solar prism attachment obviates the necessity for semi-diameter corrections to sun azimuths made on a single face.

7.09 Corrections to azimuths for curvature of path of celestial body

Referring to Fig. 7.5 in **7.05**, we see that the zenith angle of the star when it is assumed to be at $\bar{S}$ is equal to the arc $M_S A$. $\bar{S}$ is the mid-point of the chord joining S_1 and S_2 and it may be taken as lying half-way between the azimuths of S_1 and S_2, i.e. the azimuth of $\bar{S}$ is the mean of the azimuths of S_1 and S_2 (very nearly). However, we have noted that the star will be at S at the mean of the recorded times for S_1 and S_2. For S the zenith angle is equal to the arc $M_S B$. Hence to correct the arithmetic mean of the azimuths for S_1 and S_2 to the azimuth for S we must find the arc $AB(\Delta Z)$. It should be noted that the solution of the main spherical triangle PZS to give Z, using ϕ, δ and mean t, will give Z for the position S (i.e. the arc $M_S B$). We want Z for $\bar{S}$, since it is the angle between the reference mark and $\bar{S}$ which we shall use to determine the azimuth of the reference mark.

The second-order correction for curvature in azimuth

By a process similar to that used in deriving equation (7.5) in section **7.05** above, it may be shown that:

$$\Delta A'' = \sin A . \cos \phi . \sec^2 h . (\cos h . \sin \delta - 2 . \cos A . \cos \phi)$$
$$\times \{\Sigma (T_i^s - \bar{T}^s)^2 . 225/(2.n.\rho)\}, \qquad (7.10)$$

where

$\Delta A''$ = the correction in arc seconds to be applied to the azimuth computed from the mean of the observed times,
A = the azimuth of the celestial object observed,
ϕ = the latitude of the observing station, +north and −south,
h = the altitude of the celestial object observed,
δ = the declination of the celestial object observed, +north and −south;

$\{\Sigma (T_i^s - \bar{T}^s)^2 . 225/(2.n.p)\}$ is the same factor as in (7.5); as in that equation, it may be replaced by $\{\Sigma (m)/n\}$. If two pointings only are made, this factor

Astronomical corrections

may be reduced to $\{(\Delta t'')^2 \text{.arc } 1''/8\}$, where $\Delta t''$ is the time elapsed between the two pointings in seconds of time multiplied by 15 to convert it to seconds of arc.

EXAMPLE. We shall derive the curvature correction in azimuth from the data given in the example in **7.05**, using formula (7.10).

In the example, $A = 231°\ 16'$; $\phi = -45°\ 52'$; $h = 37°\ 46'$; $\delta = -51°\ 38'$. The factor $\{\Sigma(T_i^s - \overline{T}^s)^2 .225/(2.n.\rho)\}$ has the same value as in the example of **7.05**, i.e. it is equal to $+51.436''$. Thus,

$$\begin{aligned}\Delta A'' &= \sin(231°\ 16').\cos(-45°\ 52').\sec^2(37°\ 46') \\ &\quad \times \{\cos(37°\ 46').\sin(-51°\ 38') - 2.\cos(231°\ 16').\cos(-45°\ 52')\} \\ &\quad \times 51.436'' \\ &= (-0.8692) \times (+0.2516) \times 51.436'' \\ &= -11.3''\end{aligned}$$

The azimuth of the celestial object at the mean of the observed times is obtained by solving the astronomical triangle for Z the zenith angle, using the co-latitude of the observing station, the polar distance of the star and the hour angle of the star at the mean of the times. From section **7.05** the required data are

$$\omega = (90° - \phi) = (90° - 45°\ 52'\ 15'') = 44°\ 07'\ 45''$$
$$p = (90° - \delta) = (90° - 51°\ 37'\ 37'') = 38°\ 22'\ 23''$$
$$t = 83°\ 23'\ 27''.$$

We may use a formula derived from (1.15):

$$\cot Z = \frac{\cot p.\sin \omega - \cos \omega.\cos t}{\sin t}.$$

Inserting the data, $Z = 51°\ 16'\ 07''$
and azimuth $A = 231\ 16\ 07$
Curvature correction from above $= -11$
Corrected azimuth of star $= 231\ 15\ 56$

When we observe a celestial object for azimuth, we normally transfer that azimuth to the earth's surface by measuring the included angle between the mean of the pointings (horizontal circle) to the star or sun and the mean of a set of pointings to a reference object (R.O.) on the earth's surface (see **8.04**). Having computed the azimuth of the star or sun, we apply this included angle to it and so obtain the azimuth of the R.O. from the observing station. This gives us a reference line for checking ground survey work. The curvature correction in azimuth adjusts the azimuth of the celestial object, as computed from the mean of the times, to correspond with the mean of the observed horizontal circle readings on it, so that the included angle may be applied directly to that mean value to yield the azimuth of the R.O.

If, in the example of **7.05**, the mean horizontal circle reading on the R.O. had been 108° 14′ 07.5″, the included angle between star and R.O. would have been 231° 15′ 55.8″ − 108° 14′ 07.5″ = 123° 01′ 48.3″. After correcting the computed azimuth to the star for curvature as we did earlier, we obtain the azimuth of the R.O. as follows:

$$
\begin{array}{ll}
\text{Corrected azimuth of star} & = 231° \ 15′ \ 56″ \quad \text{(see above)} \\
\text{Included angle, star to R.O.} & = 123 \ \ 01 \ \ 48.3 \\
\hline
\text{Azimuth of R.O.} & = 108 \ \ 14 \ \ 07.7 \\
\text{or, to the nearest second} & = 108 \ \ 14 \ \ 08
\end{array}
$$

In some types of observation for determining azimuth, altitudes only are measured, i.e. the pointings to the celestial object are not timed (see **10.19**). However, we may still compute a curvature correction to the azimuth, if necessary, by using the change in the observed altitudes instead of the change in the times. A formula may be derived in a similar way to that in which we arrived at (7.5) in section **7.05**. It is:

$$\Delta Z'' = \cot S . \sec^2 h . (\sin h - 2 . \cot Z . \mathrm{cosec}\, 2S)$$
$$\times \{\Sigma(h_i - \bar{h})^2/(2.n.\rho)\}, \tag{7.11}$$

where

$\Delta Z''$ = the correction in seconds of arc to the zenith angle as computed from the latitude, declination and mean observed altitude,
Z = the zenith angle of the astronomical triangle,
S = the parallactic angle of the astronomical triangle,
h = the observed altitude of the celestial object,
$\Sigma(h_i - \bar{h})^2$ = the sum of the squares of the differences, in seconds of arc, between the mean altitude and the individual altitudes from which the mean was computed,
n = the number of pointings to the celestial object,
ρ = 1/arc 1″ = 206 265.

Since Z is required in (7.11) and we also need it to obtain the azimuth of the R.O., we shall compute it at this stage using the formula based upon (1.12a):

$$\tan \tfrac{1}{2} Z = \sqrt{\left(\frac{\sin(s-\zeta).\sin(s-\omega)}{\sin s . \sin(s-p)}\right)} \quad \text{where } 2s = (\zeta + \omega + p).$$

Astronomical corrections

In the example of **7.05**,

$$\omega = (90° - \phi) = (90° - 45° \; 52' \; 15'') = 44° \; 07' \; 45''$$
$$\zeta = (90° - h) = (90° - 37° \; 46' \; 15'') = 52° \; 13' \; 45''$$
$$p = (90° - \delta) = (90° - 51° \; 37' \; 37'') = 38° \; 22' \; 23''$$

and Z computes as $51° \; 16' \; 13''$, making the azimuth of the star $231° \; 16' \; 13''$.
We also need S, the parallactic angle, and it may be found from:

$$\cos S = \frac{\cos \omega - \cos \zeta . \cos p}{\sin \zeta . \sin p}$$

which gives $S = 61° \; 04' \; 12''$.

In using (7.11) the various angles may be rounded to the nearest minute:

$$S = 61° \; 04'; \quad h = 37° \; 46'; \quad Z = 51° \; 16'.$$

Thus

$$\Delta Z'' = \cot(61° \; 04') . \sec^2 (37° \; 46')$$
$$\times \{\sin(37° \; 46') - 2.\cot(51° \; 16') . \operatorname{cosec}(122° \; 08')\}$$
$$\times \{(2620)^2 + (1560)^2 + (-238)^2$$
$$+ (-3942)^2\}/(2.4.206\;265),$$

i.e.

$$\Delta Z'' = (-1.1340) \times (+15.09'')$$
$$= -17.1''$$

Then,

Zenith angle computed from mean altitude	=	$51° \; 16' \; 13''$
Curvature, or second-order correction	=	-17
Corrected zenith angle		$51 \; 15 \; 56$
Hence azimuth of star (in this case)		$231 \; 15 \; 56$
Included angle, from above, star to R.O.		$123 \; 01 \; 48$
∴ Azimuth of R.O.	=	$108 \; 14 \; 08$

This result is the same as that obtained using the times of the pointings and the hour angle, as of course it should be.

If only two pointings are made, the factor $\{\Sigma(h_i - \bar{h})^2/(2.n.\rho)\}$ reduces to $\{(\Delta h)^2 . \text{arc } 1''/8\}$, where Δh is the difference in altitude, in seconds of arc, between the two pointings.

As an example, let us take the first and last pointings of the example in **7.05**.

Astronomy for surveyors

The difference in altitude is $(38° 31' 06'' - 36° 41' 44'') = 1° 49' 22'' = 6562''$, and $\{(\Delta h)^2 .\text{arc } 1''/8\} = \{(6562'')^2 .0.000\,004\,848/8\} = 26.09''$. In (7.11), $\cot S.\sec^2 h.(\sin h - 2.\cot Z.\text{cosec } 2S)$ will be the same as before, i.e. it will be (-1.134). Hence

$$\Delta Z'' = 26.09'' \times (-1.134) = -29.6''.$$

If we now compute Z from ϕ $(-45° 52' 15'')$, $\delta(= -51° 37' 37'')$ and the mean of the two altitudes corrected for refraction $(= 37° 35' 14'')$ we find that $Z = 51° 08' 29''$. Applying the curvature correction $(-29.6'')$ to this makes $Z = 51° 07' 59.4''$ and the corrected azimuth A equal to $231° 07' 59.4''$.

The mean of the first and last horizontal circle readings on the star is $\frac{1}{2}(231° 46' 30'' + 230° 29' 28'')$ or $231° 07' 59''$, and the mean of the readings on the R.O. is, as before, equal to $108° 17' 07.5''$, making the included angle between star and R.O. equal to $(231° 07' 59'' - 108° 14' 07.5'')$ or $122° 53' 51.5''$.

Thus,

Corrected horizontal circle reading on star = $231° 07' 59.4''$
Included angle between star and R.O.   = 122 53 51.5

Azimuth of R.O.   = 108 14 07.9

This, rounded to $01''$, is the same as the previous value calculated from the mean of the four pointings on the star.

It should be noted that the need for a curvature correction may be eliminated if each pointing is treated as a separate observation with its own included angle, and an azimuth of the R.O. obtained using that included angle and a separate azimuth calculation for the time of the pointing; the mean of these azimuths will then give the azimuth of the R.O. In our example above, this would mean four separate calculations of the azimuth instead of one plus a curvature correction calculation. However, with modern electronic calculators the extra calculations present no great chore, especially if the calculator is programmable. Separate calculations done in this way can also provide a useful check on the individual pointings.

Using the four pointings of the example in **7.05**, we get:

Pointing	Hour angle (t)	Computed azimuth (A)	Included angle (L)	$A - L$
1.	82 03' 01.5''	231° 46' 29.4''	123° 32' 22.5''	108 14' 06.9''
2.	82 35 24.0	231 34 18.8	123 20 11.5	108 14 07.3
3.	83 30 28.5	231 13 26.6	122 59 18.5	108 14 08.1
4.	85 24 54.0	230 29 28.2	122 15 20.5	108 14 07.7
			Mean =	108 14 07.5

Thus, as before,   the mean azimuth of R.O. = $108°14' 08''$

Astronomical corrections

7.10 Correction to azimuth for horizontal collimation error

Suppose that the line of collimation of the telescope, instead of being accurately at right angles to the trunnion axis, is in error by a small angle c; that is to say the line of sight makes an angle $(90° - c)$ on one side and $(90° + c)$ on the other side with the axis. Thus, as in Fig. 7.7, if there were no collimation error, the line of sight would trace out the great circle $ZS'N$ as it was rotated about the trunnion axis, but if in error it would sweep out the parallel small circle LSM. Now, suppose that the star S is observed in such a telescope, and let SS' be an arc of a great circle drawn at right angles to ZN. $SS' = NM = c$, the collimation error.

If we denote by z the angle SZS', the error in azimuth, we have

$$\sin c = \sin SZ . \sin z,$$

and since c and z are both small, we may write

$$z = c . \operatorname{cosec} SZ$$
$$= c . \operatorname{cosec} \text{(zenith distance)}$$
$$z = c . \sec h \quad \text{(where } h = \text{the altitude)}. \tag{7.12}$$

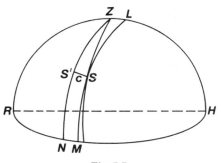

Fig. 7.7

The correction is meaningless, of course, for a single pointing. However, if the horizontal angle between two objects A and B, whose altitudes are h_A and h_B, is measured, the corrections will be $c \sec h_A$ and $c \sec h_B$, and the horizontal angle which is the difference between the circle readings will be in error by $c(\sec h_A - \sec h_B)$. If B is at the same level as the instrument (e.g. a reference mark), h_B will be $0°$ and $\sec h_B$ will be equal to 1, so that the collimation error present in the horizontal angle will now be $c(\sec h_A - 1)$. If $h_A = h_B$, the error vanishes. The sign of the correction and the value of c may be found by taking a series of L.F. and R.F. readings to a reference mark on the same level as the instrument. A comparison of the mean of all the readings with the L.F. mean and R.F. mean will give both sign and magnitude of c.

Astronomy for surveyors

The following table shows the way in which the error varies with the altitude of the star:

Error in horizontal angle between a reference mark at the same level as the instrument and objects at various altitudes, corresponding to a collimation error c (due to measurement on one face only)

Altitude of object:	0°	30°	60°	70°	80°	85°	89°
Error in horizontal angle	0	$0.15c$	c	$1.92c$	$4.76c$	$10.47c$	$56.30c$

The error is eliminated by measuring the included angle on both faces and taking the mean.

7.11 Correction to azimuth due to dislevelment of the trunnion axis

In the modern glass-circle theodolite the trunnion axis is set accurately perpendicular to the vertical axis by the manufacturer, and the bearings are completely enclosed, so that if the instrument is properly looked after there is no reason for it to get out of adjustment in this respect. If, then, the instrument is carefully levelled so that the vertical axis is made to point accurately to the zenith, the trunnion axis will remain level as the theodolite is moved in azimuth. In these circumstances, however, even if the trunnion axis is not quite at right angles to the vertical axis, the slight error introduced can be removed by taking the mean of left- and right-face readings.

In standard vernier theodolites there is usually a slipping block under one bearing of the trunnion axis to enable the latter to be adjusted so that it is perpendicular to the vertical axis.

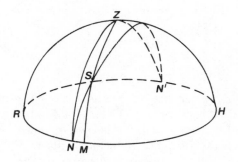

Fig. 7.8

In spite of careful initial levelling of the instrument, especially if it is mounted on a tripod, it is generally found that the plate bubble moves slightly off centre during a set of observations, and this will mean that the trunnion axis will be thrown off level. Error introduced as a result of this dislevelment (due to non-verticality of the vertical axis) cannot be cured by taking the mean of left- and

Astronomical corrections

right-face readings, and it is necessary to derive a correction to be applied to horizontal angles measured between objects at different altitudes.

If the axis of the telescope is not truly horizontal, the line of sight, when the telescope is turned about the axis, will not trace out a great circle in the sky passing through the zenith, as it should do, but will trace out a great circle inclined to the vertical. Thus, in Fig. 7.8, if NZN' denotes the great circle that would be traced out on the celestial sphere if the axis were horizontal, NSN' denotes the circle actually traced out if the axis is inclined at a small angle a. Let S be a star observed with this telescope, and draw the portion of the great circle ZSM passing through the zenith and the star.

The angle $ZNS = a$.

The actual observed altitude of the star is measured by the arc NS, whereas the true altitude is given by the arc MS.

Again, the azimuth of the star is actually measured on the circle of the horizon from the point N, whereas it should be measured from the point M; so that the error in azimuth is the angular measure of the arc MN.

In the right-angled triangle NSM, the angle $SNM = 90° - a$. Therefore, by Napier's rules, we have

$$\sin NM = \tan a . \tan MS$$

or, since both NM and a are small,

$$NM = a . \tan MS$$

that is to say, the error in azimuth = the error in level multiplied by the tangent of the altitude of the star.

Again, by Napier's rules,

$$\sin MS = \sin NS . \cos a$$

and since a is small and $\cos a$ may be taken $= 1$, it follows that we may take $MS = NS$, which means that no appreciable correction has to be made to altitude.

The following table shows the magnitude of the correction for various altitudes:

Error in azimuth corresponding to a level error a in the trunnion axis for various altitudes of object

Altitude of star	0°	30°	45°	60°	70°	80°	85°	89°
Error in azimuth	0	0.58a	a	1.73a	2.75a	5.67a	11.43a	57.30a

The older vernier and micrometer-type instruments were provided with a striding level which could be mounted on the ends of the trunnion axis for determining its dislevelment. It is not normally possible, however, to use such a level with modern glass-circle instruments, but the plate-level will serve the same purpose, even if it is lacking somewhat in sensitiveness compared with a good striding level (see Fig. 6.3).

Astronomy for surveyors

Therefore, when azimuths are being observed astronomically, the plate-level readings and approximate altitudes should be noted. The procedure is much the same as that used with graduated alidade bubble-tubes and already described in **7.07**. The plate-level is parallel to the trunnion axis, so that when the surveyor is observing, one end is on his left (L) and the other on his right (R). For each pointing, then, the L and R ends of the bubble are read against the graduations on the tube, outwards from the centre. Then the correction to the azimuth is equal to

$$+\frac{\Sigma(L) - \Sigma(R)}{n} \times v \times \tan h, \qquad (7.13)$$

where

n = the number of end readings
v = the angular value of one bubble-tube division
h = the mean altitude.

Suppose that two observations were made, one on face left and the other on face right, to a star whose mean altitude was 47° 20', as follows (one bubble-tube division = 30''):

	Bubble readings	
	L	R
F.L.	6.5	5.5
F.R.	5.0	7.0

In this case:

$$+\frac{\Sigma(L) - \Sigma(R)}{n} \times v \times \tan h = \frac{11.5 - 12.5}{4} \times 30'' \times \tan 47° 20'$$

$$= -7.5'' \times 1.085$$

$$= -8.1''.$$

Thus 8'' would have to be subtracted from the mean observed circle reading to give the correct reading.

Section 3
CONCLUSION

7.12 Summary

Corrections to observed altitudes:

(A) Astronomical: (i) Parallax (see **7.02**). Sun observations only. Maximum on horizon (9''), zero in zenith. Always added to observed altitudes. Equal to 09'' $\times \cos h$. Table in

Astronomical corrections

    text, but generally take as 6″ for all altitudes up to 70° and 0″ above 70°.
- (ii) Celestial refraction (see **7.03**). All celestial objects. Always subtracted from observed altitudes. Tables in *Star Almanac* on pages 62–64.
- (iii) Semi-diameter (see **7.04**). Sun observations only when single pointing made. Eliminated by taking mean of L.F. and R.F. readings to upper and lower limbs respectively. Mean value of semi-diameter of sun = 16.0′. Necessity for correction eliminated by use of Roelofs solar prism attachment.
- (iv) Curvature of path (see **7.05**). Applied to mean of series of L.F. and R.F. altitudes.

$$\Delta h'' = \{\cos\phi.\cos Z(\cos\phi.\cos Z.\tan h - \sin\phi)\} \times [\Sigma(m)]/n.$$

Values of m on page 68 of *Star Almanac*. Added to observed altitudes above elongation.
Zero at elongation.
Subtracted from observed altitudes below elongation; zero when body on P.V.

(B) Instrumental:
- (v) Vertical collimation (vertical circle index error) (see **7.06**). Eliminated by taking mean of L.F. and R.F. altitudes.
- (vi) Alidade bubble off centre (see **7.07**). Correction: $\{+[\Sigma(O) - \Sigma(E)]/n\} \times v$ to mean observed altitude.

Corrections to observed azimuths:

(A) Astronomical:
- (i) Semi-diameter (see **7.08**). Sun observations only when single pointing made. Eliminated by taking mean of L.F. and R.F. azimuths to left and right limbs respectively.
Correction to single observed azimuth = semi-diameter $\times \sec h$.
Eliminated by use of Roelofs solar prism attachment.
- (ii) Curvature of path (see **7.09**). Applied to value of A calculated from ϕ, δ and mean of observed hour angles, to give A for mean observed horizontal circle reading.

$$\Delta A'' = \sin A.\cos\phi.\sec^2 h$$
$$\times (\cos h.\sin\delta - 2\cos A.\cos\phi)$$
$$\times \{\Sigma(T_i^s - \bar{T}^s)^2.225\}/(2.n.\rho)$$

which reduces to
$$\Delta A'' = \{\tan A . \sin^2 \delta\} \times [\Sigma(m)]/n$$
at elongation.
If altitudes are measured, not times, use (7.11).

(B) Instrumental: (iii) Horizontal collimation (see **7.10**). Eliminated by taking mean of L.F. and R.F. azimuths.
For single-face pointings the correction to the measured included angle between A and B is $c(\sec h_A - \sec h_B)$.

(iv) Trunnion axis not level (see **7.11**). Not elminated by taking mean of L.F. and R.F. azimuths if due to vertical axis of instrument not being truly vertical. With modern instruments the plate-bubble is used to find the correction, which is equal to $\{+[\Sigma(L) - \Sigma(R)]/n\} \times v \times \tan h$.

From the foregoing it will be noted that the principle of making observations on both faces of the instrument and taking the mean should be strictly adhered to if most of the slight constructional defects of even the best theodolites are to be eliminated; but remember that even this procedure does not eliminate the error described in **7.11** above if the vertical axis is not plumb.

References

Baranov, V. N., Effect of Astronomical Refraction in the Tsinger Method. *Geodesy and Aerophotography* (English edn), No. 4, 1968, pp. 255-7.

Bomford, A. G., Small Corrections to Astronomic Observations. *Australian Surveyor,* **20,** 1964-5, pp. 199-210.

Clark, D., *Plane and Geodetic Surveying,* **2.** Constable, London, 1963.

Mackie, J. B., A Method for determining the Sensitiveness of the Plate Level of the Theodolite. *Jour. N.Z. Inst. Surveyors,* **23,** No. 220, 1961, pp. 400-5.

Saastamoinen, J., Contributions to the Theory of Atmospheric Refraction. *Bulletin Géodésique,* **105,** Sept. 1972, pp. 279-98, and **106,** Dec. 1972, pp. 383-97.

CHAPTER 8
Preparation for observations

8.01 General

Field astronomy is exacting work, and the wise surveyor will spend some time and thought in preparing for his observations. Nothing is more exasperating than to discover, after the station has been occupied, that any single part of the gear is not functioning properly or that something has been forgotten. If an assistant is available, there must be a high degree of team-work between him and the observer. While personal comfort is not always possible to obtain, consideration should be given to this aspect, since it is conducive to better observing. A properly timed programme should be prepared to suit the type of observations to be undertaken, and all the necessary calculations should be done well in advance of the commencement of operations (even if the weather conditions do not look entirely promising!).

8.02 The instrument

There may be no choice as far as the instrument is concerned, but on the other hand, if a selection does happen to be available there is no doubt that the modern glass-circle, one-second theodolite of European or Japanese manufacture is the easiest to use. In any case, whatever the instrument, it should be in good adjustment, with all motions working smoothly. The set-up should be at a height convenient for the observer, remembering that objects at relatively high altitudes may have to be sighted. If the work demands observations to objects at altitudes over, say, 45°, a set of 45° eyepiece-prisms, or even diagonal eyepieces, will be necessary. It is a good plan to allow ten or fifteen minutes for the instrument to "settle down" before the final levelling prior to the commencement of observations; this will give time for strains in the tripod legs to disappear, and for the metal of the theodolite to come to air temperature. The slow-motion (tangent) screws should be set in the middle of their runs to avoid their "running out" when a star is being tracked.

Nothing is more discomforting to both observer and booker than a cold breeze sweeping across the station; a light canvas screen rigged just up-wind of the instrument helps a great deal in this respect, or else a vehicle can be used as a wind-break if one is available (remembering that hot radiators cause turbulence of the air above them!). None of these comforts may be possible, of course, but if they can be provided they are usually very welcome—see Fig. 8.2 and 8.3, *below*.

If observations are to be made to the sun, *a suitable dark filter for the eyepiece is essential*; various filters can be obtained in different colours and densities to suit differing atmospheric conditions in ordinary (non-astro) observations, but only a properly designed sun filter should be used for sun observations.

One of the difficulties in night observing is to get the desired star into the field of view, and a useful tip is to put a small spot of luminous paint on the two external telescope sights.

Focusing is usually critical with theodolite telescopes when used for observing stars in daylight, and this remark applies also to very faint stars observed at night; thus the focusing ring should be marked in some way so that it may be pre-set accurately at infinite focus.

8.03 Illuminating gear

Glass-circle theodolites are usually supplied with internal electric illumination for reading the micrometers and making the cross-hairs visible. An external socket in the base takes a plug connected by a flexible lead to the battery-box which is held in a fixture on one of the tripod legs. Means of altering the strength of the illumination is provided by incorporating a rheostat in the battery-box; often, too, there is a small rotatable metal disc in the centre of the telescope, operated by a milled knob, for controlling the amount of light reflected from the alidade lamp on to the cross-hairs. The fainter stars can only be found and observed with reduced illumination of the graticule.

Some lighting sets have an additional flexible lead terminating in a small hand-lamp which can be used for illuminating levelling bubbles, watch, and field-book. A small electric torch will serve the same purpose, however, and lessen the load on the instrument battery. Since most microptic theodolites are already provided with plug-in lamps, it is a comparatively simple matter to make up a battery-box fitted with a suitable rheostat and supply lead and so complete the lighting set oneself. If this is done, the rheostat should be mounted high up on one of the tripod legs, and the batteries should be large capacity, longlife ones such as the Eveready No. 6. The latter are too heavy to attach to a tripod leg, but the box containing them can just as well be placed under the instrument if the supply lead is made long enough. Some manufacturers' lighting sets use several ordinary-size (Eveready No. 950 Size D) torch-cells to supply two lamps in the instrument and a third on a free lead, with the result that the current-drain is too high and the lights grow dim after a comparatively short period of use. Therefore, if a home-made supply containing larger cells is not used, it is wise to have at least one set of spare batteries available. A spare electric torch will always be found useful, too.

Before using a theodolite for star observations, it should be checked over

Preparation for observations

sometime beforehand with all its illuminating gear in place, to ensure that everything is in working order. Sometimes, for example in an instrument that has been lying idle for some time, the contact ring and wiper-arm which convey the current from the base to the alidade lamp become slightly oxidized, so that the light flickers on and off when the upper part of the theodolite is rotated. This can usually be cured by steady rotation about the vertical axis until the contacts have cleaned themselves.

Vernier instruments are not usually provided with lighting gear; the graticule may be illuminated by shining the light from a torch obliquely down the telescope tube from the objective end, although some manufacturers provide a sunshade containing a small prism which reflects the light down the tube more conveniently. The circle verniers must be read by the light of an electric torch.

8.04 Reference mark

When an observation is made by pointing to a celestial object to determine direction, all that is achieved from the subsequent calculation is the azimuth of the line of sight from the instrument to the object (which is moving) at the instant of pointing. In order to be of any practical value, this azimuth must be transferred to a line on the earth's surface between the observing station and some other fixed station. The azimuth is transferred by measuring the horizontal angle between the line to the celestial object and the line to the other fixed ground-point—the *reference mark* (R.M.) or the *reference object* (R.O.). By applying this angle in the correct sense to the computed azimuth of the star or the sun at the moment of observation, the astronomical azimuth of the R.M. is at once obtained. The azimuth of the ground reference line can then be used to find the azimuths of other ground-lines by triangulation or traverse.

It is desirable that there should be no need to refocus the telescope after pointing to the celestial body and then directing it to the R.M., and this requires that the latter should be, where practicable, about a mile away. When stellar observations are being made the R.M. should be constructed to imitate the appearance of a star as nearly as possible. This may be done by placing an electric lamp in a box or behind a screen through which a small circular hole is cut to admit the light to the observer. The face of the screen may be painted with stripes, so that it may be readily observable in the daytime. If the reference mark is not to appear larger than a star in the field of view of the telescope, the diameter of the hole must not be more than about 5 millimetres for a distance of 1 kilometre. Some observers prefer a narrow vertical slit on the screen; others use a larger hole with two cross-wires at right angles to each other against a translucent white screen illuminated from the rear. A bare $1\frac{1}{2}$-volt torch-bulb mounted in a socket at the top of a plumbed pole, about $1\frac{1}{2}$ kilometres from the instrument, makes a good mark.

Astronomy for surveyors

If, for any reason, it has not been possible to arrange for a distant R.M., a second theodolite (or level), if available, may be used as a temporary R.M. (collimator) from which a permanent line may be laid down later. The second instrument is mounted 4 or 5 metres away from the observing instrument at the same height; the two telescopes are then pointed at each other (both at infinite focus). The cross-hairs of the second instrument will be clearly seen by the observer through the main instrument telescope, and will form an excellent temporary R.M. Alignment of the telescopes is fairly critical. A white screen behind the eyepiece of the auxiliary theodolite makes the cross-hairs stand out more clearly; such a screen, illuminated by torchlight, will be necessary at night.

8.05 Timekeeping equipment

As has been remarked in Chapter 6, it is a great convenience to have at the observing station a lightweight portable radio capable of receiving time-signals. Given resonable conditions, it is then possible to check the watch used for the observations immediately before and after each set, thus reducing the chance of error due to irregular rate. This is not so important, of course, if a good electronic clock is available, since a much longer interval between checks can be tolerated, If the observer is working single-handed, a stop-watch (preferably digital electronic) is really an essential part of the equipment, in addition to the main watch or clock; even if he has a booker he may prefer to do his own timing with a stop-watch.

When more elaborate timekeeping equipment, involving a Chronocord and associated electrical apparatus, is used it is desirable to accommodate it in a lightweight tent set up close to the observing station and provided with subdued lighting. However, this is usually only required for geodetic work when more manpower and general facilities are available.

8.06 Other equipment

The only other equipment required is the following:
    Aneroid barometer (which has been compared with a mercurial barometer to determine any index error), for refraction correction.
    Thermometer for refraction correction.
    Field-book or booking forms.   Pencils.

8.07 Assistant

If an assistant is available he should be given explicit instructions beforehand on what his duties are. It is advisable for observer and assistant to have a few "dummy runs" together on various types of observation so that good team-work

Preparation for observations

can be achieved during the actual work; in this way the assistant can come to anticipate what is required of him and so be of great help to the surveyor. While making the observations the surveyor is naturally concentrating on using the instrument, and it is an advantage if the assistant keeps him "on the track" by detailing the steps at the correct time and in the correct sequence.

If no standard booking forms are available, the field-book should be ruled up and set out early, ready to receive the data.

If the observer has made a left-face observation to a faint star, he may have difficulty in finding it again on right face. In such a case the assistant can quickly work out the horizontal and vertical circle settings to enable the observer to get his second pointing without waste of time. If there are several stars in view on the first pointing, the observer should try to memorize the pattern so that when he changes face he can identify the particular star he sighted previously.

8.08 Observing programme

The need for preparing a detailed programme for an observing period cannot be too strongly emphasized. Stars must be selected so that they are in the right positions for the type of observation to be made. Times must be computed so that the observations may be commenced at the right moment. In making initial selections a star globe or a revolving planisphere can be of great help. The *Star Almanac* lists the stars in order of their Right Ascensions, and once the Local Sidereal Time corresponding to a given Local Standard Time on the evening of the observations has been established, suitable stars may be chosen without much trouble from the *Almanac*, and checked either by rough calculations or by the use of diagrams for graphical solution of the astronomical triangle (e.g. Plate XII in the 1958 edition of the War Office *Text Book of Field Astronomy*, or a device* described by L. P. Lee in the *Empire Survey Review*, No. 65, July 1947, page 123).

When the programme has been worked out it should be set down in correct chronological order in the field-book along with all the relevant data. The usual thing that happens when all this has been done is that the weather deteriorates rapidly and prevents implementation of the programme! However, the same programme can generally be used for the next few nights with alterations which are mostly of a minor character and are a consequence of the gain of sidereal on mean solar time of about $3^m\ 56^s$ in 24 hours.

Sometimes it is an advantage or even a necessity to know the azimuth of the R.O. to within a few minutes of arc before starting the observations. If

---

* In *Field and Geodetic Astronomy* (Robbins), Ministry of Defence, United Kingdom, 1976, a pair of stereographic projections, one printed in red on a transparent medium to be used as an overlay on the other which is printed in black on opaque material, are provided in a pocket of the cover.

Astronomy for surveyors

a reasonable value of the latitude of the observing station is known or can be found from a good map, the horizontal circle of the theodolite may be oriented with zero on astronomical north to within the necessary tolerance by using equation (7.3) of Chapter 7,

$$\frac{\mathrm{d}h}{\mathrm{d}t} = \cos\phi . \sin A.$$

The method of finding the azimuth A may be used with either a star or the sun, although it is somewhat easier to apply it to a star.

A stop-watch is used to time the change in altitude of the star over a period of, say, about 3 minutes. The watch is started when the star is on the intersection of the telescope cross-hairs and the altitude and horizontal circle are read; the watch is stopped about 3 minutes later with the star again on the intersection, and the circles read once more. The difference between the two altitudes ($\mathrm{d}h$) is converted into seconds of arc, and the time-interval ($\mathrm{d}t$) in seconds is multiplied by 15 to convert it to seconds of arc also. Then

$$\sin A = \frac{\mathrm{d}h}{\mathrm{d}t}.\sec\phi,$$

whence A is found. The instrument is pointed to the mean of the horizontal circle readings and the computed azimuth is then set on by using the circle setting knob. Any error in determining $\mathrm{d}h/\mathrm{d}t$ will be accentuated in higher latitudes, since it is multiplied by $\sec\phi$ which increases rapidly with ϕ. Any error in $(\mathrm{d}h/\mathrm{d}t).\sec\phi$ will give rise to increasing errors in A as the star approaches the prime vertical, i.e. when A approaches 90° or 270°.

We may test this method by using the data in the example in **7.05** of the previous chapter. Take the second and third pointings: $\mathrm{d}h = 0° 29' 58'' = 1798''$, and $\mathrm{d}t = 03^\mathrm{m} 40.3^\mathrm{s} = 3304.5''$. Hence $\mathrm{d}h/\mathrm{d}t = 0.544\,11$ and $\sin A = 0.544\,11 \times \sec(45° 52' 15'')$, so that $A = 180° + 51° 23' 38''$, or $231° 23' 38''$. The mean of the horizontal circle readings for the two pointings is $231° 23' 53''$, and since we know from previous worked examples that the circle is oriented within a minute or so of the correct value, we see that our approximate method has given quite a good result.

As an alternative to the above, a simple sun observation made on one face of the instrument by placing the intersection of the cross-hairs on the estimated position of the sun's centre, reading the altitude and horizontal circle and doing a quick calculation using the latitude and declination, will give an azimuth to a few minutes of arc.

In the present era of electronic computers, the provision of star programs is an easy matter once the appropriate computer program has been written. The repetitive type of computation needed is, in fact, ideally suited to these computers. Brief examples of the kinds of predictions possible are given below:

Preparation for observations

Azimuth, altitude and time. At 3- or 5-minute intervals of Local Standard Time, the azimuths and altitudes of selected stars can be provided, for a given period, as a computer printout. Since all azimuths and altitudes are not required, only those between suitable limits need be shown. This type of printout is especially suitable for teaching purposes, since most learners tend to be slow with their initial observations and to "lose" the star after changing the face of the instrument, particularly with daylight stars. With a tracking program available, the star can be quickly found again.

Time and azimuth for various stars at some fixed altitude. This type of programme is easily produced and is of great use in astro-fix and astrolabe work. Azimuths can be restricted so that the printout shows only those stars near the centres of the four quadrants, or near the prime vertical for balanced time and longitude work.

Altitude and time of meridian transit. This information is most useful for balanced latitude observations.

For the programs developed in the author's department, the R.A.s and declinations of many FK4 stars have been punched into cards. The programs are so written that the computer sorts through these cards and finds the stars suitable and available for the various types of observation. The only other data needed when a star program is required are latitude and longitude of the observing station and time, azimuth, altitude and magnitude limits.

8.09 List of equipment required for field astronomy

Since it is a good plan to check through one's gear before going into the field, a list of the items required is set out below (see also Fig. 8.1-8.3).

Instrument: Theodolite, tripod, plumb-bob.
    Instrument accessories: Dark glass for eyepiece, eyepiece prisms, diagonal eyepieces.
    Illuminating gear: Bulbs for theodolite, and spares; battery-box, batteries and connecting lead; spare batteries; two electric torches.
Reference mark of a suitable type to be set up.
Timekeeping equipment: Good quality electronic clock, wrist-watch and/or digital stop-watch.
    A portable transistor radio covering the medium and shortwave bands.
    A stop-watch, if the observer prefers to do the timekeeping himself and a recording device is not available.
    Recording device (e.g. Chronocord) if available and if required.

Astronomy for surveyors

Fig. 8.1 The Kern DKM 2A Theodolite and some auxiliary equipment for astronomical observations

- *A*   Theodolite (with automatic levelling on vertical circle) mounted on trivet
- *B*   Battery container with connection to theodolite
- *C*   Star globe
- *D*   Stop-watch with split-second hand, or preferably a digital stop-watch as in Fig. 6.1. (see **6.03**)
- *E*   Multi-band transistor radio set for receiving time-signals
- *F*   *Star Almanac for Land Surveyors* (H.M.S.O.)
- *G*   Pad of booking sheets
- *H*   Star atlas
- *J*   Pocket calculator

Not shown are the theodolite tripod, sun-glass, lamp-holder, eyepiece prism, thermometer, aneroid barometer and pencil.

The bare essentials for observing are *A, B, D, E, G*, thermometer, aneroid, sun-glass, lamp-holder and pencil, although *B* and the lamp-holder may be omitted from this shortened list if star work is not contemplated. (See **8.09**)

Preparation for observations

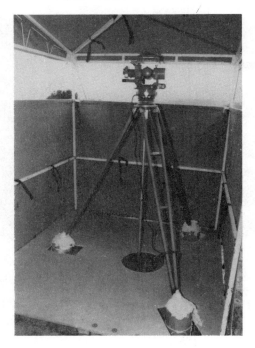

Fig. 8.2   Kern DKM 3A theodolite mounted in observing tent
The tent roof is easily removable for star observations but is kept in place during the daytime to shade the instrument from the sun. Note how the tripod legs are plastered to heavy wooden pegs for greatest stability, and how the observing platform is kept clear of these pegs.
Photo by Mr Glen Rowe, printed with the permission of the Surveyor-General, N.Z. Lands and Survey Department

Other equipment: Aneroid barometer ⎫
                  Thermometer   ⎬ for refraction correction.
                                     ⎭
 Prismatic compass and knowledge of magnetic declination for approximate determination of the direction of the meridian.
 Field-book or booking forms in pad.
 Pencils.
 Light canvas wind-break, light poles and guy ropes if expedient (Fig. 8.3).
 Light tent for timekeeping gear when this is elaborate.

Information: Observing programme entered into field-book prior to observations or as computer printout.
 Star charts (e.g. pages 60–61, but preferably larger scale).

155

Astronomy for surveyors

Fig. 8.3 Observing tent for geodetic astronomy

The observing section is on the right, the booking section on the left. Note the flaps for closing the horizontal observing gap against the wind.

Photo by Mr Glen Rowe, printed with the permission of the Surveyor-General, N.Z. Lands and Survey Department

References

Allan, A. L., The Application of Tape and Disc Recording Machines to Field Astronomy and an Observational Method arising therefrom. *Survey Review*, **XVII**, No. 133, July 1964, pp. 305-12.

Astronomisches Rechen-Institut, Apparent Places of Fundamental Stars. Heidelberg. (Based on *The FK4 Catalogue*.)

Loxton, J., A. Station Star Locator. *Survey Review*, **XXIII**, No. 177, July 1975, pp. 126-9.

Olliver, J. G., A Planisphere for Stereographic Star Charts. *Survey Review*, **XVIII**, No. 137, July 1965, pp. 134-40.

Robbins, A. R. *Field and Geodetic Astronomy*. Ministry of Defence, United Kingdom, 1976.

Vamosi, S., The Use of the Geodimeter Reflector in Geodetic Astronomy. *Canadian Surveyor*, **XXII(2)**, 1968, pp. 255-6.

CHAPTER 9

The location of objects on the celestial sphere

9.01 Introduction

In order that the surveyor may select and observe a particular star with a theodolite, it is sometimes necessary, more especially when he wishes to make the observations in daylight or evening twilight, that he should know the altitude and azimuth of the star at the given time. From the *Star Almanac* he obtains its right ascension and declination, and from these data he has to compute altitude and azimuth. In this chapter we will deal with this problem and show how, given the position of a star in one system of coordinates, we may determine its coordinates in another.

9.02 Given the latitude, longitude and time at the place of observation and the right ascension and declination of a particular star, it is required to find its altitude and azimuth

In Fig. 9.1, let P be the pole, S the star, Z the zenith, $M_N ZPM_S$ the plane of the meridian.

Draw the great circle through Z and S to intersect the horizon in A.

If we know the local mean time we can compute the corresponding sidereal time by the methods of Chapter 5. But we have seen that the right ascension of the star is the same thing as the sidereal time at the moment of the star's transit across the meridian. Consequently the difference between the sidereal

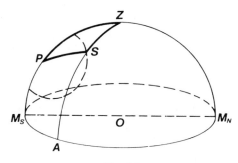

Fig. 9.1

Astronomy for surveyors

time at the instant of observation and the right ascension of the star gives the interval in sidereal time between the moments of the star's transit across the meridian and of observation—that is to say, it gives, when turned into degrees, minutes and seconds, the hour angle of the star, *SPZ*. If the sidereal time at the moment of observation is less than the right ascension of the star, the difference measures the angle *SPZ* towards the east of the meridian; if the right ascension is the less, the angle is measured toward the west.

Thus, in the spherical triangle *ZSP*, we know *ZP*, the complement of the latitude, and *SP*, the polar distance of the star which is the complement of the declination, and the included angle *ZPS*.

From these data we can compute the third side *ZS*, which is the zenith distance of the star, or the complement of the altitude, and the angle *SZP*, which determines the azimuth.

Angle $SZP = Z$ ZS = zenith distance = ζ SA = altitude = h

$ZSP = S$ ZP = co-latitude $= \omega$

$ZPS = t$ PS = polar distance $= p$.

Choice of formulae

Two sides and the included angle are available for solution of the astronomical triangle. Formula (1.15) gives

$$\cot Z = \frac{\cot p . \sin \omega - \cos \omega . \cos t}{\sin t} \qquad (9.1)$$

for finding the zenith angle Z and hence the azimuth of the star. Then formula (1.17) may be used to find the altitude h:

$$\cos \zeta = \sin h = \cos \omega . \cos p + \sin \omega . \sin p . \cos t. \qquad (9.2)$$

Often formulae based upon (1.21), (1.22) and (1.1) may, of course, be used but they tend to be more cumbersome.

EXAMPLE. At a place in New Zealand in Longitude E. $11^h\ 22^m\ 03.9^s$ and Latitude S. $45°\ 52'\ 15''$, it is required to determine the altitude and azimuth of Sirius (No. 185 in the *Star Almanac*) at $4^h\ 30^m$ p.m. L.Std.T. on 16 March 1975. R.A. Sirius is $6^h\ 44^m\ 04.3^s$ and its declination is S. $16°\ 41'\ 11''$. The Standard Meridian is 12^h E.

Given L.Std.T., 16 March 1975	=	$16^h\ 30^m\ 00.0^s$
Longitude of Standard Meridian, E.	=	12 00 00.0
Corresponding U.T.C., 16 March	=	4 30 00.0
R at 0^h U.T., 16 March (see p. 98)	=	11 31 51.2
Gain in R in $4^h\ 30^m$ (see p. 104)	=	44.4

Location of objects on celestial sphere

$$
\begin{aligned}
&\text{G.S.T.} (= \text{U.T.} + R) & &= 16^h\ 02^m\ 35.6^s \\
&\text{Longitude of observing station, E.} & &=\ 11\ \ \ 22\ \ \ 03.9 \\
\\
&\text{L.S.T. corresponding to given L.Std.T.} & &=\ 27\ \ \ 24\ \ \ 39.5 \\
&\text{R.A. Sirius} & &= -6\ \ \ 44\ \ \ 04.3 \\
\\
&\text{Hour angle of Sirius (W.)} & &=\ 20\ \ \ 40\ \ \ 35.2 \\
&\text{Hour angle of Sirius (E.)} & &=\ \ \ 3\ \ \ 19\ \ \ 24.8 \\
&t\ (\text{E.}) \text{ in arc} & &=\ 49°\ 51'\ 12'' \\
&ZP = \text{co-latitude}, & \omega &=\ 44\ \ \ 07\ \ \ 45 \\
&PS = \text{polar distance}, & p &=\ 73\ \ \ 18\ \ \ 49
\end{aligned}
$$

Applying formulae (9.2) and (9.1) above,

$$
\begin{aligned}
h &= 39°\ 30'\ 10'' \\
Z &= 108\ \ \ 23\ \ \ 08
\end{aligned}
$$

Since the hour angle of Sirius exceeds 12^h, the star must be in the east, hence for the observing station which is in the southern hemisphere, the azimuth of Sirius $= 180° - Z$, i.e. azimuth required, $A = 71°\ 36'\ 52''$.

From the above the computed altitude of Sirius (h) was $39°\ 30'\ 10''$. The observed altitude will be equal to this value *plus* the refraction correction which is $70''$ for mean values of pressure and temperature. Hence,

Altitude required $\quad h = 39°\ 30'\ 10'' + 01'\ 10'' = 39°\ 31'\ 20''$.

Note that the correction for refraction is added in this case because we are finding what the observed altitude would be from a calculated one, so that we may set the telescope at the correct elevation. In many cases it would not be necessary to include the refraction correction, since the telescope field will usually cover about $1\frac{1}{2}°$ of sky, and without the correction the star would still be near the centre of this field. However, if the observer is trying to locate a daylight star at a relatively low altitude, it would be prudent for him to allow for refraction and ensure that his telescope is correctly focused at infinity. (See **10.04**.)

9.03 **Having computed the altitude and azimuth of a star for a given time of observation, it is required to determine its approximate position at some short interval of time afterwards**

When a surveyor is preparing for daylight observations of a star, it will generally be necessary for him to take at least two readings of its position, one on each face. To give him time to read the circles and reverse the instrument before taking the second observation, he requires to know the altitude and azimuth of the star at an interval of two or three minutes after the first reading.

The computation for the second position may, of course, be made by precisely the same method that we have already used for the first, in which case ω and p will remain the same but the hour angle will be two or three minutes of time greater.

It may be slightly quicker, however, to use the differential formulae of **12.10**(b) and **10.23**(i).

In **12.10**(b) it is shown that

$$\frac{dt}{dh} = -\frac{\cos h}{\cos\phi.\cos\delta.\sin t}.$$

Thus,

$$\frac{dh}{dt} = -\frac{\cos\phi.\cos\delta}{\cos h.\operatorname{cosec} t}$$

where dh is a small change in the altitude h produced by a small change dt in the hour angle t. Then, for these small changes, we may write,

$$dh = -\left\{\frac{\cos\phi.\cos\delta}{\cos h.\operatorname{cosec} t}\right\} dt. \qquad (9.3)$$

The sign of cosec t must be watched and the west hour angle used always. From 0^h to 12^h (0°-180°), cosec t is positive, and dh is therefore negative, i.e. the altitude of the star is decreasing when the star is west of the meridian; from 12^h to 24^h (180°-360°) cosec t is negative and dh is therefore positive; the altitude is therefore increasing with increase of time when the star is east of the meridian.

In **10.23**(i) it is shown that

$$\frac{dZ}{dt} = -\sin Z.\operatorname{cosec} t.\cos S \qquad (10.26)$$

so that, for small quantities, we may write

$$dZ = -(\sin Z.\operatorname{cosec} t.\cos S)dt. \qquad (9.4)$$

In this formula, as in that for dh/dt, we must always use the west hour angle for t, so that, for $0^h < t < 12^h$, cosec t is positive and for $12^h < t < 24^h$, cosec t is negative. Further, since S may vary between 0° and 180°, we must watch the sign of $\cos S$. If $0° < S < 90°$, $\cos S$ is positive, and if $90° < S < 180°$, $\cos S$ is negative. $\sin Z$ is always positive, since Z is never greater than 180°.

EXAMPLE. To illustrate the application of formulae (9.3) and (9.4), we shall extend the scope of the example already worked in **9.02** above and compute the position of Sirius 3 sidereal minutes after 4.30 p.m. L.Std.T. From the previous work,

Location of objects on celestial sphere

$$\phi = 45° 52' 15" \qquad h = 39° 30' 10"$$
$$\delta = 16° 41' 11" \qquad t = 310° 08' 48"$$
$$Z = 108° 23' 08" \qquad S = 43° 36' 46"$$

$$S \text{ computed from } \cos S = \frac{\cos \omega - \cos \zeta . \cos p}{\sin \zeta . \sin p}.$$

3 sidereal minutes = 15 × 3 = 45' of arc.

Thus,

$$dh = -\frac{(0.6963 \times 0.9579) \times 45'}{(0.7716) \times (-1.3082)}$$

$$= +0.6608 \times 45'$$

$$= +29.74'$$

and

$$dZ = -(0.9490) \times (-1.3082) \times (0.7240) \times 45'$$

$$= +(0.8989) \times 45'$$

$$= +40.44',$$

so that, for the new altitude of Sirius:

Previously computed altitude,	$h =$	39° 30' 10"
	$dh =$	+29 44
New calculated altitude 3 sidereal minutes later	=	39 59 54
Mean refraction	=	+01 09
New observed altitude of Sirius	=	40 01 03

and for the new azimuth:

Previously computed zenith angle,	$Z =$	108° 23' 08"
	$dZ =$	+40 26
New zenith angle	=	109 03 34
New azimuth of Sirius 3 sidereal minutes later	=	70 56 26

Since four-figure working only has been used, the above values should be rounded to 40° 01' and 70° 56' respectively, which are quite adequate for locating the star.

It will be noted that the value of S was needed above to compute dZ, and since this needs another solution of the astronomical triangle, the work is

Astronomy for surveyors

increased to the stage where it may be better to re-compute the original triangle using the amended hour angle. This is particularly so when one is using a good electronic calculator with several memories.

9.04 Identification of stars

Having observed the altitude and azimuth of a star, the latitude and longitude of the observing station being known and the time of observation noted, it is required to determine its right ascension and declination.

In Fig. 9.2, let Z be the zenith point, P the pole, and S the star, as before.

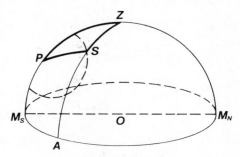

Fig. 9.2

In the spherical triangle ZSP, ZP is known, being the co-latitude; ZS, the zenith distance, is also known, as well as the angle SZP, which the vertical plane passing through the star makes with the meridian.

Thus we know two sides and the included angle, and the triangle may be solved to find SP and the angle SPZ.

The formulae to be used are similar to those of **9.02**:

$$\cot t = \frac{\cot \zeta . \sin \omega - \cos \omega . \cos Z}{\sin Z} \qquad (9.5)$$

$$\cos p = \cos \zeta . \cos \omega + \sin \zeta . \sin \omega . \cos Z. \qquad (9.6)$$

(9.5) is based upon (1.15), and (9.6) on (1.17).

The angle SPZ (t), being turned into hours, minutes and seconds at the rate of 15° for 1 hour, measures the sidereal time that will elapse before S comes to the meridian if S is to the east, or the interval of sidereal time since S was on the meridian if it is to the west.

But the right ascension of the star is the sidereal time when it is on the meridian.

Therefore, to obtain the right ascension of the star, add the time-value of the angle SPZ to the local sidereal time at the moment of observation if the star is to the east of the meridian, and subtract it if the star is to the west.

Location of objects on celestial sphere

The declination of the star is, of course, the complement of the computed polar distance $SP(p)$ if $p < 90°$, but it is equal to $p - 90°$ if $p > 90°$.

EXAMPLE. On 11 July 1975 a star of about magnitude 3 was observed at $21^h\ 58^m\ 03^s$ L.Std.T. to have an altitude of $34°\ 28'\ 00''$ and an approximate azimuth of $232°\ 10'\ 30''$. If the Latitude and Longitude of the observing station were S. $45°\ 52'\ 15''$ and E. $11^h\ 22^m\ 03.9^s$ respectively, identify the star. The Standard Meridian of the place is 12^h E. Barometer 1020 mb, temperature $-1\ °C$.

Observed altitude		=	34° 28'	00''
Refraction = 85'' × 1.05		=	−01	29
Corrected altitude,		$h =$	34 26	31
Zenith distance,		$\zeta =$	55 33	29
Azimuth		$A =$	232 10	30
Subtract 180°		=	180 00	00
Zenith angle,		$Z =$	52 10	30
Co-latitude,	$\omega = (90° - \phi) =$		44 07	45
from (9.5),		$t =$	87 17	40
		=	$5^h\ 49^m$	10.7^s
from (9.6)		$p =$	40° 42'	13''

To find the right ascension:

L.Std.T. of observation, 11 July 1975		=	$21^h\ 58^m\ 03^s$
Longitude of Standard Meridian, E.		=	12 00 00
Corresponding U.T.C. 11 July		=	9 58 03
R at 6^h U.T., 11 July		=	19 14 07.3
Gain in R in $3^h\ 58^m\ 03^s$		=	39.1
G.S.T. (= U.T. + R)		=	5 12 49.4
Longitude of observing station, E.		=	11 22 03.9
L.S.T. at time of observation		=	16 34 53.3
Hour angle (W.) of star, from above,	$t =$		5 49 10.7
∴ Right ascension = (L.S.T. − t)		=	10 45 42.6

To find the declination:

Polar distance, p from above	=	40° 42' 13''
Declination, $\delta = (90° - p)$	=S.	49° 17' 47''

Astronomy for surveyors

On looking through the catalogue of stars on pages 26-53 in the *Star Almanac*, we find one, No. 292 μ Velorum, which has a right ascension of $10^h\ 45^m\ 42.7^s$ and declination of S. $49°\ 17'\ 43''$ on the date concerned; there are no other stars in the Almanac which have coordinates near the calculated values, so that it is most likely that μ Velorum was the star observed, especially as it was of about magnitude 3, and the *Star Almanac* contains all stars brighter than magnitude 4.

Another approach in identifying stars is to use the measured rate of change of altitude with time as explained by H. F. Rainsford. The method is based upon equation (7.3) of Chapter 7:

$$\frac{dh}{dt} = \cos\phi.\sin A.$$

From this we get Rainford's equation:

$$\sin A = +\frac{\Delta h}{15.0411\Delta t.\cos\phi}, \tag{9.7}$$

where Δh is in seconds of arc and Δt in seconds of mean solar time if a mean solar watch or chronometer is used. For a star in the east (rising) Δh is positive and for a west star Δh is negative. Then the hour angle and right ascension are found from

$$\cot t = (\sin\phi.\cos A - \cos\phi.\tan h)/\sin A \tag{9.8}$$

and

$$\text{R.A.} = \text{L.S.T.} - t.$$

The declination is found from

$$\sin\delta = \sin\phi.\sin h + \cos\phi.\cos h.\cos A. \tag{9.9}$$

(9.8) is based on (1.15) and (9.9) on (1.17).

EXAMPLE. At a place in Latitude N. $30°\ 51'\ 10''$ and Longitude E. $1^h\ 55^m\ 47.0^s$ on 2 May 1975 a star of about magntiude 4 was observed in the east to have an altitude of $26°\ 57'\ 09''$ at L.Std.T. $22^h\ 07^m\ 27.0^s$; $2^m\ 37.6^s$ later, by a mean-time chronometer, the star's altitude was $27°\ 26'\ 48''$. Both altitudes have been corrected for refraction. The Standard Meridian of the place is 2^h east. Identify the star.

Here

$$\Delta h = 0°\ 29'\ 39'' = 1779''$$

and

$$\Delta t = 0^h\ 02^m\ 37.6^s = 157.6^s$$

Location of objects on celestial sphere

Thus from (9.7) above,

$$A = 60° \ 57' \quad \text{(nearest minute of arc);}$$

from (9.8),

$$t = -5^h \ 10^m \ 24.5^s$$

and from (9.9),

$$\delta = +37° \ 15'.$$

Mean observed L.Std.T.	=	$22^h \ 08^m \ 45.8^s$
Longitude Standard Meridian, E.	=	2 00 00
U.T.		20 08 45.8
R at 18^h U.T., 2 May	=	14 40 06.7
Gain in R in $2^h \ 08^m \ 45.8^s$	=	21.2
G.S.T.	=	10 49 13.7
Longitude of observing station, E.	=	1 55 47.0
L.S.T.	=	12 45 00.7
t, from above	=	−5 10 24.5
L.S.T. − t = R.A.	=	17 55 25.2

From pages 44 and 45 of the *Star Almanac* the star is clearly identified as No. 493 θ Herculis.

In using equations (9.8) and (9.9), due attention must be paid to sign conventions: north latitudes are positive and south latitudes are negative; north declinations are positive and south declinations negative. To check this, the student should apply formulae (9.8) and (9.9) to the values in the first example of star identification given earlier.

Reference

Rainsford, H. F., *Survey Adjustments and Least Squares*. Constable & Co., London, 1968.

EXERCISE

At a place in Latitude 41° 12′ 40″ S. and Longitude $11^h \ 39^m \ 34^s$ E. on the evening of 15 January 1977, the altitude and azimuth of a second-magnitude star was obtained through a small break in the clouds. From the following data, find the star's R.A. and declination:

Observed altitude of star: 44° 03′ 32″
Barometer: 1029 mb; thermometer: 15.6 °C

Astronomy for surveyors

Azimuth of star: 131° 21′ 55″
R at U.T. 6^h, 15 January 1977: 7^h 38^m 18.6^s
L.Std.T. of observation: 20^h 50^m 51^s
Std. Meridian is 12^h E.

                              Ans.   R.A. 8^h 44^m 07^s
                                            Dec. S. 54° 37′ 31″
                                        (The star is δ Velorum, No. 238)

CHAPTER 10

The determination of azimuth

Section 1

GENERAL

10.01 Azimuth defined

Azimuth is a term which tends to be somewhat loosely used to denote the direction of a given line with reference to some other datum direction; this datum direction is sometimes true north or true south, or something close to these, or even some quite arbitrary direction selected to suit the needs of a particular project. However, in this book we want to be sure we know just what we mean when we refer to the azimuth of a line, so some kind of definition is necessary.

We shall therefore select true (astronomical) north as our datum direction, and express all azimuths in terms of it, on the full circle $0°$ to $360°$. A line bearing true north from the observer will have azimuth $0°$ (or $360°$ which is the same thing); one bearing true east will have azimuth $90°$; south, $180°$, and west, $270°$. If the term "bearing" is used it will have the same meaning as azimuth, unless qualified in some way.

We have seen from the earlier chapters that solution of the astronomical triangle for direction will give us the zenith angle Z. This angle is not always the azimuth of the celestial body in terms of our definition. From Fig. 10.1(a) for the northern hemisphere it will be seen that the angle Z, measured from the elevated (north) pole, gives the azimuth of S, the celestial body, if S lies to the east of the meridian; if S lies to the west of the meridian, its azimuth is given by $(360° - Z)$. Fig. 10.1(b) shows the situation in the southern hemisphere: when S is to the east of the meridian, its azimuth is $(180° - Z)$, and when to the west $(180° + Z)$.

10.02 Need for astronomical azimuth

The determination of the azimuth of a line is a very important and common operation for the surveyor. In most cases he will commence his survey from previously established points, the records of which will give him an initial azimuth to carry forward. Situations do arise, however, when the absence of such old points, or their unreliability, compels him to set up his own initial

Astronomy for surveyors

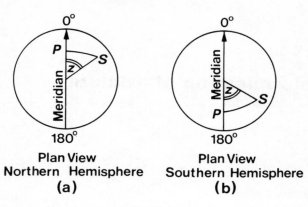

Fig. 10.1

station and determine the azimuth of the first line radiating from it by observations to the sun or stars.

More usually the land surveyor will wish to determine the azimuth of a line by astronomical methods in order to check its bearing as brought forward by other means. For example, a long traverse run to locate a title boundary will accumulate errors, particularly in rough terrain, and if periodical bearing checks cannot be obtained from connections to triangulation stations or from cross bearings, they must be got from astronomical observations. Local bearings, however, are often in terms of the meridian of the origin of a "circuit" or limited area whose survey points are coordinated on a plane projection (as is the case in New Zealand). The astronomical azimuth of the line must therefore be converted to the local, or circuit, bearing by application of the convergence between the meridians of the circuit origin and the observing station, before a comparison of the bearings may be made. Methods of calculating the convergence are to be found in textbooks on Geodesy, but for the convenience of students, formulae, of sufficient accuracy for most purposes, and rules for application are contained in Appendix III.

A situation that quite often arises, particularly in rural areas where recent cadastral surveys are sparse, is one where a single old mark only can be found. In such a case a reasonably accurate azimuth, in terms of the local survey projection or circuit, is needed to enable the surveyor to locate other nearby marks shown on existing survey plans. A quick sun observation in these circumstances will often save him much time, provided he has with him the *Star Almanac*, a pocket calculator, a sun-glass (filter) for the theodolite eyepiece, and moderately accurate local time on his wrist-watch; a knowledge of the latitude of the place to a minute or so of arc is also necessary, but this can usually be obtained from a map which the surveyor would normally have with him.

Determination of azimuth

10.03 Reference mark

The setting-up and use of reference marks has been dealt with in **8.04**.

When booking the results of an observation for azimuth a diagram should always be included to show the relative positions of R.M., celestial body and meridian in plan view. This information helps to remove any difficulty in deciding later how to apply the included angle between body and R.M. in finding the azimuth of the latter, although the mean horizontal circle readings alone should usually make this clear. Figure 10.2 shows such a simple diagram.

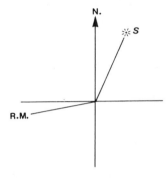

Fig. 10.2

10.04 Star observations in daylight

It is often a very great convenience to the surveyor to be able to make his observations for azimuth to stars in the daytime, and certain stars are bright enough to render this possible if the observations are done in the late afternoon. They can also be done in the early morning, of course, but with the disadvantage that the time available is rapidly curtailed by increasing daylight.

The relative brightness of a star is indicated by its *magnitude* which is expressed by astronomers as a number; the larger the number, however, the fainter is the star. With the naked eye, under good atmospheric conditions at night, it is possible for persons with exceptionally keen eyesight to see stars down to about magnitude 6, and these are reckoned to be 1/100th as bright as stars of magnitude 1. Now $2.51^5 = 100$, and when we wish to compare the brightness of two stars whose magnitudes are known we arrive at the brightness ratio by raising 2.51 to the power of the difference of the magnitudes. Thus if m_1 and m_2 are their magnitudes, the ratio is given by $2.51^{(m_2-m_1)}$. For example, if the stars have magnitudes 1 and 6, we get $2.51^{(6-1)} = 100$. so that the first is 100 times brighter than the second; again, if $m_1 = 3$ and $m_2 = 1$, the ratio is 2.51^{-2}, or 1/6.3. The brightest star in the heavens is Sirius (α Canis Majoris) whose magnitude is -1.6; comparing this with a star of the first magnitude, we find that it is $2.51^{2.6}$ or 10.9 times as bright. The faintest star listed in the *Star Almanac* is τ Octantis with a magnitude of 5.6; this star is therefore $2.51^{-7.2}$, or 1/754th of the brightness of Sirius.

Stars suitable for daylight observations are marked with "d" following their numbers on the left-hand side of the even-numbered pages of the *Star Almanac* in the star catalogue section (see p. 56).

As these stars cannot be seen with the naked eye in daylight it is necessary to compute the positions of those selected for observation before directing the telescope to them. The method of computing a star's altitude and azimuth, knowing latitude, longitude, R.A. and declination, is given in Chapter 9. To enable us to set the telescope in the direction of the star, we must know the azimuth of the R.M. within a few minutes of arc; this may be found from the magnetic bearing if a good compass is available, or (better) from a quick, single-face sun observation. Those unfamiliar with the appearance of a star as seen through a telescope in daylight sometimes have difficulty in finding it in the field of view. The background is a pale blue against which the star appears as a small but intensely white pinpoint; as mentioned in Chapter 8, *focusing of the telescope (at infinite focus) is critical and should be checked on a distant object before setting the circle readings and looking for the star.*

Some readers will wonder why surveyors want to observe stars in daylight when the sun is usually available. There are two main reasons for this:

(i) It is easier to observe the star's pinpoint of light on the cross-hair than to make tangency with the hair against the disc of the sun. Considerable practice is required to get a moving star exactly on the intersection of the cross-hairs and even more to get both horizontal and vertical cross-hairs simultaneously tangential to the sun's disc; the eye has difficulty in focusing at one instant on the two points of tangency, and there will always be an element of doubt about the pointing. As mentioned in Chapter 7, however, the Roelofs solar prism attachment does much towards eliminating this difficulty, but it is a somewhat expensive accessory unless the surveyor is required to make regular sun observations.

(ii) Certain errors of observation and position may be minimized by observing pairs of north–south and east–west stars in a way not possible with the sun.

Daylight star observations have been used with advantage by surveyors working in Antarctica during the summer season of "midnight sun", when daylight conditions prevail for 24 hours a day. Although the sun is always available in these circumstances, it is never very far above the horizon, and measured altitudes are liable to considerable refraction corrections. Under the conditions of extremely low temperatures and large expanses of ice, there appear to be anomalies in corrections computed from the ordinary refraction tables, indicating that sun altitudes are not altogether reliable.

Determination of azimuth

Section 2

FIRST METHOD

AZIMUTH BY EQUAL ALTITUDES OF A STAR

10.05 General

This type of observation is not one in common use; it has the advantage, however, of requiring the minimum of equipment. Neither star tables nor knowledge of the latitude and longitude of the observing station are necessary, nor are any but the simplest calculations needed. The main disadvantage is the length of time required to complete the observation, and the chance of its being ruined by poor weather conditions. Furthermore, a change in atmospheric refraction during the period covered by the observations may introduce errors.

10.06 Principle

To define the meridian, we have to determine the direction of the celestial pole, and the simplest method is probably that of observing a star at equal altitudes. If the circle in Fig. 10.3 represents the circular path of a star round the pole, the problem is to determine the direction of the centre P of this circle. Suppose that the star is observed at S, and then, keeping the angle of elevation of the telescope unchanged, the observer waits until he sees the star again at H at the same altitude.

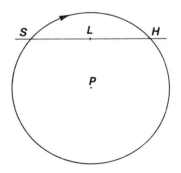

Fig. 10.3

Clearly the point L, midway between S and H, will be vertically above the pole P, and all that the observer has to do to get his true meridian is to bisect the angle between S and H. Nothing could be simpler in principle, but certain precautions are necessary to get good results.

In the first place, when fixing either the point S or H, we are really marking the point of intersection of the horizontal line with the circle. Now, we can fix

Astronomy for surveyors

the intersecting point of two lines most accurately when the two lines are at right angles, and so the best position for the line *SH* is when it passes somewhere near *P*. As the star takes 24 sidereal hours to complete its circle round the pole, this would mean that the second observation would be made about 12 hours after the first. This would be often impossible and generally inconvenient. If, on the other hand, the line *SH* is taken too near the top of the circle, the star is moving so rapidly in a horizontal direction that it is not possible to secure good intersections.

10.07 Observing procedure

Two simple observations *S* and *H*, such as we have just described, would not be sufficient to enable instrumental errors to be eliminated, and so in practice at least four observations are made, as illustrated in Fig. 10.4.

They are made as follows:

(i) Intersect the R.M. on left face and note the horizontal circle reading.
(ii) Intersect the star on L.F. in the position S_1 and note both horizontal and vertical circle readings, making sure that the alidade bubble is centred before reading the vertical circle. Note the plate bubble readings *L* and *R* (see **7.11**).

Since both horizontal angle and altitude are required, the star must be observed on the intersection of the cross-hairs of the telescope. Skill in doing this comes only with practice, and the best way is to preset the horizontal hair a little ahead of the star; the vertical hair is then kept on the star with the horizontal slow-motion screw until the star reaches the intersection.

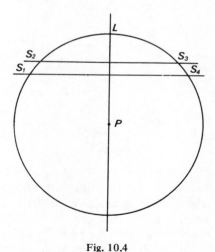

Fig. 10.4

Determination of azimuth

(iii) Without delay change to right face and repeat the procedure of (ii) for the star at S_2.

(iv) Keeping the instrument on R.F., wait (several hours) until the star is approaching S_3; 3 or 4 minutes before it reaches S_3, check the vertical circle reading and ensure that it is the same as for S_2, with the alidade bubble central. Without altering the telescope inclination, keep the vertical cross-hair on the star with the horizontal slow-motion screw until it reaches the intersection of the cross-hairs in position S_3. Note the horizontal circle reading and plate bubble readings.

(v) Change to L.F., set the telescope to the same inclination as for S_1 (with alidade bubble central), and intersect the star as described in (iv). Note the horizontal circle reading and the plate bubble readings.

(vi) Intersect the R.M. on R.F. and note the horizontal circle reading.

The direction midway between S_2 and S_3 should, of course, if there are no errors, coincide with that midway between S_1 and S_4. This will not usually be the case, but the mean of the two results is taken and instrumental errors are largely eliminated.

10.08 Calculation of azimuth of R.M.

If a, b, c and d represent the horizontal circle readings on the star at S_1, S_2, S_3 and S_4 respectively, corrected by means of the plate bubble readings (see 7.11), then $(a + b + c + d)/4$ represents the circle reading corresponding to true north or south (0° or 180°) according to whether the observer is in the northern or southern hemisphere. The difference between this circle reading and the mean reading on the R.M. will give the angle between the R.M. and the meridian, and application of this in the correct sense to 0° or 180° will give the true bearing (azimuth) of the R.M.

For example, suppose the mean circle reading on the R.M. was 46° 30′, and a, b, c, d were 84°, 85°, 122° and 123° respectively, then

$$(a + b + c + d)/4 = 103° \ 30'$$

and

103° 30′ − 46° 30′ = 57° 00′,   the angle between the meridian and the R.M.

If the observer were in the southern hemisphere, the azimuth of the R.M. would then be 180° − 57° = 123°. On the other hand, if the observer were in the northern hemisphere, the azimuth of the R.M. would be 360° − 57° = 303°. (The fact that the circle reading on the R.M. was less than the mean of the four readings on the star indicates that the R.M. was anticlockwise from the meridian.)

The accuracy of the method may be further increased by taking a number of settings at say $10'$ or $20'$ intervals in altitude, the instrument being reversed for half the readings on each side to eliminate collimation and horizontal axis errors.

The polar distances of the stars are not absolutely constant, as the theory of the method assumes, but undergo very slight changes during the year, which are tabulated in the *Star Almanac*. In the course of 6 or 8 hours, however, the alteration never amounts to more than a small fraction of a second of arc, and therefore need not be considered. An unknown error may be introduced by changes in atmospheric refraction during the considerable interval of time that must separate the first and second sets of observations. Nevertheless, the method will give results which are sufficiently accurate for checking ordinary traverse bearings.

Section 3

SECOND METHOD

AZIMUTH BY OBSERVING A CIRCUMPOLAR STAR NEAR ELONGATION

10.09 Principle of method

In Fig. 10.5 let P be the celestial pole, Z the zenith of the observer, E, M_N and W the east, north and west points of the observer's horizon. ZPM_N therefore represents the plane of the meridian. The aspect is one from a hypothetical point in space looking down upon the celestial sphere from the north. A similar diagram could be drawn to show the aspect looking from the south, but the E and W points would then have to be interchanged and the direction of the star's movement reversed.

The small circle with P as centre represents the path of a circumpolar star S. The vertical plane passing through the zenith of the observer and the star traces out the circle ZSB on the celestial sphere. This will be the circular path swept out by the telescope of a theodolite when the telescope, after being directed

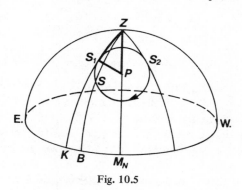

Fig. 10.5

Determination of azimuth

to the star, is turned in a vertical plane about the transverse axis. As the star moves from the position S shown in the figure, this vertical plane will make a greater and greater angle with the plane of the meridian ZPM_N until the star arrives at the position S_1, where the vertical circle ZS_1K, swept out by the telescope, is a tangent to the circular path of the star. This is the point where the vertical plane containing the star makes its greatest angle with the plane of the meridian. At this point the star is said to be *at elongation*, and clearly, its motion being vertical, it is in a favourable position for observation of its azimuth, because its horizontal movement is so slight for some time before and after it arrives at S_1. There will be a corresponding point S_2 in the path of the star to the west of the celestial pole, and the points S_1 and S_2 are referred to as the points of eastern and western elongation respectively.

It is clear from the figure that the points S_1 and S_2 will always be at a greater altitude than the celestial pole P, but the smaller the circle of the star's path, or the greater the declination of the star, the more nearly will the altitude of S_1 and S_2 approach that of P.

Now, if an *Almanac* star is selected for observation, we shall know its declination, and the polar distance PS_1 is the complement of the declination. If, in addition, we know the latitude of the place of observation (from a map or previous observation), then in the right-angled spherical triangle ZPS_1, we shall know PS_1, and ZP which is the complement of the latitude. Hence, by Napier's rules, we can compute the angle PZS_1. We have

$$\sin PS_1 = \sin ZP . \sin PZS_1$$

or

$$\sin PZS_1 = \cos \text{declination} \times \sec \text{latitude}$$

i.e.

$$\sin Z = \cos \delta . \sec \phi. \qquad (10.1)$$

This calculation gives us the angle that the star at S_1 makes with the meridian, and hence its azimuth. Thus, if we measure the angle that the star at S_1 makes with some reference mark, the azimuth of the R.M. is determined.

The method so far indicated would require the direction of the star to be measured at the exact moment of elongation. But we have set it down as a general principle that at least two observations should be made, one with face left and the other with face right, and it becomes important to inquire what error in azimuth will be made if sufficient time is taken to obtain two readings.

On making the necessary calculations, it will be found that, for a place in latitude $30°$, the azimuths of stars at different polar distances will not alter by $5''$ after the moment of elongation until the times given in Table 10.1 have elapsed.

As there will be a corresponding and nearly equal period before elongation, it follows that for a star whose polar distance is $10°$ there will be a total time

Astronomy for surveyors

Table 10.1

Polar distance of star	Time after moment of elongation before azimuth changes by 5″
10°	3 min 33 sec
15	3 07
20	2 35
30	2 11

of about 7 minutes during which its motion is so nearly vertical that the total change of azimuth in that period is not more than 5″. For a star whose polar distance is 30°, the corresponding period is $4\frac{1}{3}$ minutes.

If, then, the surveyor is not seeking to determine the azimuth of the R.M. to better than 15 or 20 seconds, it will be sufficiently accurate for him to take two observations of the star, one with face left and the other with face right, not at the exact moment of elongation, but one just before and the other probably just after elongation. The time required to read the circle, reverse face, and set the telescope again on the star should not be more than 2 or 3 minutes, so that there should be time to get both observations within the period we have just calculated to cover a total azimuth change of not more than 5″. The nearer the star is to the pole, the greater the length of time available for the observations.

The method requires, of course, that the local time be known, but it is simple to carry out in practice, provided the selected star is readily identifiable.

The average value of the angle that the star makes with the meridian, as determined by two observations in this way, is clearly always a little less than the angle at elongation. In order to get the most accurate results with this method, it is better not to use the formula for the star at elongation at all, but to get one or more careful sets of two observations (one L.F. and one R.F.) of the star *near* elongation, observing the altitude of the star at each measurement. In Fig. 10.6, let S represent the star moving in its circular path

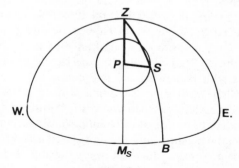

Fig. 10.6

Determination of azimuth

round the pole P, Z the zenith, ZSB the vertical circle passing through the zenith and the star. Then, in the triangle ZPS, if the altitude (h) of the star is measured, the values of ZS ($90° - h$), ZP ($90° - \phi$) and PS (p, the polar distance of the star) are known, and Z, the zenith angle, may be calculated. In the diagram,

$$PM_S = \phi = \text{latitude}$$
$$PZ = \omega = \text{co-latitude} = (90° - \phi)$$
$$PS = p = \text{polar distance of star} = (90° - \delta)$$
$$BS = h = \text{altitude of star}$$
$$ZS = \zeta = \text{zenith distance of star} = (90° - h).$$

From formula (1.2) we get

$$\cos Z = \frac{\cos p - \cos \zeta . \cos \omega}{\sin \zeta . \sin \omega}$$

$$= \frac{\sin \delta - \sin h . \sin \phi}{\cos h . \cos \phi}$$

i.e.
$$\cos Z = \sin \delta . \sec h . \sec \phi - \tan h . \tan \phi. \qquad (10.2)$$

Note that the sun will elongate when observed from places within the Tropics at dates when its polar distance is less than the co-latitude of the observing station.

10.10 Observing procedure

(i) Intersect R.M. on L.F. and read horizontal circle.
(ii) Intersect star on L.F. (see **10.07**(ii); read horizontal and vertical circles (alidade bubble central), also plate bubble L and R.
(iii) Change to R.F. without delay and repeat procedure of (ii).
(iv) Intersect R.M. on R.F. and read horizontal circle.

The difference between the mean circle readings on R.M. and star gives the mean included angle between R.M. and star; the mean of the vertical circle readings gives the mean altitude. Acceptance of these mean values assumes, of course, that the star was moving in a straight line, which we know to be incorrect (*vide* curvature corrections in **7.05** and **7.09**). However, provided there are only 2 to 3 minutes of time between the L.F. and R.F. pointings, the curvature correction is negligible for all ordinary purposes.

The mean azimuth derived from two or three such sets should be within 6 or 7 seconds of the true value if the observations are done with a good glass-circle

Astronomy of surveyors

theodolite reading to 1″. If these sets are balanced by a similar number to a star of about the same polar distance near elongation on the other side of the pole, the mean azimuth from all sets should be within 3″ of the true value.

A worked example of a single set is given in **10.13** below.

10.11 Calculation of the time of elongation

In order to prepare for these observations, it will generally be necessary for the surveyor to work out beforehand the time at which the star will elongate. In Fig. 10.5 the angle ZPS_1 measures, when turned into time, the sidereal time that must elapse before the star at S_1 comes on to the meridian. But when the star is on the meridian the sidereal time is given by its right ascension. Thus the sidereal time when the star is at S_1 is equal to

R.A. of star − the hour angle of the star, ZPS_1.

Of course, the sidereal time when the star is at S_2 is equal to

R.A. of star + the hour angle of the star, ZPS_2.

This sidereal time has then to be turned into mean solar time by the methods we have discussed in Chapter 5.

EXAMPLE. To find the Local Standard Time of western elongation of α Pictoris on 23 April 1975 at a place in Latitude 45° 52′ 15″ S., Longitude $11^h\ 22^m\ 03.9^s$ E. The Standard Meridian of the place is 12^h E.

From the *Star Almanac*: R.A. of α Pictoris = $6^h\ 47^m\ 55.5^s$

Dec. of α Pictoris = S.61° 55′ 16″

(α Pictoris is No. 186 on pages 32 and 33 in the *Star Almanac*—see pages 56, 57 herein.)

In the right-angled triangle which exists at elongation,

$$\cos t = \tan p . \cot \omega$$

i.e.

$$\cos t = \cot \delta . \tan \phi. \qquad (10.3)$$

So

$$\cos t = \cot 61° 55′ 16″ . \tan 45° 52′ 15″$$

$$\therefore t\ (\text{arc}) = 56° 38′ 12″$$

and

Determination of azimuth

	t(time) =	3ʰ 46ᵐ 32.8ˢ	
R.A. α Pictoris	=	6 47 55.5	
t + R.A. = L.S.T.	=	10 34 28.3	23 April
Longitude, E.	=	11 22 03.9	
∴ G.S.T.	=	23 12 24.4	
U.T. + R at U.T. 6ʰ	=	20 02 39.4	23 April
Sidereal interval after U.T. 6ʰ	=	3 09 45.0	
Convert to mean solar interval	=	−31.1	
Mean solar interval after U.T. 6ʰ	=	3 09 13.9	23 April
i.e. U.T. of western elongation	=	9 09 13.9	23 April
Standard longitude, E.	=	12 00 00	
L.Std.T. of western elongation	=	21 09 13.9	23 April

Note. If the L.Std.T. of *eastern* elongation had been required, we would have had to *subtract* t from the R.A. to get the L.S.T. of the event, and then convert this L.S.T. to the corresponding L.Std.T., as above. If this is done, we get $13^h \ 37^m \ 22.5^s$, but this time is close to midday and the star, magnitude 3.3, would not have been observable.

10.12 Azimuth, altitude and hour angle at elongation

We have seen in **10.11** that, at elongation,

$$\cos t = \cot \delta . \tan \phi. \qquad (10.3)$$

From Napier's rules, using the same right-angled triangle, we can deduce formulae for finding the zenith angle (Z) and altitude h, when the declination of the star (δ) and the latitude (ϕ) are known. They are:

$$\sin Z = \cos \delta . \sec \phi \qquad (10.1)$$

$$\sin h = \operatorname{cosec} \delta . \sin \phi. \qquad (10.4)$$

10.13 Worked example: *Observation of a star near western elongation for azimuth*

Observer: J. B. M.
Instrument: Wild T2 with diag. eyepiece prism

Barometer: 999 mb
Thermometer: 9.5 °C
Plate bubble: 1 division = 20″

Astronomy for surveyors

Star observed: No. 186 α Pictoris
R.A.: $6^h\ 47^m\ 55.5^s$
Dec.: S. 61° 55′ 16″
Date: 1975 April 23
Place: Arts Block, Pillar F,
    University of Otago,
    Dunedin, N.Z.
Latitude: 45° 52′ 15″ S.
Longitude: $11^h\ 22^m\ 03.9^s$ E.
R.M.: Red light, 1 Cobden St.

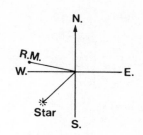

Face	Object	Horizontal circle	Vertical circle	Altitude	Plate bubble L	R
L	R.M.	275° 20′ 54″				
L	Star	226 48 01	35° 53′ 34″	54° 06′ 26″	1.0	2.2
R	Star	46 47 33	323 49 01	53 49 01	3.4	3.4
R	R.M.	95 20 52				

Azimuth correction from plate bubble readings

$$= \{(\Sigma L - \Sigma R)/n\} \times 20'' \times \tan h$$
$$= \{(4.4 - 5.6)/4\} \times 20'' \times 1.374$$
$$= -08''.$$

For horizontal angle between R.M. and star:

Mean horizontal circle reading on star	=	226° 47′ 47″
Plate bubble correction, as above	=	−08
Corrected horizontal circle reading on star	=	226 47 39
Mean horizontal circle reading on R.M.	=	275 20 53
Mean angle between star and R.M., clockwise from star	=	48 33 14

For corrected observed altitude:

Mean observed altitude	=	53° 57′ 44″
Correction for refraction = 42″ × 0.99 $(r_0 \times f)$	=	−42
Corrected observed altitude	=	53 57 02

Calculation of azimuth of R.M.:

$$\cos Z = \sin \delta . \sec h . \sec \phi - \tan h . \tan \phi.$$

Determination of azimuth

A formula such as this may be approached in two ways:

(i) If the Latitude and the Declination are *both* north or *both* south, they can both be taken as positive. In this example both are south, so the formula gives:

$$\cos Z = \sin(61° 55' 16'').\sec(53° 57' 02'').\sec(45° 52' 15'')$$
$$-\tan(53° 57' 02'').\tan(45° 52' 15'')$$
$$= +0.736\,977$$

and

$$Z = 42° 31' 31''.$$

Since the place is in the southern hemisphere and the star was observed west of the meridian, reference to **10.01** and Fig. 10.1(b) shows that the azimuth of the star when observed was equal to $(180° + Z)$, or $222° 31' 31''$.

(ii) The convention that northern Latitudes and Declinations are positive and southern ones are negative may be used. In this case, then,

$$\cos Z = \sin(-61° 55' 16'').\sec(53° 57' 02'').\sec(-45° 52' 15'')$$
$$-\tan(53° 57' 02'').\tan(-45° 52' 15'')$$
$$= -0.736\,977$$

and

$$Z = 137° 28' 29''.$$

At first sight this may appear to be incompatible with the value obtained for Z in (i) above, but the convention we have used relates the astronomical triangle to the north celestial pole, and we need to refer to Fig. 10.1(a), star west of the meridian. We see that the azimuth of the star in this case is equal to $(360° - Z)$, or $222° 31' 31''$, the same value as we found in (i) above, Therefore,

Azimuth of star	$= 222° 31' 31''$
Mean angle from star to R.M., clockwise from star	$= 48\ \ 33\ \ 14$
∴ Azimuth of R.M.	$= 271\ \ 04\ \ 45$

Note. The L.Std.T. of western elongation of this star was worked out in **10.11** above as $21^h\ 09^m\ 13.9^s$ to give us the knowledge of when to prepare for the observation, Equations (10.1) and (10.4) allow us to compute the azimuth at elongation, i.e. where to look for the star at the appropriate time.

In the above example, the mean L.Std.T. of the two pointings on the star was $21^h\ 13^m\ 23.9^s$, or $04^m\ 10^s$ (mean solar time) after elongation.

Astronomy for surveyors

Instead of measuring the altitude of the star, we may note the time, using a digital stop-watch, when the star is bisected by the vertical cross-hair near the centre of the field of view. The star near elongation will usually be moving so slowly in azimuth that the observer must put the cross-hair on it with the horizontal slow-motion screw and start the watch at the moment of bisection. The hour angle of the star can then be found from knowledge of the accurate time of the observation and the R.A. of the star, and the astronomical triangle solved for Z, using this value of t and the known values of Latitude and Declination.

In the example given above,

L.Std.T. of observation, already given	$= 21^h\ 13^m\ 23.9^s$
Longitude of Standard Meridian, E.	$=\ 12\ \ \ 00\ \ \ 00$
U.T.C. of observation, 23 April 1975	$=\ \ \ 9\ \ \ 13\ \ \ 23.9$
DUT1	$=\ \ \ \ \ \ \ \ \ \ \ \ \ \ \ \ +0.3$
U.T.1 of observation, 23 April	$=\ \ \ 9\ \ \ 13\ \ \ 24.2$
R at U.T. 6^h, 23 April	$=\ 14\ \ \ 02\ \ \ 39.4$
ΔR in $3^h\ 13^m\ 24.2^s$	$=\ \ \ \ \ \ \ \ \ \ \ \ \ +31.8$
G.S.T.	$=\ 23\ \ \ 16\ \ \ 35.4$
Longitude of observing point, E.	$=\ 11\ \ \ 22\ \ \ 03.9$
L.S.T.	$=\ 10\ \ \ 38\ \ \ 39.3$
R.A. star	$=\ \ \ 6\ \ \ 47\ \ \ 55.5$
t (time)	$=\ \ \ 3\ \ \ 50\ \ \ 43.8$
t (arc)	$=\ 57°\ 40'\ 57''$
$p (= 90° - \delta)$	$=\ 28\ \ \ 04\ \ \ 44$
$\omega (= 90° - \phi)$	$=\ 44\ \ \ 07\ \ \ 45$

$$\cot Z = \frac{\cot p . \sin \omega - \cos \omega . \cos t}{\sin t}$$

from which Z computes as $42°\ 31'\ 31''$, the same value as was obtained earlier.

10.14 The effect of an error in the latitude

In the preceding calculations we require to know the declination of the star and the latitude of the place of observation. The declination is found with negligible error from the *Star Almanac*, but it is very likely that the latitude will not be known with the same precision. From **10.09** above,

Determination of azimuth

$$\sin Z = \cos \delta . \sec \phi \qquad (10.1)$$

or

$$\sin Z . \cos \phi = \cos \delta .$$

If we differentiate this with respect to ϕ, keeping δ constant, we have:

$$\cos Z . \frac{dZ}{d\phi} . \cos \phi + (-\sin \phi) . \sin Z = 0$$

i.e.

$$\frac{dZ}{d\phi} . \cos Z . \cos \phi = \sin Z . \sin \phi .$$

Therefore

$$\frac{dZ}{d\phi} = \frac{\sin Z . \sin \phi}{\cos Z . \cos \phi} = \tan Z . \tan \phi \qquad (10.5)$$

or

$$\frac{dZ}{d\phi} = \tan \phi . \frac{\sin Z}{\sqrt{(1 - \sin^2 Z)}}$$

$$= \tan \phi . \frac{\cos \delta}{\sqrt{(\cos^2 \phi - \cos^2 \delta)}}$$

Thus the change produced in Z by a small change (error) in ϕ is zero if $\phi = 0$ (since $\tan 0° = 0$), i.e. at places on the equator where stars elongate on the horizon. Since we have noted in Chapter 7 that it is not advisable to observe stars at altitudes less than 15°, we would not use this method for such places. When $\phi = 90°$, $dZ/d\phi = \infty$, but zenith and pole coincide, so that elongation is impossible, and in any case azimuth at the poles does not have any meaning. $dZ/d\phi$ is also ∞ if $\phi = \delta$, i.e. if $\omega = p$, for in this case the path of the star passes through the zenith, and there is no position of elongation.

Table 10.2 gives the values of the error in azimuth compared with the error in latitude, as calculated by the preceding formula, for various values of ϕ and δ.

Table 10.2   Ratio of error in azimuth to a small error in latitude

Dec. of star observed	In Lat. 20°	In Lat 30°	In Lat 40°	In Lat 50°
60°	0.22	0.40	0.70	1.47
70	0.14	0.24	0.40	0.75
80	0.06	0.10	0.19	0.41

Astronomy for surveyors

In the cases tabulated for $\phi = 20°$, $30°$ and $40°$, an error in latitude of, say, $5''$ will produce an error in azimuth of less than $5''$, the tabulated ratios being all less than 1. The error in azimuth may, however, be much greater than the error in latitude if the star observed has a declination approaching the value of the latitude. This is seen in the last column of the table, where $\phi = 50°$ and $\delta = 60°$ and the error in azimuth $= 1.47$ times the error in latitude.

In any given latitude, the error is least when the star selected is nearest to the pole. From the formula, $dZ/d\phi = 0$ if $\delta = 90°$. This and other considerations, as we have seen, all point to the desirability of choosing a star for observation as near to the celestial pole as possible.

10.15 Daylight observations of elongating stars

Reference to the list of stars in the *Star Almanac* will show that some of those marked with "d" (see **10.04**) have small enough polar distances to make them suitable for observation near elongation in daylight as well as at night.

10.16 Circum-elongation observation of stars

By making an equal number of timed pointings on each face of the theodolite to a star as it is approaching and leaving the position of elongation, we can get a very good determination of azimuth. Each such pointing can be reduced, or corrected by a small amount to give it the direction it would have had when the star was at elongation, and the mean of all these corrected values will give an accurate value of the star's azimuth, free of any horizontal collimation error in the instrument because of the equal number of measures on each face.

To keep the computations reasonably simple, the pointings should be made within the period extending from 15 minutes before to 15 minutes after elongation. Altitudes are not measured.

Development of reduction formulae

By using the relationships

$$\cos p = \cos \zeta . \cos \omega + \sin \zeta . \sin \omega . \cos Z$$

and

$$\cos \zeta = \cos \omega . \cos p + \sin \omega . \sin p . \cos t$$

we can obtain

$$\cos Z . \sin \zeta = \cos p . \sin \omega - \cos \omega . \sin p . \cos t \qquad (10.6)$$

and from the sine rule

$$\sin Z . \sin \zeta = \sin p . \sin t. \qquad (10.7)$$

Determination of azimuth

If the triangle is right-angled at S (i.e. if the star is at elongation) the sine rule gives

$$\sin Z_0 = \frac{\sin p}{\sin \omega} \quad \text{which, from Napier's rule,}$$

$$= \frac{\cos p . \cos t_0}{\cos \omega} \tag{10.8}$$

where $Z_0 = Z$ at elongation, and $T_0 = t$ at elongation.

Also from Napier's rule we can get

$$\cos Z_0 = \sin t_0 . \cos p. \tag{10.9}$$

Now, by multiplying (10.6) by (10.8), LHS by LHS, and RHS by RHS, and subtracting the product of (10.7) by (10.9) obtained in the same way, we can reduce to

$$\sin \zeta . \sin(Z_0 - Z) = \cos p . \sin p . 2 . \sin^2 \tfrac{1}{2}(t_0 - t). \tag{10.10}$$

Further manipulation leads to

$$\frac{\sin(Z_0 - Z)}{\sin Z} = \cos p . \frac{2 . \sin^2 \tfrac{1}{2}(t_0 - t)}{\sin t} = y, \quad \text{say.}$$

By expanding the numerator of the LHS and dividing by the denominator, we get

$$\sin Z_0 . \cot Z - \cos Z_0 = y \tag{10.11}$$

in which Z_0 is a constant.

Thus Z is a function of y, or $Z = F(y)$. We can now expand $F(y)$ as a power series in y, using Maclaurin's Theorem:

$$F(y) = f(0) + y . f'(0) + \frac{y^2}{2!} f''(0) + \ldots,$$

where

$f'(0)$ is the first derivative of $F(y)$ when $y = 0$, and

$f''(0)$ is the second derivative of $F(y)$ when $y = 0$.

Differentiating (10.11) gives

$$\frac{dZ}{dy} = -\frac{\sin^2 Z}{\sin Z_0} \tag{10.12}$$

and

$$\frac{d_2 Z}{dy^2} = 2 . \sin Z . \cos Z . \frac{\sin^2 Z}{\sin^2 Z_0}. \tag{10.13}$$

From (10.11), when $y = 0$, we get

$$\sin Z_0 . \cot Z = \cos Z_0, \quad \text{hence} \quad \cot Z = \frac{\cos Z_0}{\sin Z_0} = \cot Z_0.$$

Thus, when $y = 0$, $Z = Z_0$, and from (10.12),

$$\frac{dZ}{dy} = -\sin Z_0 \quad \text{and} \quad \frac{d_2 Z}{dy^2} = \sin 2Z_0.$$

Summarizing,

$$f(0) = Z_0$$
$$y.f'(0) = y.(-\sin Z_0)$$
$$\frac{y^2}{2!} f''(0) = \frac{y^2}{2}.(\sin 2Z_0).$$

Thus,

$$F(y) = Z = Z_0 - \sin Z_0 . y + \sin 2Z_0 . \frac{y^2}{2} + \ldots$$

and

$$(Z - Z_0) = -\sin Z_0 . y + \sin 2Z_0 . \frac{y^2}{2} + \ldots \quad (10.14)$$

But

$$y = \frac{\cos p}{\sin t} . 2.\sin^2 \tfrac{1}{2}(t_0 - t) \quad \text{(see above)}$$

and $2.\sin^2 \tfrac{1}{2}(t_0 - t)/\sin 1''$ is equal to the m of the "Table for Circum-meridian Observations" on page 68 of the *Star Almanac*.

Thus, if we divide (10.14) all through by $\sin 1''$, we get

$$(Z - Z_0)'' = -\frac{\sin Z_0 . \cos p}{\sin t} . m + \frac{\sin 2Z_0 . \cos^2 p}{\sin^2 t . \sin 1''} . 2.\sin^4 \tfrac{1}{2}(t_0 - t)$$

or

$$(Z_0 - Z)'' = P.m - Q.2.\sin^4 \tfrac{1}{2}(t_0 - t), \quad (10.15)$$

where

$$P = \frac{\sin Z_0 . \cos p}{\sin t}; \quad Q = \frac{\sin 2Z_0 . \cos^2 p}{\sin^2 t . \sin 1''}$$

and $(Z_0 - Z)''$ is in seconds of arc.

By using (10.15) we can reduce the horizontal circle reading of an observation done some time away from elongation to the value it would have had at

Determination of azimuth

elongation, and if the time away from elongation were not too long, we could cut (10.15) down to

$$(Z_0 - Z)'' = P.m.$$

As an illustration, let us look at the previously worked example in **10.13**. Here,

$(t_0 - t) = 4^m\ 10^s$; see beginning of *Note* in **10.13**.
$Z_0\quad = 42°\ 31'\ 55''$; see (10.1).
$p\quad = 28°\ 04'\ 44''$.
$t\quad = 57°\ 40'\ 57''$; see hour angle computation in *Note*, **10.13**.

Since $(t_0 - t) = 4^m\ 10^s$, from the table on page 68 of the *Star Almanac* we find $m = 34''$. Then

$$(Z_0 - Z)'' = P \times 34'' = 0.706 \times 34'' = 24''.$$

This means that the mean horizontal circle reading must be increased by 24" to coincide with the azimuth of western elongation (since the star would be observed slightly short of elongation). Hence the corrected circle reading at elongation would be (226° 47' 39" + 24"), or 226° 48' 03". The angle between R.M. and star at elongation would therefore be (275° 20' 53" −226° 48' 03") = 48° 32' 50". But Z_0 is 42° 31' 55", hence the azimuth of the star at western elongation is 222° 31' 55". By adding the angle 48° 32' 50" to this, we get 271° 04' 45" as the azimuth of the R.M., a figure which agrees with that previously obtained.

Reduction of multiple pointings

Formula (10.15) allows us to make a series of pointings straddling elongation, and thus strengthen the value of the azimuth obtained. From the total number of pointings made to the star we take the mean horizontal circle reading; then by taking the differences between the time of elongation and the times of the individual pointings, we get a series of $(t_0 - t)$ values for each of which we can obtain the m value from the *Star Almanac*. If the mean of these m values is taken and multiplied by factor P, we obtain $(Z_0 - Z)''$, the correction to be applied to the mean horizontal circle reading to bring it to the value it would have had if the star were at elongation. The azimuth of the R.M. is then found in the same way as in the example of reduction of a single observation above.

It will have been noted that, in factor P of (10.15), $\sin t$ and not $\sin t_0$ appears in the denominator. This is not a constant, and strictly it should be worked out for each pointing. However, provided the circum-elongation pointings do not extend more than 15 minutes of time on either side of elongation, we may substitute $\sin t_0$ for $\sin t$ and also neglect the Q term of (10.15). Some attention should also be paid to getting a reasonably well-balanced set of pointings on each side of elongation. If the pointings are not particularly well distributed,

Astronomy for surveyors

t should be found for each, (10.15) applied to each pointing individually, and a mean taken of the computed azimuths of the R.M. The Q term of (10.15) should be included if the pointings extend more than 15 minutes on either side of elongation, which would be rare unless clouds intervened.

Worked example: *Circum-elongation observation for azimuth*

Date: 1975 May 9
Place: Arts Block, Pillar F,
   University of Otago,
   Dunedin, N.Z.
Latitude: 45° 52′ 15″ S.
Longitude: $11^h\ 22^m\ 03.9^s$ E.
Observer: J. B. M.
Booker: C. Wratt
Instrument: Wild T2 No. 76 004
Plate bubble: 1 division = 20″
Star observed: FK4 264 at
   western elongation
R.A. star: $6^h\ 42^m\ 01.0^s$
Dec. star: 80° 47′ 43″ S.

R.M.: Red light,
   1 Cobden Street
Chronometer:
   Mercer No. 27 959
Error on U.T.C. =
   $12^h\ 00^m\ 18.5^s$ fast

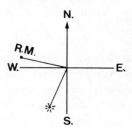

Face	Object	Horizontal circle	Chronometer	Stop-watch	Chronometer time of obs.	Plate bubble L	R
L	R.M.	271° 08′ 37″					
L	Star	193 19 43	$21^h\ 26^m\ 20^s$	12.3^s	$21^h\ 26^m\ 07.7^s$	3.2	2.3
L	Star	193 19 55	21 28 00	17.3	21 27 42.7	3.5	2.1
L	R.M.	271 08 32					
R	R.M.	91 08 41					
R	Star	13 20 34	21 33 05	16.2	21 32 48.8	3.5	2.1
R	Star	13 20 35	21 34 40	16.9	21 34 23.1	3.4	2.2
R	R.M.	91 08 41					
R	R.M.	91 08 39					
R	Star	13 20 28	21 38 15	10.8	21 38 04.2	3.8	2.1
R	Star	13 20 26	21 40 05	18.2	21 39 46.8	3.7	2.0
R	R.M.	91 08 40					
L	R.M.	271 08 34					
L	Star	193 19 45	21 44 10	12.7	21 43 57.3	3.7	2.0
L	Star	193 19 31	21 45 45	10.6	21 45 34.4	3.8	2.0
L	R.M.	271 08 32					
					Sum	28.6	16.8

Note. There are as many pointings on the R.M. as there are on the star.

Mean horizontal circle reading on R.M.   = 271° 08′ 37.0″
Mean horizontal circle reading on star   = 193   20   07.1
Approximate altitude of star   = 46° 40′

Determination of azimuth

Computation of azimuth
Plate bubble correction:

$$\{(\Sigma L - \Sigma R)/n\} \times v \times \tan h = \{(28.6 - 16.8)/16\} \times 20'' \times 1.06$$
$$= (11.8/16) \times 20'' \times 1.06$$
$$= +15.6''$$

Mean horizontal circle reading on star	$= 193° \; 20' \; 07.1''$
Plate bubble correction, from above	$= +15.6$
Corrected horizontal reading on star	$= 193 \; \; 20 \; \; 22.7$

Time of elongation:
Hour angle at elongation $t_0 = \cos^{-1}(\cot \delta . \tan \phi)$
$ = 80° \; 23' \; 01''$
$ = 5^h \; 21^m \; 32.1^s$

L.S.T. at transit	R.A. =	$6 \;\; 42 \;\; 01.0$
L.S.T. at western elongation	=	$12 \;\; 03 \;\; 33.1$
Longitude of observing station, E.	=	$-11 \;\; 22 \;\; 03.9$
G.S.T. (= U.T. + R)	=	$24 \;\; 41 \;\; 29.2$
U.T. + R at U.T. 6^h, 9 May 1975	=	$21 \;\; 05 \;\; 44.2$
Sidereal interval after U.T. 6^h	=	$3 \;\; 35 \;\; 45.0$
ΔR	=	-35.3
Mean solar interval after U.T. 6^h	=	$3 \;\; 35 \;\; 09.7$
U.T.1 of elongation in west	=	$9 \;\; 35 \;\; 09.7$

Individual times of pointings and mean value of m:

Chronometer fast on U.T.C.	$= 12^h \; 00^m \; 18.5^s$
DUT1 (= U.T.1 − U.T.C.)	$= 0.3$
Chronometer fast on U.T.1	$= 12 \;\; 00 \;\; 18.2$

			From Star Almanac
Chronometer	U.T.1	$t_0 - t$	m
$21^h \; 26^m \; 07.7^s$	$9^h \; 25^m \; 49.5^s$	$9^m \; 20.2^s$	$171''$
27 42.7	27 24.5	7 45.2	118
32 48.8	32 30.6	2 39.1	14
34 23.1	34 04.9	1 04.8	2
38 04.2	37 46.0	2 36.3	13
39 46.8	39 28.6	4 18.9	37
43 57.3	43 39.1	8 29.4	142
45 34.4	45 16.2	10 06.5	201
			8)698
			Mean m 87.3 $= m_0$

Astronomy for surveyors

Calculation of azimuth:
From (10.15),

$$P = \frac{\sin Z_0 . \cos p}{\sin t_0}$$

(assuming $t = t_0$ for reasonable distribution of pointings).

$Z_0 = \sin^{-1}(\cos \delta . \sec \phi) = 13° \ 16' \ 54''$
$p = (90° - \delta) \qquad = \ 9 \ 12 \ 17$
$t_0 = \text{(from above)} \qquad = 80 \ 23 \ 01$

From which

$$P = +0.230.$$

Thus

$$P \times m_0 = 0.230 \times 87.3'' = 20.1''.$$

Mean corrected horizontal circle reading on star	$= 193° \ 20' \ 22.7''$
$P.m_0$	$= \qquad \quad 20.1$
∴ Mean circle reading on star at western elongation	$= 193 \ 20 \ 42.8$
Mean horizontal circle reading on R.M.	$= 271 \ 08 \ 37.0$
Angle between star at western elongation and R.M.	$= \ 77 \ 47 \ 54.2$
Azimuth of star at western elongation $= 180° + Z_0$	$= 193 \ 16 \ 54$
∴ Azimuth of R.M.	$= 271 \ 04 \ 48$

Section 4

THIRD METHOD

EX-MERIDIAN OBSERVATIONS OF THE SUN OR A STAR FOR AZIMUTH BY MEASURING ALTITUDES AND HORIZONTAL ANGLES

10.17 Principle of method

We have already discussed a particular case of this method in the preceding section where we were dealing with stars near elongation. A celestial object will not elongate, however, if its polar distance is greater than the co-latitude of the observing station, but even so, by observing its altitude near certain optimum positions when it is well away from the meridian and at the same time measuring the angle between it and the R.M., we can still compute the azimuth of the latter and get satisfactory results. The formulae to be used are those we have already noted at the end of **10.09**. ϕ and δ are known, and

Determination of azimuth

h is measured, so that Z may be computed. (Note, however, that in the formulae δ and ϕ may have different signs.)

10.18 Best position of star or sun for observations

Since the three quantities used in the calculation for Z are ϕ, h and p, we must examine the effects that small errors in these will have on Z. We can best do this by making use of the differential calculus.

(i) *The effect of an error in latitude on Z*

One of the basic formulae connecting the four quantities concerned is:

$$\cos p = \cos \zeta . \cos \omega + \sin \zeta . \sin \omega . \cos Z \quad \text{(see 10.09)}.$$

Since

$$p = 90° - \delta, \quad \zeta = 90° - h \quad \text{and} \quad \omega = 90° - \phi,$$

we have

$$\sin \delta = \sin h . \sin \phi + \cos h . \cos \phi . \cos Z. \tag{10.16}$$

Differentiating this with respect to ϕ, keeping δ and h constant, we have

$$0 = \sin h . \cos \phi + \cos h . \cos \phi \left(-\sin Z \frac{dZ}{d\phi} \right) + \cos h . \cos Z (-\sin \phi)$$

so that

$$\frac{dZ}{d\phi} = \frac{\sin h . \cos \phi - \cos h . \cos Z . \sin \phi}{\cos h . \cos \phi . \sin Z}$$

$$= \frac{1}{\cos \phi} \left\{ \frac{\sin h . \cos \phi}{\cos h . \sin Z} - \frac{\cos Z . \sin \phi}{\sin Z} \right\}.$$

But, from (10.16) above,

$$\cos Z = \frac{\sin \delta - \sin h . \sin \phi}{\cos h . \cos \phi}$$

and substituting this in the right-hand term of the bracket, we get

$$\frac{dZ}{d\phi} = \frac{1}{\cos \phi} \left\{ \frac{\sin h . \cos \phi}{\cos h . \sin Z} - \frac{\sin \phi}{\sin Z} \left(\frac{\sin \delta - \sin h . \sin \phi}{\cos h . \cos \phi} \right) \right\}$$

$$= \frac{1}{\cos \phi} \left(\frac{\sin h - \sin \phi . \sin \delta}{\cos h . \sin Z . \cos \phi} \right). \tag{10.17}$$

But from basic formula (1.2)

$$\cos \zeta = \cos \omega . \cos p + \sin \omega . \sin p . \cos t$$

or
$$\sin h = \sin\phi.\sin\delta + \cos\phi.\cos\delta.\cos t$$
i.e.
$$\sin h - \sin\phi.\sin\delta = \cos\phi.\cos\delta.\cos t.$$

Substituting this in (10.17) above,
$$\frac{dZ}{d\phi} = \frac{1}{\cos\phi}\left(\frac{\cos\phi.\cos\delta.\cos t}{\cos h.\sin Z.\cos\phi}\right) = \frac{\cos\delta.\cos t}{\cos h.\sin Z.\cos\phi}$$
$$= \frac{\sin p.\cos t}{\sin\zeta.\sin Z.\cos\phi}$$

But from the sine formula (1.1),
$$\frac{\sin p}{\sin Z.\sin\zeta} = \frac{1}{\sin t}.$$

Therefore,
$$\frac{dZ}{d\phi} = \frac{1}{\sin t}.\frac{\cos t}{\cos\phi} = \cot t.\sec\phi. \tag{10.18}$$

Also from the sine formula,
$$\cos\phi = \sin\omega = \frac{\sin S.\sin\zeta}{\sin t} = \frac{\sin S.\cos h}{\sin t};$$

$\sec\phi$ is therefore equal to
$$\frac{\sin t}{\sin S.\cos h}.$$

Thus, substituting in (10.18) and putting $\cot t = \cos t/\sin t$,
$$\frac{dZ}{d\phi} = \frac{\cos t}{\sin t}.\frac{\sin t}{\sin S.\cos h} = \cos t.\text{cosec}\, S.\sec h. \tag{10.19}$$

(10.19) tells us, again, that t should be near 6^h, and, further, that the parallactic angle S should be near $90°$ to keep cosec S near its minimum value of 1. (6 hours) so that $\cot t$ is close to zero, and ϕ small so that $\sec\phi$ is close to its minimum value of 1. As long, then, as we do not use the method in high latitudes without ensuring that we have a good knowledge of the latitude or without balancing the observations east and west, we can keep $dZ/d\phi$ small by observing the sun or star when its hour angle is as close to 6^h as possible.

(10.19) tells us, again, that t should be near 6^h, and, further, that the parallactic angle S should be near $90°$ to keep cosec S near its minimum value of 1. and that h should be small to keep $\sec h$ near its minimum value of 1. If $S = 90°$,

Determination of azimuth

the body is at elongation, and for t to be near 90° at the same time, the body would have to be a close circumpolar star at elongation. Such a star in low latitudes would also have ϕ and h small and would be the ideal body to observe, were it not for the fact that refraction effects are uncertain below about 15° altitude. However, we are not concerned with elongation observations in this section, so we conclude that, to minimize errors due to doubtfully known latitudes in determining azimuth from other ex-meridian observations, we must have the hour angle of the body near 6^h, the parallactic angle as near to 90° as possible, and the altitude as low as possible, consistent with keeping refraction errors small. The parallactic angle S is a maximum when the body is on the Prime Vertical, so it appears that the best position for observations to a non-elongating body, if $dZ/d\phi$ is to be kept small, lies between the Prime Vertical and the point where the hour angle (east or west) is 6^h, with the body so selected as to make the altitude greater than 15° in this position.

(ii) *The effect of an error in the altitude h on Z*

By differentiating the equation

$$\sin\delta = \sin h.\sin\phi + \cos h.\cos\phi.\cos Z \quad (10.16)$$

with respect to h, keeping δ and ϕ constant, we get

$$0 = \cos h.\sin\phi + (-\sin h).\cos\phi.\cos Z$$

$$+ \cos h.\left(-\sin Z \frac{dZ}{dh}\right).\cos\phi$$

and on simplifying this in a way similar to that for $dZ/d\phi$ in **10.18**(i) above, we get

$$\frac{dZ}{dh} = \cot S.\sec h \quad (10.20)$$

$$\frac{dZ}{dh} = \cos S.\csc t.\sec\phi. \quad (10.21)$$

Examination of these results will show that S should be as near 90° as possible, t should be close to 6^h and h and ϕ should be low, so that the conditions for keeping error in Z small are the same as those found in **10.18**(i).

(iii) *The effect of an error in declination on Z*

As far as the stars are concerned, their declinations change very slowly and by very small amounts during the year. Their declinations are listed in the *Star Almanac* and can always be obtained to the nearest $1''$ of arc, so that there is unlikely to be much effect on Z due to any small residual errors in them. The sun's declination, however, is changing quite rapidly in comparison, and, therefore, inaccuracy in the knowledge of the time of the observation may produce a significant error in the value of the declination interpolated from the *Star Almanac**. Furthermore, the tabulated values of the sun's declination in the *Star Almanac* are given to $0.1'$ only, so that they may be in error by nearly $03''$, and possibly more after the interpolation.

If we differentiate the formula

$$\sin\delta = \sin h.\sin\phi + \cos h.\cos\phi.\cos Z \qquad (10.16)$$

with respect to δ, keeping h and ϕ constant, we get

$$\cos\delta = 0 - \cos h.\cos\phi.\sin Z.dZ/d\delta,$$

so that

$$\frac{dZ}{d\delta} = \frac{-\cos\delta}{\cos h.\cos\phi.\sin Z}.$$

But

$$\cos h = \frac{\sin t.\cos\delta}{\sin Z} \quad \text{(from sine formula),}$$

so that

$$\frac{dZ}{d\delta} = \frac{-\cos\delta.\sin Z}{\sin t.\cos\delta.\cos\phi.\sin Z} \qquad (10.22)$$

$$= -\sec\phi.\operatorname{cosec} t.$$

This result shows that ϕ should be small and t as close to $90°$ or 6^h as possible, and is in line with previous conditions deduced in **10.18**(i) and (ii). We have no control over ϕ for any particular observation (except, of course, by not using the method if errors due to uncertain latitude might be large), but if we observe the sun when it is between the Prime Vertical and the point where its hour angle is 6^h (east or west), with altitude close to but not less than $15°$, we shall minimize the effect of errors in ϕ, h and δ. It is by no means always possible to do this—for example, in mid-latitudes in winter when the sun does not cross the P.V. above the horizon, and is also below the horizon 6 hours before and after upper transit. Under these circumstances, the best one can do is to observe it as far away from transit as possible but not at altitudes less than $15°$.

---

* If possible, the time of a sun observation should be obtained to within 2 or 3 minutes, for this reason.

Determination of azimuth

10.19 Observing procedure

(a) *For stars*

This is exactly the same as that described in steps (i) to (iv) of **10.10**. Since both horizontal angle and altitude are required, the star must be observed on the intersection of the cross-hairs; the best method of doing this is to preset the horizontal cross-hair a little ahead of the star and keep the vertical hair on it with the horizontal slow-motion screw until it reaches the intersection.

If one such set, consisting of L.F. on R.M., L.F. on star, R.F. on star, R.F. on R.M., is balanced by another to a second star at about the same altitude and about the same distance from the meridian but on the other side of it, the mean of the two sets should yield an azimuth of the R.M. within $10''$ of the true value (assuming a 1-second instrument is used). If several sets are taken on both sides of the meridian the accuracy will be improved, but the results will probably not be as good as those obtained from similarly balanced elongation observations, since the star is moving much more quickly in azimuth, thus making accurate intersection more difficult.

(b) *For the sun*

Again the observing routine is the same as in steps (i) to (iv) of **10.10**. However, since we cannot observe the sun's centre on the intersection of the cross-hairs, we must get the sun into one quadrant of the field of view so that both hairs touch the edge of the sun's disc simultaneously. This is a difficult operation, as has already been mentioned in **10.04**(i), and even after much practice few observers would guarantee to get perfect tangency between both cross-hairs and the sun's disc at every observation. The Roelofs solar prism attachment is the only real answer to this difficulty (see **7.04**).

Having observed the sun in, say, the upper left quadrant on the left face, the instrument is quickly changed to right face and the sun observed in the diagonally opposite quadrant, in this case the lower right one. If the time-interval between the observations on the two faces is short (2-3 minutes), there will be negligible curvature error in taking the mean of the two altitudes as the altitude of the sun's centre at the instrument azimuth given by the mean of the two horizontal circle readings. The need for semi-diameter corrections is thus eliminated by taking these means, *vide* **7.04** and **7.08**. Figure 10.7 shows the positions of the sun in the field of view for the two pointings described above. The arrangement of the cross-hairs is one common to many modern theodolites, but there is no difficulty in applying the principle to other types of graticule as long as the two images are disposed symmetrically about the centre of the field of view in such a way as to eliminate the need for semi-diameter corrections in both azimuth and altitude (but see **7.08**).

The observing routine is, then:
(i) Intersect the R.M. on L.F. and read horizontal circle.

Astronomy for surveyors

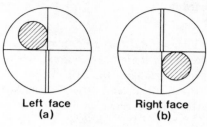

Fig. 10.7

(ii) Observe the sun in the upper left quadrant, as in Fig. 10.7(a). This is best done by pre-setting the horizontal cross-hair a little ahead of the sun and maintaining tangency with the vertical hair on the sun's apparent right limb, by means of the horizontal slow-motion screw, until tangency with the horizontal hair occurs at the apparent lower limb. Read the horizontal and vertical circles (alidade bubble central), also the plate bubble L and R. Note the approximate L.Std.T. of observation (which is needed to compute the sun's declination). Always book the sun's position *as it appears in the field of view*; in this instance it is entered thus: ⌖

(iii) Change to R.F. without delay and observe the sun in the lower right quadrant, as in Fig. 10.7(b). Check alidade bubble, and read circles and plate bubble as in (ii) above. Note approximate L.Std.T. of observation. Book sun's position as: ⌖

(iv) Intersect R.M. on R.F. and read horizontal circle.

There is, of course, no reason why the sun should not be observed in the upper right and lower left quadrants on L.F. and R.F. respectively, and in fact it is good practice to follow the first set with a second in which the sun is observed in these quadrants.

The only way to balance sun observations of this kind (to reduce errors introduced by uncertain knowledge of latitude) is to do the first group in the morning when the sun is in the east and then to wait until the sun has reached the appropriate position in the afternoon, in the west, for the balancing group. Even so, changes in atmospheric refraction conditions between the times of the two groups may introduce slight errors, especially at altitudes near or below 15°.

The accuracy of azimuths obtained from observation of the sun by this method is difficult to estimate because many factors are involved. For example, if $\phi = 46°$, $h = 16°$, $p = 102°$, $S = 34\frac{1}{2}°$ for an observation, and an error of 5″ is made in measuring the sun's altitude because tangency was not perfect with the horizontal cross-hair, the error in Z due to this cause alone would be of the order of 7.5″. However, with care in the work, a single set (one pointing to the sun on each face) with a 1″ instrument should give an azimuth which

Determination of azimuth

is within 15" of the true value. By taking several balanced east-west sets when the sun is in the optimum positions (see **10.18**), the accuracy will be considerably improved.

10.20 Calculation of the sun's declination

The values of the sun's declination at 6-hourly intervals during the year for U.T. are given in the columns headed "Dec." on pages 2-25 of the *Star Almanac*. An interpolation table for the sun is also given on page 73. An example will serve to illustrate the method of finding the declination.

EXAMPLE. Find the sun's declination at $9^h\ 01^m\ 30^s$ L.Std.T. on 20 February 1975 at a place whose Standard Meridian is 12^h E.

We must first find the U.T. corresponding to the given L.Std.T. so that we may enter the sun table in the *Star Almanac*:

Given L.Std.T., 20 February 1975	=	$9^h\ 01^m\ 30^s$
Longitude of Standard Meridian, E.	=	12 00 00
Corresponding U.T., 19 February	=	21 01 30
Sun's declination at U.T. 18^h, 19 February	=	S. 11° 18.5'
Change in $3^h\ 01^m\ 30^s$ (table on p. 73 of *S.A.*)	=	−02.7
Sun's declination at $21^h\ 01^m\ 30^s$ U.T., 19 February	=	S. 11 15.8
	=	S. 11° 15' 48"

On page 5 of the *Star Almanac* for 19 February 1975, U.T. 18^h, the sun's declination is given as S. 11° 18.5'; between this tabulated value and the next at U.T. 0^h on 20 February, the change in the six hours is given at the right of the "Dec." column as 54, representing 05.4' (as may be verified by taking the difference between the two tabulated values). It is obvious from the table that the sun's southerly declination is decreasing, so the change in the 6 hours we are concerned with is −05.4'. Now, entering the table on page 73 of the *Star Almanac*, "Interpolation Table for Sun", with 54 as the tabular 6-hourly difference, and the time-difference of $3^h\ 00^m$ (the nearest value to the $3^h\ 01^m\ 30^s$ interval), we find that the change in this interval is 27, i.e. −02.7'. This is applied to the declination at U.T. 18^h to give the declination at U.T. $21^h\ 01^m\ 30^s$.

Astronomy for surveyors

10.21 Worked examples: *Observations for azimuth by the altitude method*

(a) *Using the sun*

Place: Arts Block, Pillar F,
   University of Otago,
   Dunedin, N.Z.
Latitude: 45° 52′ 15″ S.
Longitude: $11^h\ 22^m\ 03.9^s$ E.
Observer: G. Arthur
Instrument: Wild T2, No. 128 992

Plate bubble: 1 division = 20″
Object observed: sun
R.M.: Red light, 1 Cobden Street
Booker: C. Wratt
Standard Meridian: 12^h E.

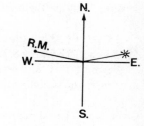

(i) Morning observation, 20 February 1975
   Barometer: 1019 mb. Temperature: 20 °C

Face	Object	Horizontal circle	Vertical circle	Altitude	Plate bubble L	Plate bubble R	Approx. L.Std.T.
L	R.M.	348° 11′ 10″					
L	☉	151 07 30	59° 45′ 37″	30° 14′ 23″	2.9	2.0	$9^h\ 00^m$
R	☉	331 10 40	300 09 12	30 09 12	4.4	0.3	9 03
R	R.M.	168 11 11					

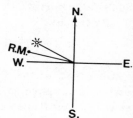

(ii) Afternoon observation, 26 February 1975
   Barometer: 1017 mb. Temperature: 23 °C

Face	Object	Horizontal circle	Vertical circle	Altitude	Plate bubble L	Plate bubble R	Approx. L.Std.T.
L	R.M.	271° 05′ 37″					
L	☉	289 22 15	59° 43′ 22″	30° 16′ 38″	2.0	3.0	$16^h\ 36^m$
R	☉	109 29 57	299 21 47	29 21 47	2.8	2.1	16 38
R	R.M.	91 05 41					

Computations

(i) Morning observation, 20 February 1975
   Plate bubble correction:

$$\{(\Sigma L - \Sigma R)/n\} \times v \times \tan h = \{(7.3 - 2.3)/4\} \times 20'' \times 0.582$$
$$= +15''.$$

Determination of azimuth

Mean horizontal circle reading on sun	=	151° 09' 05"
Plate bubble correction	=	+15
Corrected horizontal circle reading on sun	=	151 09 20
Mean horizontal circle reading on R.M.	=	348 11 11
Angle, sun to R.M., clockwise from sun	=	197 01 51

Sun's altitude:

Mean observed altitude of sun		=	30° 11' 48"
Refraction correction ($r_0 = 100, f = 0.97$)		=	−01 37
Parallax correction ($9'' \times \cos h$)		=	+08
Corrected observed altitude,	$h =$		30 10 19

Sun's declination:
This was computed in **10.20**, $\delta =$ S. 11° 15' 48"

Three sides of the astronomical triangle are now known:

Co-latitude $= \omega = (90° - \phi) = 90° - 45° 52' 15'' = 44° 07' 45''$
Polar distance $= p = (90° - \delta) = 90° - 11\ 15\ 48 = 78\ 44\ 12$
Zenith distance $= \zeta = (90° - h) = 90° - 30\ 10\ 19 = 59\ 49\ 41$

A suitable formula is:

$$\cos Z = (\cos p - \cos \omega . \cos \zeta)/\sin \omega . \sin \zeta, \quad \text{see (1.1)}.$$

Substituting the above values in this gives $Z = 105° 57' 06''$.
From Fig. 10.1(b), with the sun in the east (morning):

Azimuth of sun's centre when observed $= 180° - Z =$		74° 02' 54"
Angle, sun to R.M., clockwise from sun	=	197 01 51
∴ Azimuth of R.M.	=	271 04 45

(ii) Afternoon observation, 26 February 1975
   Plate bubble correction:

$$\{(4.8 - 5.1)/4\} \times 20'' \times 0.573 = -01''.$$

Mean horizontal circle reading on sun	=	289° 26' 06"
Plate bubble correction	=	−01
Corrected horizontal circle reading on sun	=	289 26 05

Astronomy for surveyors

Mean horizontal circle reading on R.M.	=	271° 05′ 39″
Angle, sun to R.M., anticlockwise from sun	=	18 20 26

Sun's altitude:

Mean observed altitude of sun	=	29° 49′ 13″
Refraction correction ($r_0 = 101″, f = 0.96$)	=	−01 37
Parallax correction ($9″ \times \cos h$)	=	+08
Corrected observed altitude, $h =$		29 47 44

Sun's declination:

Approximate L.Std.T.	=	$16^h 37^m$	26 February
Longitude Standard Meridian	=	12 00	
Approximate U.T.	=	4 37	26 February

Sun's declination at U.T. 0^h, 26 February	=	S. 9° 01.9′
Change in $4^h 37^m$	=	−04.4
Sun's declination at time of observation	=	S. 8° 57.5′
i.e. $\delta =$		S. 8° 57′ 30″

Thus,

$$\omega = 44° 07′ 45″, \quad p = 81° 02′ 30″, \quad \zeta = 60° 12′ 16″.$$

Using the formula in (i) above,

$$Z = 109° 25′ 29″.$$

From Fig. 10.1(b), with the sun in the west (afternoon):

Azimuth of sun's centre when observed $= 180° + Z =$		289° 25′ 29″
Angle sun to R.M., anticlockwise from sun	=	18 20 26
∴ Azimuth of R.M.	=	271 05 03
From (i) and (ii) above, mean azimuth of R.M.	=	271° 04′ 54″

(b) *Using the stars*

Place: Arts Block, Pillar F,
   University of Otago,
   Dunedin, N.Z.
Latitude: 45° 52′ 15″ S.
Longitude: $11^h 22^m 03.9^s$ E.
Observer: G. Arthur

Instrument: Wild T2, No. 128 992
Plate bubble: 1 division = 20″
Object observed: Star
R.M.: Red light, 1 Cobden Street
Booker: G. Arthur
Standard Meridian: 12^h E.

Determination of azimuth

(i) Star in East. 10 March 1975
   Barometer: 1020 mb.
   Temperature: 7.5 °C
   Star: S.A. No. 464,
   η Ophiuchi: R.A. $17^h\ 08^m\ 58.3^s$,
   Dec.: S. 15° 41′ 44″

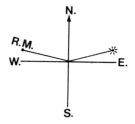

Face	Object	Horizontal circle	Vertical circle	Altitude	Plate buble L	R	Approx. L.Std.T.
L	R.M.	271° 04′ 45″					
L	Star	79 07 47	57° 50′ 44″	32° 09′ 16″	2.4	3.4	$02^h\ 42^m$
R	Star	258 23 46	302 47 02	32 47 02	1.3	4.5	02 44
R	R.M.	91 04 44					

(ii) Star in West. 10 March 1975
   Barometer: 1019 mb. Temperature: 7.5 °C
   Star: No. 156 110, Smithsonian
   Astrophysical Observatory Catalogue:
   R.A. $10^h\ 36^m\ 21.9^s$, Dec.: S. 13° 15′ 34″.

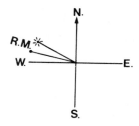

Face	Object	Horizontal circle	Vertical circle	Altitude	Plate bubble L	R	Approx. L.Std.T.
L	R.M.	271° 04′ 42″					
L	Star	293 14 49	51° 40′ 09″	38° 19′ 51″	3.5	2.4	$03^h\ 19^m$
R	Star	112 26 49	307 47 14	37 47 14	2.2	3.8	03 21
R	R.M.	91 04 44					

Computations
(i) Star in East
   Plate bubble correction:

$$\{(3.7 - 7.9)/4\} \times 20'' \times 0.636 = -13''$$

Mean horizontal circle reading on star	=	78° 45′ 47″
Plate bubble correction	=	−13
Corrected horizontal circle reading on star	=	78 45 34
Mean horizontal circle reading on R.M.	=	271 04 45
Angle, star to R.M., clockwise from star	=	192 19 11

Star's altitude:
Mean observed altitude of star	=	32° 28′ 09″
Refraction correction: ($r_0 = 91″, f = 1.02$)	=	−01 33
Corrected observed altitude of star $h =$		32 26 36

(Note no parallax correction for a star; see **7.02**)

Formula:

$$\cos Z = (\cos p - \cos \omega . \cos \zeta)/\sin \omega . \sin \zeta$$

$$p = 74° 18′ 16″, \quad \omega = 44° 07′ 45″, \quad \zeta = 57° 33′ 24″.$$

$$\therefore Z = 101° 14′ 23″.$$

From Fig. 10.1(b), since star is in the east,
its azimuth is given by (180° −Z)	=	78° 45′ 37″
But angle, star to R.M., clockwise from star	=	192 19 11
∴ Azimuth of R.M.	=	271 04 48

(ii) Star in West
Plate bubble correction:
$\{(5.7 - 6.2)/4\} \times 20″ \times 0.783 = -02″$

Mean horizontal circle reading on star	=	292° 50′ 49″
Plate bubble correction	=	−02
Corrected horizontal circle reading on star	=	292 50 47
Mean horizontal circle reading on R. M.	=	271 04 43
Angle, star to R.M., anticlockwise from star	=	21 46 04

Star's altitude:
Mean observed altitude of star	=	38° 03′ 33″
Refraction correction ($r_0 = 74″, f = 1.02$)	=	−01 15
Corrected observed altitude of star $h =$		38 02 18

Using formula in (i) above:

$$p = 76° 44′ 26″, \quad \omega = 44° 07′ 45″, \quad \zeta = 51° 57′ 42″.$$

$$\therefore Z = 112° 50′ 48″.$$

From Fig. 10.1(b), since star is in the west,
its azimuth is given by (180° +Z)	=	292° 50′ 48″
But angle, star to R.M., anticlockwise from star	=	21 46 04

Determination of azimuth

∴ Azimuth of R.M.   =   271° 04′ 44″

From (i) and (ii) above, mean azimuth of R.M.   =   271° 04′ 46″

The above worked examples were based upon actual observations done in the southern hemisphere. To help readers in the northern hemisphere, the following "fictional" observation is computed; for convenience, bubble, refraction and parallax corrections are included in the data given.

EXAMPLE. At a place whose Longitude is $0^h\ 35^m\ 31.8^s$ E., Latitude 14° 54′ 12″ N., the sun was observed on 24 November 1975 at a corrected altitude of 26° 18′ 38″ in the east at approximately $7^h\ 35^m$ a.m. L.Std.T. The corrected mean horizontal circle reading on the sun's centre was 58° 24′ 27″ and that on the R.M. was 329° 07′ 36″. The Standard Meridian of the country is at 0^h. What was the azimuth of the R.M.?

Computation

Corrected mean horizontal circle reading on sun   =   58° 24′ 27″
Mean horizontal circle reading on R.M.   =   329 07   36

Angle, sun to R.M., anticlockwise from sun   =   89 16   51

Sun's declination:
Approximate L.Std.T. of observation = $7^h\ 35^m$ 24 November 1975
Longitude of Standard Meridian   = 0 00

Approximate U.T. of observation   = 7 35   24 November 1975
Sun's declination at U.T. 6^h, 24 November 1975   = S. 20° 25.6′
Change in $1^h\ 35^m$ (table on page 73 of *S.A.*)   =   +0.8

Declination of sun when observed   = S. 20° 26.4′
                                       i.e. δ =   S. 20° 26′ 24″

In the astronomical triangle, now based upon the elevated north celestial pole,

$$p = 90° - \delta = 110°\ 26'\ 24'' \quad \text{(north polar distance)}$$
$$\omega = 90° - \phi = 75\ 05\ 48$$
$$\zeta = 90° - h = 63\ 41\ 22$$

Formula, as before:

$$\cos Z = (\cos p - \cos \omega . \cos \zeta)/\sin \omega . \sin \zeta.$$

From this,

Astronomy for surveyors

$$Z = 122° \ 19' \ 35''.$$

From Fig. 10.1(a), since the sun was in the east,
is azimuth is given by Z itself   =   122° 19′ 35″
Angle, sun to R.M., anticlockwise from sun   =   89 16 51

∴ Azimuth of R.M.   =   33 02 44

Section 5

FOURTH METHOD

EX-MERIDIAN OBSERVATIONS OF THE SUN OR A STAR FOR AZIMUTH BY MEASURING THE HOUR ANGLE AND THE HORIZONTAL ANGLE BETWEEN THE OBJECT AND THE R.M.

10.22 Principle of method

If we have the means of finding accurately the Local Standard Time when an Almanac star or the centre of the sun comes on to the vertical cross-hair, we can find its hour angle. In the case of a star, we must convert the L.Std.T. to the corresponding Local Sidereal Time (*vide* **5.13**), and then from the formula

$$H.A. = t = L.S.T. - R.A.$$

we can find t. Alternatively, if we have a timekeeper which keeps L.S.T. and whose error on L.S.T. is known, we can use this to time the observation and so avoid having to convert the L.Std.T. to L.S.T. In the case of the sun, we must convert the L.Std.T. of the observation to U.T. and use the formula

$$G.H.A. \ Sun = U.T. + E$$

to obtain the G.H.A. Sun; from this the local angle (L.H.A.) of the sun may be found by application of the longitude of the observing station. So, with a knowledge of the latitude and longitude of the observing station, the declination of the celestial body and its hour angle, we can solve the astronomical triangle (Fig. 10.8) for Z by using the formula

$$\cos p = \cos \omega . \cos \zeta + \sin \omega . \sin \zeta . \cos Z \quad (\text{see } (1.2))$$

in which $(\cos p . \cos \omega + \sin p . \sin \omega . \cos t)$ is substituted for $\cos \zeta$, and $(\sin t . \sin p / \sin Z)$ for $\sin \zeta$ to give

$$\cot Z = \cot p . \sin \omega . \operatorname{cosec} t - \cos \omega . \cot t \qquad (10.23)$$

or, in terms of ϕ and δ,

$$\cot Z = \tan \delta . \cos \phi . \operatorname{cosec} t - \sin \phi . \cot t. \qquad (10.24)$$

Determination of azimuth

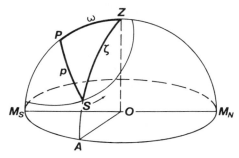

Fig. 10.8

This type of observation does not involve the somewhat difficult operation of getting the moving star on the intersection of the cross-hairs (or the sun in a quadrant), since we are not required to measure its altitude. In the case of a star we need only to note the instant when it is on the vertical cross-hair near the centre of the field of view; in the case of the sun, the instant when its left or right limb touches the vertical hair is required. As with other observations, we observe on both faces of the instrument to eliminate some of the instrumental errors; with the sun we observe to the left limb on one face and to the right limb on the other to eliminate semi-diameter correction as well. If there is only a short interval (2-3 minutes) between the L.F. and R.F. observations there will be negligible curvature error in accepting the mean of the azimuths as correct for the mean of the times. It is most important, of course, that we use a reliable timekeeper whose rate is known, since t is found from the time of the observation. The disadvantage of the method compared with the previous one in which altitudes are measured is that we need additional timekeeping equipment.

10.23 Best position of star or sun for the observations

To find the best position of the celestial body we shall investigate by means of the differential calculus the effects on Z of small errors in $t, \phi,$ and δ.

(i) *The effect of an error in t*
One formula for the computation of Z is (10.24)

$$\cot Z = \cot p . \cos \phi . \operatorname{cosec} t - \sin \phi . \cot t.$$

Multiply all through by $\sin t$, and we get

$$\cot Z . \sin t = \cot p . \cos \phi - \sin \phi . \cos t. \qquad (10.25)$$

Now differentiate this with respect to t, keeping p and ϕ constant:

Astronomy for surveyors

$$\cot Z.\cos t + \left(-\operatorname{cosec}^2 Z.\frac{dZ}{dt}\right).\sin t = 0 - \sin\phi.(-\sin t),$$

i.e.
$$\cot Z.\cos t - \operatorname{cosec}^2 Z.\frac{dZ}{dt}.\sin t = \sin\phi.\sin t$$

so that

$$dZ/dt = \frac{\cot Z.\cos t - \sin\phi.\sin t}{\operatorname{cosec}^2 Z.\sin t} = \left\{\begin{array}{l}\text{the change in } Z \text{ produced} \\ \text{by a small change (error)} \\ \text{in } t\end{array}\right.$$

i.e.
$$dZ/dt = \sin Z(\cos Z.\cot t - \sin\phi.\sin Z)$$
$$= \sin Z\left(\cos Z.\frac{\cos t}{\sin t} - \sin\phi.\sin Z.\frac{\sin t}{\sin t}\right)$$
$$= \sin Z.\operatorname{cosec} t(\cos Z.\cos t - \sin\phi.\sin Z.\sin t).$$

But
$$\cos Z.\cos t - \sin\phi.\sin Z.\sin t = \cos Z.\cos t - \sin Z.\sin t.\cos\omega$$

and from formula (1.4),
$$\cos S = -(\cos Z.\cos t - \sin Z.\sin t.\cos\omega).$$

Therefore,
$$dZ/dt = -\sin Z.\operatorname{cosec} t.\cos S. \qquad (10.26)$$

Also, since
$$-\sin Z.\operatorname{cosec} t = \frac{-\sin Z}{\sin t} = \frac{-\sin p.\sin t}{\sin \zeta}.\frac{1}{\sin t}$$
$$= -\cos\delta.\sec h,$$
$$dZ/dt = -\cos\delta.\sec h.\cos S. \qquad (10.27)$$

From (10.26) and (10.27), therefore, we see that to keep dZ/dt as small as possible, we must try to keep S, δ and t near 90°, and both h and Z small. This means selecting a star very close to the pole and observing it when the parallactic angle and the hour angle are both close to 90°, i.e. at elongation, when its azimuth is changing least. In these circumstances, provided that the altitude is not too great, Z will be least affected by a small error in t. Since, however, the maximum value for $(\cos S.\cos\delta)$ is 1, $(\cos S.\cos\delta.\sec h)$ will always be small for close circumpolar stars, provided the latitude—and therefore the altitude—is not too high, so that they may be observed at any time. Time observations for azimuth on close circumpolar stars are considered as a special case in Section 6 of this chapter. Failing a close circumpolar star, however, one should select a circumpolar star, the closer to the pole the better, and observe it near elongation.

As far as the sun is concerned, we can keep $(\cos\delta.\sec h.\cos S)$ as small as possible by observing when h is small and S is large. S is a maximum when the sun is on the Prime Vertical. The sun will cross the P.V. above the horizon only when it has moved on to the same side of the celestial equator as the elevated pole; as it moves closer to the elevated pole, the longer will be the period, in mid-latitudes, between the times of crossing the P.V. and setting, as summer advances (in either hemisphere). If it is observed during this period, S will be near its maximum value and t will be reasonably close to 6^h, with cosec t near its minimum value. Since we do not need to measure altitudes and are therefore not concerned with refraction, we can observe at less than $15°$ above the horizon as long as we do not make our pointing too close to the horizon to have trouble with "shimmer" or lateral refraction. When the sun is on the side of the celestial equator away from the elevated pole, it should be observed as far away from transit as possible, but at a sufficiently high altitude to avoid the effects of shimmer.

If it is necessary to use stars which do not elongate, they should be observed between the times when they are on the P.V. and when their hour angles are 6^h.

(ii) *The effect of an error in ϕ*

Using the formula (10.25) of **10.23**(i) above,

$$\cot Z.\sin t = \cot p.\cos\phi - \sin\phi.\cos t$$

and differentiating with respect to ϕ, keeping p and t constant, we get

$$-\cosec^2 Z.\frac{dZ}{d\phi}.\sin t = \cot p.(-\sin\phi) - \cos\phi.\cos t$$

so that

$$\frac{dZ}{d\phi} = \frac{\cot p.\sin\phi + \cos\phi.\cos t}{\cosec^2 Z \sin t}$$

or

$$\frac{dZ}{d\phi} = \sin Z\left(\frac{\cot p.\sin\phi.\sin Z + \cos\phi.\cos t.\sin Z}{\sin t}\right)$$

But since

$$\frac{\sin Z}{\sin t} = \frac{\sin p}{\sin \zeta} \quad \text{(sine rule)},$$

$$\frac{dZ}{d\phi} = \sin Z \left(\frac{\cot p . \sin\phi . \sin p}{\sin \zeta} + \frac{\cos\phi . \cos t . \sin p}{\sin \zeta} \right)$$

$$= \sin Z \left(\frac{\cos p . \sin\phi}{\cos h} + \frac{\cos\phi . \cos t . \sin p}{\cos h} \right)$$

$$= \frac{\sin Z}{\cos h} (\cos p . \cos\omega + \sin p . \sin\omega . \cos t) = \frac{\sin Z}{\cos h} . \sin h$$

$$= \sin Z . \tan h. \tag{10.28}$$

Thus, if there is uncertainty in the latitude, in order to minimize any consequent error in Z, we must keep the altitude as low as is practicable, and observe when Z is small. In middle latitudes in midwinter the sun transits at low altitudes, and it is feasible, therefore, to make the observation for azimuth near its transit (only if latitude is in doubt, however). At such a time the sun will be moving across the field of view at its fastest, so that it is best to preset the telescope to point near the meridian and to time the transit of the limbs over the vertical cross-hair with stop-watch and an electronic clock (see the first example in **10.25**).

(iii) *The effect of an error in* δ

Since an accurate knowledge of time is essential in this method of determining azimuth, the declination of the celestial body observed will not be in doubt, and therefore we do not need to consider the effect of errors in it.

10.24 Observing procedure

(a) *For stars*
  (i) Intersect the R.M. on L.F. and note the horizontal circle reading.
  (ii) Get the star into the field of view of the telescope on L.F. and pre-set the vertical cross-hair a little ahead of it so that it will cross the hair near the centre. As soon as the star comes on to the hair, note the time (by using one of the methods of **6.03**). Read the horizontal circle, the plate bubble L and R, and the approximate altitude.

Note. If the observation is being made to a star near elongation, it will be better to bring the cross-hair on to the star by using the horizontal slow-motion screw, than to wait for the star to transit the hair.
  (iii) Change to R.F. without delay and repeat the procedure in (ii) above.
  (iv) Intersect the R.M. on R.F. and note the horizontal circle reading.

As for all astronomical observations, the instrument should be firmly set up and carefully levelled before commencing.

Another set consisting of steps (i) to (iv) above should then be made to

Determination of azimuth

another star at approximately the same altitude and zenith angle on the other side of the meridian and the mean of the two computed azimuths accepted; this will help to reduce errors arising from small inaccuracies in the assumed latitude and clock correction. An additional one or two such balanced pairs of sets will improve the accuracy still further.

(b) *For the sun*

The steps are the same as in (a) above, except that in (ii) the time of transit of the sun's left limb across the vertical cross-hair in the L.F. position is noted, and in (iii) the time of transit of the right limb across the hair in the R.F. position. But see Appendix II for another method described by G. G. Bennett (1974): the sun's limb that is leaving the cross-hair which is silhouetted against the bright disc is more easily timed across the hair than the limb which is approaching an unilluminated hair invisible because of the dark filter. Therefore the passage of that single limb is timed on each face and the semi-diameter correction applied. The general availability of accurate time-signals and digital "quartz" watches arguably makes this relatively simple observation the best one nowadays for sun azimuths. See (c) below. The azimuth is calculated from each individual pointing and the mean taken, rather than by calculating it from the mean of the two pointings and times.

(c) *Accuracy*

Because the altitude of the celestial body is not required in this method, three advantages accrue:

(i) The observer can concentrate on the transit of the star or sun's limb across a single cross-hair—the vertical one—without the added stress of having to get the star exactly on the intersection of the hairs, or both hairs simultaneously tangential to the sun's disc.
(ii) No correction for celestial refraction is required, and a source of unknown error is therefore eliminated.
(iii) Consequent upon (ii) above, the star or sun may be observed at altitudes considerably below $15°$.

With these advantages, and an accurate knowledge of the clock error on local time or U.T., the hour angle method will yield somewhat better results than that using altitudes.

10.25 Worked examples: *Observations for azimuth by the hour angle method*

(A) *Using the sun*
Place: Arts Block, Pillar F,
   University of Otago,
   Dunedin, N.Z.
   Latitude: $45° 52' 15'' $ S.

Longitude: $11^h 22^m 03.9^s$ E.
Observer: M. D. Body
Booker: J. B. M.
Standard Meridian: 12h E.

Astronomy for surveyors

Instrument: Wild T2, No. 76015
Plate bubble: 1 division = 20″
Object observed: sun
R.M.: red light, 1 Cobden Street

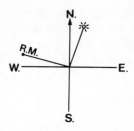

(i) Midwinter observation, sun near meridian, altitude low (see end of (ii), **10.23**). Date: 26 June 1974. Chronometer fast on U.T.C.: $12^h\ 00^m\ 14.8^s$. DUT1 = $+0.2^s$.

Face	Object	Horizontal circle	Chronometer	Stop-watch	Chronometer time of obs.	Plate bubble L	Plate bubble R	Approx. altitude
L	R.M.	167° 13′ 07″						
L	☉+	260 21 00	$12^h\ 22^m\ 40^s$	12.0^s	$12^h\ 22^m\ 28.0^s$	2.8	1.7	20° 46′
L	☉-	260 21 00	12 24 55	9.1	12 24 45.9	2.8	1.7	
L	R.M.	167 13 07						
R	R.M.	14 01 39						
R	☉+	105 38 54	12 28 45	7.0	12 28 38.0	2.5	2.4	20 46
R	☉-	105 38 54	12 31 05	9.6	12 30 55.4	2.7	2.2	
R	R.M.	14 01 45						

Computations
Time of transit:

Sun's L.H.A. at transit	=	$24^h\ 00^m\ 00^s$	
Longitude of observing station	=	11 22 03.9	
Sun's G.H.A. at local transit (= U.T. +E)	=	12 37 56.1	
U.T. +E at U.T. 0^h 26 June 1974	=	11 57 23.3	
Difference between G.H.A. at transit and at U.T. 0^h	=	0 40 32.8	
E at U.T. 0^h 26 June	=	11 57 23.3	
Change in E in $0^h\ 40^m\ 32.8^s$	=	−0.4	
∴ E at local transit	=	11 57 22.9	
But U.T. +E (= G.H.A.) at local transit (see above)	=	12 37 56.1	
Difference = U.T. at local transit	=	0 40 33.2	
Longitude of Standard Meridian, E.	=	12 00 00	
∴ L.Std.T. of transit	=	12 40 33.2	

First group of L.F. readings:
Plate bubble correction:
$$\{(5.6 - 3.4)/4\} \times 20'' \times \tan 20°\ 46' = +04''$$

Determination of azimuth

Mean horizontal circle reading on sun	=	260° 21′ 00″
Plate bubble correction	=	+04
Corrected horizontal circle reading on sun	=	260 21 04
Mean horizontal circle reading on R.M.	=	167 13 07
Angle, sun to R.M., anticlockwise from sun	=	93 07 57

Sun's declination:

Sun's declination at U.T. 0^h, 26 June 1974	= N.	23° 22.8′
Change in 0^h 40^m 33.2^s	=	0.0
Sun's declination at time of observation	= N.	23° 22′ 48″
and p =		133° 22′ 48″

Sun's hour angle:

Mean chronometer time of observation	=	12^h 23^m 37.0^s
Chronometer fast on U.T.C.	=	12 00 14.8
∴ U.T.C. of observation	=	0 23 22.2
DUT1	=	+0.2
∴ U.T.1 of observation	=	0 23 22.4
But U.T.1 of sun's local transit	=	0 40 33.2
Difference = sun's hour angle at observation	=	0^h 17^m 10.8^s E.
i.e. t =		4° 17′ 42″ E.
and $\omega = (90° - \phi)$ =		44° 07′ 45″

Formula for solution:

$$\cot Z = \frac{\cot p . \sin \omega - \cos \omega . \cos t}{\sin t} = -13.576\,593$$

and

$$Z = 175° \, 47′ \, 15″$$

Since the observation was made before transit, the sun was to the east of the meridian and Fig. 10.1(b) applies. Hence the azimuth of the sun's centre at the time of observation was $(180° - Z)$,

i.e. A =		4° 12′ 45″
But, from above, angle, sun to R.M., anticlockwise from sun	=	93 07 57
∴ Azimuth of R.M.	=	271 04 48

Astronomy for surveyors

(The angle between sun and R.M. was taken anticlockwise from the sun, and must therefore be subtracted from 364° 12' 45" to give the azimuth of the R.M.)

Second group of R.F. readings:
Plate bubble correction:

$$\{(5.2 - 4.6)/4\} \times 20'' \times \tan 20° 46' = +01''$$

Mean horizontal circle reading on sun =	105° 38' 54"
Plate bubble correction =	+01
Corrected horizontal circle reading on sun =	105 38 55
Mean horizontal circle reading on R.M. =	14 01 42
Angle, sun to R.M., anticlockwise from sun =	91 37 13

Sun's declination:
This will not have changed appreciably in the 6 minutes since the first group of readings, and may therefore be taken as N. 23° 22' 48".

and $p =$ 113° 22' 48"

Sun's hour angle:

Mean chronometer time of observation =	$12^h \ 29^m \ 46.7^s$
Chronometer fast on U.T.C. =	12 00 14.8
∴ U.T.C. of observation =	0 29 31.9
DUT1 =	+0.2
∴ U.T.1 of observation =	0 29 32.1
But U.T.1 of sun's local transit =	0 40 33.2
Difference = sun's hour angle at observation =	0 11 01.1 E.
i.e. $t =$	2° 45' 16" E.
and ω, from above =	44° 07' 45"

Using the same formula as before,

$$\cot Z = -21.182\,973$$

and $Z = 177° 17' 50''$,

i.e. $A =$	2° 42' 10"
But angle, sun to R.M., anticlockwise from sun =	91 37 13
∴ Azimuth of R.M. =	271 04 57

Determination of azimuth

Thus, from first group, F.L. azimuth of R.M.	=	271° 04' 48"
and from second group, F.R. azimuth of R.M.	=	271 04 57
Hence mean azimuth of R.M.	=	271 04 53

Note. A glance at the observation data at the beginning of the example will show that the vertical cross-hair of the telescope was set slightly ahead of the sun, and the contacts between it and the apparent right and left limbs were timed with stop-watch and chronometer as the sun moved across the field of view. The horizontal circle setting therefore remained unaltered during the observation of the sun.

As deduced in **10.23**, the above type of observation is appropriate when there is some doubt about the latitude of the observing station. This can be illustrated quite effectively by re-working the example using 45° 50' S. as the latitude, instead of the correct value 45° 52' 15" S. The difference in azimuth produced by the error of 02' 15" in the latitude amounts to only 03". However, the longitude of the station must be known accurately, and the timing must be done carefully. An error of 1 second of time in the longitude produces an error of 17" in the azimuth in the above example.

(ii) Ex-meridian observations, sun near Prime Vertical in east and west.

Place: Arts Block, Pillar F,
  University of Otago,
  Dunedin, N.Z.
Latitude: S. 45° 52' 15"
Longitude: E $11^h 22^m 03.9^s$
Observer: G. Arthur

Instrument: Wild T2, No. 128 992
Plate bubble: 1 division = 20"
R.M.: Red light, 1 Cobden Street
Object observed: sun
Booker: C. Wratt

(a) Morning observation. 20 February 1975. Chronometer fast on U.T.C.: $13^h 01^m 26.1^s$; DUT1 = $+0.6^s$.

Face	Object	Horizontal circle	Chronometer	Stop-watch	Chronometer time of obs.	Plate bubble L R	Approx. altitude
L	R.M.	348° 11' 03"					
L	☉	173 03 56	$8^h 06^m 05^s$	10.5^s	$8^h 05^m 54.5^s$	2.7 2.6	11° 00'
R	☉	352 08 01	8 08 20	15.2	8 08 04.8	2.8 2.4	
R	R.M.	168 11 13					

(b) Afternoon observation. 26 February 1975. Chronometer fast on U.T.C.: $12^h 01^m 56.6^s$; DUT1 = $+0.5^s$.

Astronomy for surveyors

Face	Object	Horizontal circle	Chronometer	Stop-watch	Chronometer time of obs	Plate bubble L R	Approx. altitude
L	R.M.	271° 05′ 44″					
L	☉	281 35 23	17ʰ 17ᵐ 05ˢ	7.7ˢ	17ʰ 16ᵐ 57.3ˢ	2.0 2.9	22° 55′
R	☉	100 31 26	17 19 35	8.4	17 19 26.6	2.9 2.1	
R	R.M.	91 05 49					

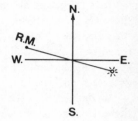

Computations

(a) Morning observation, 20 February 1975
 Plate bubble correction:

$\{(5.5 - 5.0)/4\} \times 20'' \times \tan 11° 00' = 00''$.

Mean horizontal circle reading on sun	=	172° 35′ 59″
Plate bubble correction	=	00
Corrected horizontal circle reading on sun	=	172 35 59
Mean horizontal circle reading on R.M.	=	348 11 08
Angle, sun to R.M., clockwise from sun	=	175 35 09

Mean chronometer time of observation, 20 February 1975	=	8ʰ 06ᵐ 59.7ˢ
Chronometer fast on U.T.C.	=	13 01 26.1
U.T.C. of observation, 19 February	=	19 05 33.6
DUT1	=	+0.6
U.T.1 of observation, 19 February	=	19 05 34.2

Sun's declination:

Sun's declination at U.T. 18ʰ, 19 February	= S.	11° 18.5′
Change in 1ʰ 05ᵐ 34.2ˢ	=	−01.1
Sun's declination at time of observation	= S.	11° 17.4′
	i.e. δ = S.	11° 17′ 24″

Sun's L.H.A.:

U.T.1 of observation, from above, 19 February	=	19ʰ 05ᵐ 34.2ˢ
E at U.T. 18ʰ, 19 February	=	11 46 05.7
ΔE in 1ʰ 05ᵐ 34.2ˢ	=	+0.3

Determination of azimuth

G.H.A. sun (= U.T. + E) at time of observation	=	6	51	40.2
Longitude of observing station, E.	=	11	22	03.9
L.H.A. sun at time of observation	=	18	13	44.1
i.e. sun's east hour angle, t	=	5	46	15.9
or t, arc, E.=		86°	33'	58"

Using formula:
$$\cot Z = \tan\delta . \cos\phi . \operatorname{cosec} t - \sin\phi . \cot t,$$

where $\delta = 11°\ 17'\ 24"$; $\phi = 45°\ 52'\ 15"$; $t = 86°\ 33'\ 58"$.

(δ and ϕ are both south, so we may treat them both as positive.)
From the formula, then,
$$Z = 84°\ 30'\ 21".$$

From Fig. 10.1(b), with the sun in the east (morning),

the azimuth of the sun's centre when observed = (180° − Z)	=	95°	29'	39"
Angle, sun to R.M., clockwise from sun	=	175	35	09
∴ Azimuth of R.M.	=	271	04	48

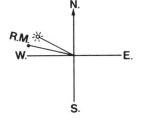

(b) Afternoon observation, 26 February 1975
Plate bubble correction:

$$\{(4.9 - 5.0)/4\} \times 20" \times \tan 22°\ 55' = 00".$$

Corrected mean horizontal circle reading on sun	=	281°	03'	25"
Mean horizontal circle reading on R.M.	=	271	05	47
Angle, sun to R.M., anticlockwise from sun	=	9	57	38
Mean chronometer time of observation, 26 February 1975	=	17^h	18^m	12.0^s
Chronometer fast on U.T.C.	=	12	01	56.6
U.T.C. of observation	=	5	16	15.4
DUT1	=			+0.5
U.T.1 of observation, 26 February	=	5	16	15.9

Sun's declination:

Sun's declination at U.T. 0^h, 26 February	= S.	9° 01.9'
Change in $5^h\ 16^m\ 15.9^s$	=	−05.1
Sun's declination at time of observation	= S.	8° 56.8'
	i.e. δ = S.	8° 56' 48"

Sun's L.H.A.:

U.T.1 of observation, from above, 26 February	=	$5^h\ 16^m\ 15.9^s$
E at U.T. 0^h, 26 February	=	11 46 54.0
ΔE in $5^h\ 16^m\ 15.9^s$	=	+02.1
G.H.A. sun (= U.T. $+E$) at time of observation	=	17 03 12.0
Longitude of observing station, E.	=	11 22 03.9
L.H.A. sun at time of observation	=	4 25 15.9
or t, arc =		66° 18' 58"

Using formula:

$$\cot Z = \tan \delta . \cos \phi . \operatorname{cosec} t - \sin \phi . \cot t,$$

where $\delta = 8°\ 56'\ 48"$; $\phi = 45°\ 52'\ 15"$; $t = 66°\ 18'\ 58"$ (δ and φ both south and therefore both taken as positive).

From the formula,

$$Z = 101°\ 02'\ 31".$$

From Fig. 10.1(b), with sun in the west (afternoon),

the azimuth of the sun's centre when observed = 180° +Z	=	281° 02' 31"
Angle, sun to R.M., anticlockwise from sun	=	9 57 38
∴ Azimuth of R.M.	=	271 04 53
From (ii) (a) and (ii) (b) above, mean azimuth of R.M.	=	271° 04' 51"

(B) *Using the stars*

Place: Arts Block, Pillar F,
   University of Otago,
   Dunedin, N.Z.
Latitude: 45° 52' 15" S.
Longitude: $11^h\ 22^m\ 03.9^s$ E.
Observer: G. Arthur

Instrument: Wild T2, No. 128 992
Plate bubble: 1 division = 20"
R.M.: Red light, 1 Cobden Street
Objects observed: stars
Booker: G. Arthur

Standard Meridian: 12^h E.

Determination of azimuth

(i) Star in East, 10 March 1975. Star No. 464 η Ophiuchi, R.A.: $17^h\ 08^m\ 58.3^s$; Dec.: S. $15°\ 41'\ 44''$. Chronometer fast on U.T.C.: $12^h\ 02^m\ 45.7^s$; DUT1 = $+0.5^s$.

Face	Object	Horizontal circle	Chronometer	Stop-watch	Chronometer time of obs.	Plate bubble L R	Approx. altitude
L	R.M.	271° 04' 40"					
L	Star	93 47 27	$01^h\ 25^m\ 15^s$	8.9^s	$01^h\ 25^m\ 06.1^s$	3.2 2.7	18° 45'
R	Star	273 08 18	01 29 00	12.1	01 28 47.9	2.1 4.0	
R	R.M.	91 04 43					

(ii) Star in West. 10 March 1975. Star No. 212 Procyon. R.A.: $7^h\ 38^m\ 02.0^s$; Dec.: N. $5°\ 17'\ 10''$. Chronometer fast on U.T.C.: $12^h\ 02^m\ 45.8^s$; DUT1 = $+0.5^s$.

Face	Object	Horizontal circle	Chronometer	Stop-watch	Chronometer time of obs.	Plate bubble L R	Approx. altitude
L	R.M.	271° 04' 43"					
L	Star	285 30 16	$02^h\ 07^m\ 00^s$	9.7^s	$02^h\ 06^m\ 50.3^s$	3.4 2.5	7° 14'
R	Star	105 06 56	02 09 00	4.8	02 08 55.2	2.6 3.4	
R	R.M.	91 04 41					

Computations
(i) Star in East
    Plate bubble correction:

$$\{(5.3 - 6.7)/4\} \times 20'' \times \tan 18°\ 45' = -02''.$$

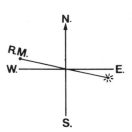

Mean horizontal circle reading on star	=	93° 27' 53"
Plate bubble correction	=	−02
Corrected horizontal circle reading on star	=	93 27 51
Mean horizontal circle reading on R.M.	=	271 04 42
Angle, star to R.M., clockwise from star	=	177 36 51

Star's L.H.A.:
Mean chronometer time of observation,
   10 March 1975   = $01^h\ 26^m\ 57.0^s$
Chronometer fast on U.T.C.   =   12 02 45.7

U.T.C. of observation, 9 March   =   13 24 11.3
DUT1   =   + 0.5

Astronomy for surveyors

U.T.1 of observation, 9 March	=	13	24	11.8
R at U.T. 12^h, 9 March (see p. 98)	=	11	06	13.7
ΔR in $1^h\ 24^m\ 11.8^s$	=			+13.8
G.S.T. of observation	=	0	30	39.3
Longitude of observing station, E.	=	11	22	03.9
L.S.T. of observation	=	11	52	43.2
R.A. of star	=	17	08	58.3
L.H.A. of star, t	=	18	43	44.9
t(E.)	=	5	16	15.1
t(E.), arc	=	79°	03′	46″

Formula:

$$\cot Z = \tan\delta.\cos\phi.\operatorname{cosec} t - \sin\phi.\cot t,$$

where $\delta = 15°\ 41'\ 44''$; $\phi = 45°\ 52'\ 15''$; $t = 79°\ 03'\ 46''$. δ and ϕ are both south, so we treat them both as positive.

From the formula,

$$Z = 86°\ 32'\ 01''.$$

From Fig. 10.1(b), since star is in the east,

its azimuth = $180° - Z$	=	93°	27′	59″
Angle, star to R.M., clockwise from star	=	177	36	51
∴ Azimuth of R.M.	=	271	04	50

(ii) Star in West
Plate bubble correction

$\{(6.0 - 5.9)/4\} \times 20'' \times \tan 7°\ 14' = 00''$

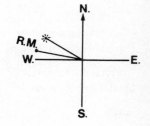

Corrected mean horizontal circle reading on star	=	285°	18′	36″
Mean horizontal circle reading on R.M.	=	271	04	42
Angle, star to R.M., anticlockwise from star	=	14	13	54
Star's L.H.A.:				
Mean chronometer time of observation, 10 March 1975	=	02^h	07^m	52.8^s

Determination of azimuth

Chronometer fast on U.T.C.	=		$12^h\ 02^m\ 45.8^s$
U.T.C. of observation, 9 March	=		14 05 07.0
DUT1	=		+0.5
U.T.1 of observation	=		14 05 07.5
R at U.T. 12^h, 9 March	=		11 06 13.7
ΔR in $2^h\ 05^m\ 07.5^s$	=		+20.6
G.S.T. of observation	=		01 11 41.8
Longitude of observing station, E.	=		11 22 03.9
L.S.T. of observation	=		12 33 45.7
R.A. of star	=		7 38 02.0
L.H.A. of star,	t	=	4 55 43.7
	t (arc) =		73° 55' 55"

Formula:

$$\cot Z = \tan\delta . \cos\phi . \operatorname{cosec} t - \sin\phi . \cot t,$$

where $\delta = +5°\ 17'\ 10"$; $\phi = -45°\ 52'\ 15"$; $t = 73°\ 55'\ 55"$. In this case, the declination is north and the latitude south, so the signs are different, as shown.

From the formula,

$$Z = 74°\ 41'\ 19".$$

With positive declinations north and negative latitudes south, the astronomical triangle is based on the north celestial pole, hence Fig. 10.1(a) applies and with the star in the west,

Azimuth of the star = (360° − Z)	=	285° 18' 41"
Angle, star to R.M. anticlockwise from star	=	14 13 54
∴ Azimuth of R.M.	=	271 04 47
From (i) and (ii) above, mean azimuth of R.M.	=	271° 04' 49"

The above worked examples were based upon actual observations done in the southern hemisphere. To help readers who live in the northern hemisphere, the following "fictional" observation is computed:

EXAMPLE. At a place whose Longitude is $4^h\ 50^m\ 00^s$ W., Latitude 48° 51' 25" N., Star No. 239 ι Cancri was observed in the west at $10^h\ 17^m\ 18.4^s$ p.m. L.Std.T. on 29 May 1975. The Standard Meridian of the place is 5^h W. DUT1 was $+0.3^s$. If the mean corrected horizontal circle reading on the star

Astronomy for surveyors

was 268° 15′ 22″ and that on the R.M. was 354° 06′ 32″, what was the azimuth of the R.M.?

Computation
R.A. of star No. 239: $8^h\ 45^m\ 12.9^s$; Dec.: N. 28° 51′ 02″.

Corrected mean horizontal circle reading on star	=	268° 15′ 22″	
Mean horizontal circle reading on R.M.	=	354 06 32	
Angle, star to R.M., clockwise from star	=	85 51 10	

Star's L.H.A.:

L.Std.T. of observation, 29 May 1975	=	$22^h\ 17^m\ 18.4^s$	
Longitude of Standard Meridian, W.	=	5 00 00	
U.T.C. of observation, 30 May	=	3 17 18.4	
DUT1	=	+0.3	
U.T.1 of observation, 30 May	=	3 17 18.7	
R at U.T. 0^h, 30 May	=	16 27 32.8	
ΔR in $3^h\ 17^m\ 18.7^s$	=	+32.4	
G.S.T. of observation	=	19 45 23.9	
Longitude of observing station, W.	=	4 50 00	
L.S.T. of observation	=	14 55 23.9	
R.A. of ι Cancri	=	8 45 12.9	
L.H.A. ι Cancri, $\quad t$	=	6 10 11.0	
t (arc)	=	92° 32′ 45″	

Formula:
$$\cot Z = \tan\delta \cdot \cos\phi \cdot \operatorname{cosec} t - \sin\phi \cdot \cot t,$$

where $\delta = 28°\ 51′\ 02″$; $\phi = 48°\ 51′\ 25″$; $t = 92°\ 32′\ 45″$. Here δ and ϕ are both north, so we regard them both as positive.

From formula, then,
$$Z = 68°\ 22′\ 53″.$$

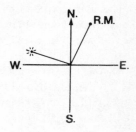

Determination of azimuth

From Fig. 10.1(a), since star is in the west,
its azimuth = (360° −Z)   =   291° 37′ 07″
Angle, star to R.M., clockwise from star   =   85 51 10

∴ Azimuth of R.M.   =   17 28 17

Section 6

FIFTH METHOD

AZIMUTH FROM TIME OBSERVATIONS UPON A CLOSE CIRCUMPOLAR STAR

10.26 Principle of method (see also **10.23**(i))

The method about to be described is one which will yield results of higher precision than those already discussed in this chapter; it is a method which can be employed for determining geodetic azimuths, provided a suitable procedure is adopted. The principle lies in measuring a series of angles between a close circumpolar star and the R.M., noting the time at which each pointing is made to the star. No altitudes need be measured, and as the time may be found, with sufficient precision for ordinary work, from simple observations of an electronic clock or digital watch, the method is easy to apply.

In the northern hemisphere the star α Ursae Minoris (Polaris) is a very convenient one for the purpose; being a star of magnitude 2.1 which lies in an easily recognizable constellation, it can be readily found, and it is within 1° of the north celestial pole. In the southern hemisphere, unfortunately, there are no stars near the pole sufficiently bright to be easily found without first calculating their positions. The best star for the purpose is σ Octantis, which is within 1° of the south pole. It is, however, of magnitude 5.5, and in order to find the star it is necessary to know beforehand the approximate azimuth of the R.M. This may be determined from a quick observation of the sun or a known star, noting time and horizontal angle between body and R.M., as described earlier in this chapter. Since σ Octantis describes a very small circle round the pole, its approximate altitude may be found by adding $\{p$ (in minutes) $\times \cos t\}$ to the latitude, remembering that when t, the hour angle, lies between 6^h and 18^h, $\cos t$ is negative and the star's altitude will be less than that of the pole. Its azimuth may also be found approximately by adding $\{p$ (in minutes) $\times \sin t \times \sec \phi\}$ to 180° (for σ Octantis), again remembering that when t lies between 12^h and 24^h $\sin t$ is negative and the star's azimuth will be east of the pole. The same formulae may be used, if necessary, to find the approximate altitude and azimuth of Polaris, if $\{p \sin t.\sec \phi\}$ is added to 0° in the case of the azimuth. The hour angle, t, is found, of course, from the formula L.S.T = R.A. $+t$, in which L.S.T. and R.A. are known.*

In Fig. 10.9, let P be the celestial pole around which moves the circumpolar star S in a small circle. Let Z be the zenith. Then in the spherical triangle ZPS, $ZP = \omega = (90° - \phi)$, the co-latitude; $PS = p$, the polar distance of the star; $ZPS = t$, its hour angle, and $PZS = Z$, the zenith angle of the star.

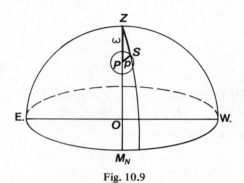

Fig. 10.9

From formula (1.16) in Chapter 1,

$$\cot p.\sin \omega = \cot Z.\sin t + \cos \omega.\cos t,$$

so that

$$\tan Z = \frac{\sin t}{\cot p.\sin \omega - \cos \omega.\cos t}$$

or

$$\tan Z = \frac{\sin t}{\cot p.\cos \phi - \sin \phi.\cos t}$$

$$= \frac{\sin t.\sec \phi}{\cot p - \tan \phi.\cos t}$$

$$= \frac{p \sin t.\sec \phi}{p \cot p - p \tan \phi.\cos t}$$

and, since p is small,

$$\tan Z = \frac{p \sin t.\sec \phi}{1 - p \tan \phi.\cos t}.$$

* A small but useful planisphere for predicting the azimuth and altitude of σ Octantis (and B Octantis) has been devised by Dr G. G. Bennett of the Surveying Department, University of New South Wales, and is available at a small charge from that Department.

Determination of azimuth

Dividing out, we get,

$$\tan Z = p \sin t \cdot \sec \phi + p^2 \sin t \cdot \sec \phi \cdot \tan \phi \cdot \cos t$$
$$+ p^3 \sin t \cdot \sec \phi \cdot \tan^2 \phi \cdot \cos^2 t + \ldots$$

Since p is small, we may neglect the terms containing p^3 and higher powers, and write

$$\tan Z = \sec \phi (p \sin t + p^2 \sin t \cdot \tan \phi \cdot \cos t).$$

Since Z is also small, there will be negligible error in taking

$$Z = \sec \phi (p \sin t + p^2 \sin t \cdot \tan \phi \cdot \cos t).$$

To find Z in seconds of arc, divide through by $\sin 1''$, i.e.

$$Z'' = \sec \phi \cdot \{p'' \sin t + (p'')^2 \sin t \cdot \tan \phi \cdot \cos t \cdot \sin 1''\}. \qquad (10.29)$$

The hour angle, t, in time, is found from

$$t = \text{L.S.T.} - \text{R.A.}$$

To determine this we must know both the L.M.T. and the longitude of the observing station, or the L.Std.T. and the longitude of the Standard Meridian. Thus, we require to know the R.A. and declination of the star—both of which we can get from the *Star Almanac*—the time of the observation, and also the latitude and longitude of the place, obtainable from previous observations or a good map.

Since the stars we are considering are at approximately the same elevation as the pole, the higher the latitude of the observing station, the larger will be the effect of any error due to non-verticality of the vertical axis of the theodolite. It is therefore essential that the ends of the plate bubble should be read during the observation and a correction made for any such error that may be present. For locations close to the equator, Polaris and σ Octantis will very near the horizon, and although we are not required to measure altitudes, the method becomes unreliable owing to lateral refraction and absorption of light in the long path through the earth's atmosphere.

Tables for finding both azimuth and latitude from observations to Polaris (altitude and time) are given on pages 56-59 of the *Star Almanac*. An explanation of the tables is also given on page xi, and a worked example on page xvi.*

10.27 Observing routine

(a) *Simple observations*

The simplest type of observation would follow the routine detailed overleaf.

---

* See Appendix I (**AI.03.05**) for examples of the use of Pole Star tables.

Astronomy for surveyors

(i) Intersect the R.M. on left face and read the horizontal circle.
(ii) Point to the star on left face and set the vertical cross-hair *very slightly* ahead of it, so that the star will make its own intersection near the centre of the field of view. Note the time of intersection, read the plate bubble left and right, and read the horizontal circle; note the approximate altitude.
(iii) Change to right face and repeat (ii) above.
(iv) Return to the R.M. on right face and read the horizontal circle.

Use the mean of the observed local times, corrected if necessary for clock error, to compute the star's mean hour angle; and the mean of the differences between horizontal circle readings on star and R.M., corrected for any dislevelment, for the mean included angle between star and R.M.

The mean azimuth from several such observations taken one after the other would yield a more accurate result; if there is any tendency for the instrument to go off level as indicated by the plate bubble readings, it should be re-levelled between sets of observations.

(b) *Multiple observations using a change of horizontal plate zero*

Another routine which will yield still more precise results is as follows (assuming that a glass-circle instrument is used, and omitting detail):

(i) Intersect R.M. on L.F. with horizontal circle set near zero.
(ii) Intersect star on L.F. as in (ii) above and note the time.
(iii) Repeat (ii), still on L.F.
(iv) Intersect R.M. on L.F.

In returning to the R.M. the instrument is moved in the opposite direction. Face is now reversed, the setting on the R.M. increased by 30°, and steps (i) to (iv) above repeated (but on R.F.). The routine is continued in this way, with further L.F. sets on 60° and 120° settings, and R.F. sets on 90° and 150°. The plate bubble is, of course, read on all star pointings. Each separate set is used as the basis of a computation by formula (10.29), and the mean of the six values of the azimuth taken. The number of sets may be increased by increasing the number of zeros used, but there should always be an equal number on L.F. and R.F., distributed evenly around the horizontal circle. The number of computations involved makes for rather tedious work, but the fact that the azimuth from each set is calculated separately shows up any blunders that may have occurred, immediately the results are compared.

(c) *Multiple observations for use with curvature correction*

If a series of readings is taken as quickly as possible in the following sequence: L.F. on R.M., L.F. on star, R.F. on star, R.F. on R.M., R.F. on R.M., R.F. on star, L.F. on star, L.F. on R.M., L.F. on R.M., L.F. on star, R.F. on star, R.F.

Determination of azimuth

on R.M. with, of course, times noted at star intersections, horizontal circle read and plate bubble positions noted, the amount of calculation may be considerably shortened by taking means and using a curvature correction. For close circumpolar stars such as Polaris and σ Octantis, it may be shown that the formula for curvature correction in azimuth (formula (7.10) of Chapter 7) reduces to

$$-\Delta Z'' = \tan Z \frac{1}{n} \Sigma \left(\frac{2 \sin^2 \tfrac{1}{2}(\Delta t)}{\sin 1''} \right),$$

where

$\Delta Z''$ = curvature correction to Z in seconds of arc,
Z = computed zenith angle of the star at the mean of the n observed times,
n = the number of pointings to the star,
Δt represents the successive differences between the mean of all the times and each individual time (sidereal).

The true mean zenith angle is always smaller than the zenith angle corresponding to the mean hour angle, hence the negative sign for $\Delta Z''$. The true mean azimuth of the star is calculated from $(Z - \Delta Z'')$, i.e. it will be $180° + (Z - \Delta Z'')$ or $180° - (Z - \Delta Z)$ for σ Octantis west or east of the south pole respectively; $0° + (Z - \Delta Z'')$ or $360° - (Z - \Delta Z'')$ for Polaris east or west of the north pole respectively. The mean of all the angles measured between the star and R.M. is then applied in the correct sense to this mean azimuth to give the true azimuth of the R.M.

Values of $\dfrac{2 \sin^2 \tfrac{1}{2}(\Delta t)}{\sin 1''}$ (usually called "m") corresponding to successive values of Δt are given on page 68 of the *Star Almanac*.

However, see the end of **7.09**. Each pair of pointings may be computed separately without too much work if an electronic calculator is available. This allows a check on the accuracy of each pair.

10.28 **Worked example**: *Simple observation to σ Octantis for azimuth by hour angles*

Date: 1974 August 19
Place: Arts Block, Pillar F,
    University of Otago,
    Dunedin, N.Z.
Latitude: 45° 52′ 15″ S.
Longitude: $11^h 22^m 03.9^s$ E.
Observer: J. B. M.
Booker: M. D. Body

Star observed: σ Octantis
R.A.: $20^h 45^m 34.9^s$
Dec.: S. 89° 03′ 16″
R.M.: Red light, 1 Cobden Street

Astronomy for surveyors

Instrument: Wild T2, No. 76 015
Plate bubble: 1 division = 20″
Chronometer: fast on L.Std.T. by 12.6^s
DUT1 = +0.1^s
Standard Meridian: 12^h E.

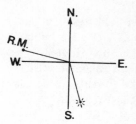

Face	Object	Horizontal circle	Chronometer	Stop-watch	Chronometer time of obs.	Plate bubble L R	Approx. altitude
R	R.M.	91° 04′ 25″					
R	Star	359 02 29	20^h 20^m 00^s	+19^{m}02.1^s	20^h 39^m 02.1^s	2.0 3.5	46° 32′
L	Star	179 03 00	20 20 00	+21 47.4	20 41 47.4	4.2 1.3	
L	R.M.	271 04 26					

Computation
Plate bubble correction:

$$\{(6.2 - 4.8)/4\} \times 20'' \times \tan 46° 32' = +07''$$

Mean horizontal circle reading on star	=	179° 02′ 45″
Plate bubble correction	=	+07
Corrected horizontal circle reading on star	=	179 02 52
Mean horizontal circle reading on R.M.	=	271 04 26
Angle, star to R.M., clockwise from star	=	92 01 34

Star's L.H.A.:

Mean chronometer time of observation, 19 August 1974	=	20^h 40^m 24.8^s
Chronometer fast on U.T.C.	=	12 00 12.6
U.T.C. of observation, 19 August	=	8 40 12.2
DUT1	=	+0.1
U.T.1 of observation	=	8 40 12.3
R at U.T. 6^h, 19 August	=	21 48 50.4
ΔR in 2^h 40^m 12.3^s	=	+26.3
G.S.T. of observation	=	6 29 29.0
Longitude of observing station, E.	=	11 22 03.9
L.S.T. of observation	=	17 51 32.9
R.A. of σ Octantis	=	20 45 34.9

Determination of azimuth

L.H.A. of σ Octantis, t = $21^h\ 05^m\ 58.0^s$
i.e. t, arc = $316°\ 29'\ 30''$

Formula:

$$Z'' = \{p''.\sin t + (p'')^2 .\sin t.\cos t.\tan \phi.\sin 1''\}.\sec \phi,$$

where $p = 00°\ 56'\ 44'' = 3404''$; $t = 316°\ 29'\ 30''$; $\phi = 45°\ 52'\ 15''$.

$p''.\sin t =$	$-2343.5''$
$(p'')^2 .\sin t.\cos t.\tan \phi.\sin 1'' =$	-28.9
sum =	-2372.4
sum $\times \sec \phi =$	-3407.3
=	$-0°\ 56'\ 47.3''$

This must be applied to the reference direction* 180 00 00

∴ Azimuth of star when observed = 179 03 12.7
Angle, star to R.M., clockwise from star = 92 01 34

∴ Azimuth of R.M. = 271 04 47

Section 7

CONCLUSION

10.29 Summary of methods described

The following methods of determining azimuth have been described in this chapter:

(i) By observations to a star at equal altitudes.
(ii) By observations to stars near elongation (altitudes measured).
(iii) By circum-elongation observations to stars, in which hour angles are measured.
(iv) By ex-meridian observations to stars or the sun, in which altitudes are measured.
(v) By ex-meridian observations to stars or the sun, in which hour angles are used.
(vi) By observations to a close circumpolar star at any hour angle.

* See **10.02**.

10.30 Choice of method

Method (i) of **10.29** above is seldom used, but it has the advantage of not requiring star or other tables, and computations are simple; its main disadvantage is the long period required to complete the observation.

The most accurate method is that described in (vi)—observations to a close circumpolar star at any hour angle but ideally near elongation, particularly if multiple observations are used, as in **10.27**(b). This method should be confined to latitudes from about 5° to 55°, unless a micrometer eyepiece and first-order instrument are available, when the latitude may be extended to 70° if the star is observed near elongation.

Of slightly lesser accuracy is the circum-elongation type of observation using hour angles, followed by the less precise method in which stars are observed near elongation, again using hour angles. Unless considerable care is exercised in measuring the times in these hour-angle methods, however, the accuracy suffers. Observations should be balanced east and west of the meridian, and the altitudes confined to the range 5°–55°.

Descending in order of accuracy, we come to observations to circumpolar stars near elongation where altitudes are measured; to observations to east and west stars at low altitudes, using hour angles; and finally to sun observations by hour angles and by altitudes.

Determination of azimuth in very low latitudes is probably best done by observations to east and west stars at low altitudes, by hour angles, provided time is accurately known.

In latitudes above 70° the accurate determination of azimuth presents considerable difficulty and is beyond the scope of this book. For those who are interested, however, a geodetic method is described in **4.45** of *Geodesy* by Bomford (Oxford, Clarendon Press, 1975).

References

Adams, L. P., Astronomical Position and Azimuth by Horizontal Directions: a Rigorous Solution. *Survey Review*, **XXI**, No. 159, Jan. 1971, pp. 2–9.

Anon., Observing Azimuth with the Roelofs Solar Prism. *Canadian Surveyor*, **XVI**(2), 1962, pp. 105–6.

Bennett, G. G., A Solar Ephemeris using the HP9801A Calculator and its Application to Azimuth Determinations. *Australian Surveyor*, **26**, No. 1, March, 1974.

Bennett, G. G., Sun Observations for Azimuth. *Australian Surveyor*, **26**, No. 1, March 1974.

Berry, R. M., Azimuth by Solar-altitude Observation: How good is it? *Surveying and Mapping*, **18**, No.3, July–Sept. 1958, pp. 311–13.

Bhattacharji, J. C., A Practical Method of Determining Geodetic Azimuth and Deviation of the Vertical. *Empire Survey Review*. **15**, No. 112, April 1959, pp. 71–8.

Determination of azimuth

Black, A. N., A Note on Azimuth Determination. *Empire Survey Review*, **12**, No. 89, July 1953, pp. 121-6.

Capon, L. B., Latitude, Longitude and Azimuth. *Australian Surveyor*, **15**, No. 2, June 1954, pp. 87-90. (Also correspondence in **15**, No. 3, 4, and 5, and in **16**, No. 1, 2, and 6.)

Cox, C. L., Daylight Star Observations in Canterbury. *Jour. N.Z. Inst. Surveyors*, **20**, No. 197, May 1950, pp. 186-202.

Ezenwere, O. C., Azimuth and Latitude from Observed Altitude: a Rigorous Solution. *Survey Review*, **XXIII**, No. 178, Oct. 1975, pp. 184-6.

Fallon, N. R., Two Short Notes on Field Astronomy. *Empire Survey Review*, **14**, No. 104, April 1957, pp. 89-90.

Ghosh, S. K., Determination of Azimuth and Latitude from Observations of a Single Unknown Star by a New Method. *Empire Survey Review*, **12**, No. 87, Jan. 1953, pp. 17-26. (See also correspondence in the same journal, **12**, No. 89, July 1953, p. 143, **12**, No. 91, Jan. 1954, p. 237, and in **13**, No. 96, April 1955, p. 95; an article by A. Gougenheim in **12**, No. 94, Oct. 1954, pp. 342-9; and another by D. A. Tarczy-Hornoch in **13**, No. 99, Jan. 1956, pp. 212-19.)

Husti, G. J., A Method of Determining Latitude and Azimuth Simultaneously by Star Altitudes. *Survey Review*, **XXIV**, No. 184, April 1977, pp. 66-70.

Johns, R. K. C., Azimuth Determination by Astronomic Transit. *Canadian Surveyor*, **XIII(6)**, 1959.

Kivioja, I. A., Precise Astro-azimuth Observations by Using Mercury Levelling for a Theodolite. *Tech. Papers 41st Annual Meeting ACSM* 1981, pp. 100-4.

Kryukov, Yu. A., The Accuracy of Initial Sides and Laplace Azimuths. *Geodesy and Aerophotography* (English edn), 1967, pp. 157-9.

Murthy, V. N. S., Simultaneous Determination of Latitude, Azimuth and Time by Observations to a Pair of Stars. *Empire Survey Review*, **15**, No. 111, Jan. 1959, pp. 15-22. (See also correspondence in the same journal, **15**, No. 114, Oct. 1959, p. 190, and **15**, No. 117, July 1960, p. 339, and **15**, No. 115, Jan. 1960, p. 243.)

Opie, B. P., The Accuracy of Circum-elongation Observations for Azimuth. *Survey Review*, **XVIII**, No. 137, July 1965, pp. 107-19; also **XVIII**, No. 138, Oct. 1965, pp. 171-5.

Pavlov, F. F., An Astronomical Method of carrying forward an Azimuth in a Single Coordinate Zone. *Geodesy and Aerophotography* (English edn), 1965, pp. 9-10; also *ibid.*, pp. 307-8 of 1966 (article by Kozlov).

Poling, A. A., Astronomical Azimuths for Local Control. *Surveying and Mapping*, **XXVII**, No. 4, Dec. 1967, pp. 645-8.

Pring, R. W., Some Notes on Astronomy as Applied to Surveying. *Empire Survey Review*, **11**, No. 85, July 1952, pp. 309-18. (See also correspondence in the same journal, **12**, No. 89.)

Robbins, A. R., Azimuth Control in Canadian Latitudes. *Canadian Surveyor*, **XXIII(4)**, 1969, pp. 389-98.

Seaborg, H. J., Astronomic Azimuths in Alaska and Canada. *Surveying and Mapping*, **8**, No. 2, April-June 1948, pp. 50-8.

Smith, A. F., Criteria for the Determination of Azimuth by Observing Stars at Elongation. *Empire Survey Review*, **16**, No. 121, July 1961, pp. 136-9. (See also correspondence in the same journal, **16**, No. 123, Jan. 1962, p. 239.)

Szpunar, W., Method of Determining the Azimuth on the Basis of Observations of Pairs of Stars in Symmetrical Elongations. *Bulletin Géodésique,* **82**, Dec. 1966, pp. 317-27.

Thornton-Smith, G. J., The Curvature Correction in Precise Azimuth Observations upon a Close Circumpolar Star at any Hour Angle. *Empire Survey Review,* **11**, No. 80, April 1951, pp. 65-9. (See also the same journal, **11**, No. 82, Oct. 1951, p. 190.)

Thornton-Smith, G. J., An Azimuth Observation in the Almucantar. *Empire Survey Review,* **12**, No. 94, Oct. 1954, pp. 362-72.

Thornton-Smith, G. J., Latitude, Longitude and Azimuth from Two Stars. *Empire Survey Review,* **13**, No. 97, July 1955, pp. 124-32.

Thornton-Smith, G. J., Semi-graphic Almucantar Azimuths. *Empire Survey Review,* **13**, No. 102, Oct. 1956, pp. 375-80.

Thornton-Smith, G. J., A Circum-almucantar Observation for Azimuth. *Empire Survey Review,* **15**, No. 112, April 1959, pp. 79-86.

Vajcekjan, V. I., Some Problems of High Precision Determination of Astronomical Azimuths (in Russian). *Tr. Centr. NII Geod. Aeros"emki i Kartogr.,* **233**, 1980, 40-66.

Wuddah-Martey, E. E. L., Geodetic Azimuth from Meridian Transits by Interpolation. *Survey Review,* **XX**, No. 158, Oct. 1970, pp. 376-84.

EXERCISES

Students are advised to draw a diagram of the visible celestial hemisphere as an aid to solving astronomical problems. Care should be taken when constructing such diagrams to place the body observed on the correct side of the equator in relation to the elevated pole. For example, in the southern hemisphere, a star or the sun with a northerly declination has a polar distance greater than $90°$, so the designation of the declination in the *S.A.* must be carefully noted.

1. At a place in Latitude $30°$ N., prove that the azimuth of a circumpolar star having a declination of S. $60°$ when at western elongation is $215° 15' 52''$, and that the hour angle of the star is then $70° 31' 43''$. Find the time that elapses before the azimuth is diminished by $5''$.

$$Ans. \quad 3^m\ 34^s.$$

2. At a place in Latitude $30°$ S., prove that the azimuth of a circumpolar star having a declination of S. $60°$ when at western elongation is $215° 15' 52''$, and that the hour angle of the star is then $70° 31' 43''$. Find the time that elapses before the azimuth is diminished by $5''$.

$$Ans. \quad 2^m\ 11^s.$$

3. In Latitude $37°$ S., the sun's declination being S. $14°$, show that when its east hour angle is 3^h its azimuth is $72° 14' 39''$.

4. Compute the azimuth of a star having a declination of S. $75°$ when at eastern elongation, at a place in Latitude $30°$ S.

$$Ans. \quad 162° 36' 39''.$$

Determination of azimuth

5. Demonstrate that if two circumpolar stars A and B are in the same vertical at some instant on the east of the meridian, A being above B, they will later be simultaneously on the vertical making the same angle on the west of the meridian, B being then above A.

6. The corrected observed zenith distance of the sun on the afternoon of 17 March at a place in Latitude $34° 56'$ S. is $62° 19'$. If the sun's declination is S. $1° 28'$, compute its azimuth, to the nearest minute of arc, at the time of observation.

Ans. $289° 20'$.

7. Determine the difference of azimuth of the sun at its rising in midwinter and midsummer, also the difference (expressed in mean solar time) in the lengths of the days at these two times, assuming the latitude of the place to be $30°$ N., and the greatest declination of the sun $\pm 23° 27'$. Disregard corrections for refraction and parallax.

Ans. Diff. of azimuth = $54° 43'$.
Diff. of day-length = $3^h 52^m$.

8. In Latitude $30° 18'$ S., Longitude $123° 40'$ E., the following sun observation was taken at $4^h 45^m$ p.m. L.M.T.

Face	Altitude	Object	Bearing to sun	Bearing to R.M.
L	$22° 28' 30''$	☉	$258° 43' 30''$	$357° 46' 00''$
R	22 04 30	☉	258 51 30	357 46 00

The sun's declination for the day concerned, at U.T. 6^h, was S. $20° 15.9'$, increasing $03.1'$ during the next 6 hours. Find the true bearing of the R.M. to the nearest minute of arc. The temperature was $9°$ C and the atmospheric pressure 1009 mb.

Ans. $357° 44'$.

9. If the R. A. and declination of a star are $6^h 21^m 30^s$ and S. $52° 37' 00''$ respectively, find its true bearing and altitude at eastern elongation, also the L.M.T. of elongation, on the night of 29/30 October. The latitude of the place is $31° 00' 00''$ S., the longitude $8^h 00^m 00^s$ W., and R at U.T. 6^h on 30 October is $2^h 30^m 57.1^s$.

Ans. Bearing $134° 54' 07''$.
Altitude $40° 24' 16''$.
L.M.T. $23^h 39^m 36^s$, 29 October.

10. In Latitude $25° 58' 00''$ N., Polaris was observed at its eastern elongation, its declination for the date concerned being N. $88° 44' 20''$. Compute the azimuth of the star.

Ans. $1° 24' 10''$.

11. During the night of 9 November, in Latitude $45° 52' 08''$ S. and Longitude $11^h 22^m 04^s$ E., β Capricorni (No. 562) is observed near the prime vertical in

the west for azimuth. The L.Std.T. (meridian 12^h E.) of the observation is 10^h 13^m 07.4^s p.m. The mean horizontal circle reading on the star is $270°$ $44'$ $10''$ and on the R.M. is $28°$ $09'$ $16''$. The value of R at U.T. 6^h on 9 November is 3^h 10^m 46.3^s. What is the true azimuth of the R.M.? R.A. and declination of β Capricorni are 20^h 18^m 57.3^s and S. $14°$ $53'$ $50''$ respectively.

Ans. $33°$ $22'$ $47''$.

12. On 14 June observations are made to Polaris to determine azimuth. Using the following data, find the true bearing of the R.M.:

Latitude: $23°$ $51'$ $12''$ N.; longtiude: $70°$ $31'$ $15''$ W.
R.A. Polaris: 1^h 57^m 24.3^s; Dec. Polaris: N. $89°$ $05'$ $16''$.

Mean horizontal circle reading on Polaris: $298°$ $16'$ $09''$.
Mean horizontal circle reading on R.M.: $0°$ $00'$ $38''$.
R at U.T. 0^h, 15 June: 17^h 30^m 13.6^s.
L.Std.T. (meridian 5^h W.) of observation: 9^h 33^m 42^s p.m.

Ans. $62°$ $05'$ $59''$.

13. In Latitude $29°$ $31'$ $10''$ S. and Longitude $30°$ $17'$ $18''$ E., find the L.Std.T. of western elongation of σ Octantis on the night of 24/25 October and also its azimuth and true altitude at this time. R.A. of σ Octantis is 20^h 30^m 54^s and declination S. $89°$ $05'$ $56''$. R at U.T. 18^h on 24 October is 2^h 09^m 39.7^s. The Standard Meridian is 2^h E.

Ans.   L.Std.T. 0^h 17^m 20.3^s on 25 October.
Azimuth $181°$ $02'$ $08''$.
Altitude $29°$ $31'$ $25''$.

CHAPTER 11

The determination of latitude

Section 1

GENERAL

11.01 The need for astronomical latitude

We have already seen in Chapter 10 that a knowledge of the latitude is necessary for most methods of determining azimuth. Again, if the surveyor wishes to find his local time by means of astronomical observations, he will need to know his latitude. If an accurate map of the area in which he is working is available, he may be able to scale off the latitude with sufficient accuracy for his purpose, but often he will want the best value he can get, and this will mean that he has to make special observations of the type to be described in this chapter.

When commencing work in an isolated area not covered by existing survey, the surveyor will normally establish an initial, or datum station in a suitable locality, and he will wish to determine its latitude and longitude astronomically, as well as an initial azimuth from it.

11.02 General principles

In Chapter 3 it was demonstrated that the altitude of the celestial pole at any place is equal to the latitude of the place. If there were stars situated exactly at the two celestial poles, it would be very simple to observe the altitude of the elevated one at any time during the hours of darkness (since it would be stationary), make the appropriate correction for refraction, and immediately deduce the latitude.

While no such stars exist we do know that Polaris is very close to the north celestial pole and σ Octantis to the south pole, and it would seem feasible to measure the altitudes of these and determine latitude by making suitable corrections. This is, in fact, a practicable method, as we shall see in a later section of this chapter.

Although no stars mark the positions of the celestial poles in the heavens, we know from our *Star Almanac* tables the accurate angular distances of many stars from the poles (their polar distances, which are the complements of their declinations). If, then, we were to measure the altitudes of such stars *when they were on the meridian*, we would be able to deduce the altitude of the pole, and hence the latitude, by application of their polar distances.

Astronomy for surveyors

In Fig. 11.1, O is the position of the observer on the earth's surface and Z is his zenith; OP is parallel to the polar axis, and OQ is parallel to the celestial equator (parallel because P and Q are infinitely distant from the earth). If the star S_1 were observed on the meridian, at upper transit, its altitude would be the angle S_1OM_S; but its polar distance $(90° - \delta)$ is equal to the angle POS_1, hence the altitude of the pole, i.e. the latitude, is given by the measured altitude S_1OM_S minus angle POS_1. On the other hand, if the star were observed at S_2, the point of lower transit, its altitude would be equal to the angle S_2OM_S, and the latitude would be given by $S_2OM_S + POS_2$, where $POS_2 = POS_1 = p$, the polar distance of the star.

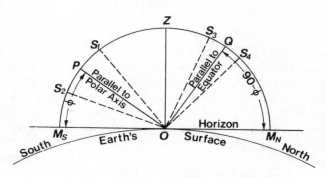

Fig. 11.1 Diagram in the plane of the meridian

The stars we have so far considered transit the meridian on the same side of the zenith as the pole; stars which are closer to the celestial equator will transit on the opposite side of the zenith, and their meridian altitudes will be measured in the direction opposite to that of the pole. However, we know that the altitude of Q, on the celestial equator, i.e. angle M_NOQ, must be equal to $(90° - \phi)$ or ω, the co-latitude. Hence, if a star is observed on the meridian at S_3, its altitude will be S_3OM_N; but QOS_3 is its declination, therefore S_3OM_N (measured) minus QOS_3 (from tables) $= (90° - \phi)$, from which ϕ may be found.

S_1, S_2 and S_3 represent the positions of stars which lie in the same celestial hemisphere as the elevated pole, i.e. they have declinations of the same "name" as the latitude of the place. Since Fig. 11.1 is drawn for the southern hemisphere, the declinations of stars crossing the meridian at S_1, S_2 and S_3 will be south. A star which transits the meridian at S_4 with altitude M_NOS_4 will have a northerly declination, but $M_NOQ = (90° - \phi) = M_NOS_4 +$ declination QOS_4, hence ϕ may be found.

The diagram for a place in the northern hemisphere is very similar to that in Fig. 11.1; the only change is the transposition of the words "South" and "North", and the letters M_S and M_N.

When determining latitude by meridian (or circum-meridian) altitudes, the student should always draw the diagram to guide him in the final computation;

Determination of latitude

indeed, he should do this before working through the calculation for *any* type of observation as an aid to clear thinking and avoidance of blunders.

Most of the commonly used methods for finding latitude depend upon the measurement of altitude, and the question of the effect of atmospheric refraction must therefore come up for consideration. Single (i.e. unbalanced) observations made to sun or stars at altitudes below $15°\text{-}20°$, where the effect of refraction is considerable, will be unreliable, and even at higher altitudes there will still be an element of doubt. We shall see, however, that a considerable degree of precision may be obtained by pairing observations at reasonably high altitudes in such a way that this difficulty is almost entirely removed.

The latest generation of glass-circle, one-second theodolites incorporates automatic vertical circle index levelling, and this is a considerable advantage during altitude observations since it relieves the surveyor of the need to level, or read an alidade bubble-tube.

Section 2

APPROXIMATE LATITUDE BY OBSERVATION OF THE SUN OR A STAR AT MERIDIAN TRANSIT

11.03 Principle of the method

This has already been discussed in **11.02**. Since the transit of a celestial body over the meridian is instantaneous, it will be impossible to measure its altitude on each face of the theodolite at this moment and so eliminate instrumental errors. However, if we know the longitude of the observing station and the correct local time, we can calculate the time of transit of the sun or an *Almanac* star, observe its altitude on one face of the instrument at this time, apply the necessary corrections, and obtain a good approximation of the latitude (probably within $20''$) with a modern glass-circle instrument reading to $01''$.

If we are uncertain of the local time or longitude to within, say, 5 minutes, it is better, after calculating the time of transit, to start observing the body at least a quarter of an hour before this time on the watch (longer if the uncertainty is greater), following it with the horizontal cross-hair until it reaches its maximum altitude (at upper transit) or minimum (at lower transit).

Alternatively, if the approximate direction of the meridian can be found (e.g. use of compass and application of known magnetic declination), the observer may start following the celestial body from a few minutes before it reaches this direction until its altitude is a maximum (or minimum).

For a single-face observation of this kind to a star, the corrections to be applied will be those for refraction and vertical collimation; a sun observation will also require these, and in addition, corrections for parallax and semi-diameter.

Astronomy for surveyors

11.04 Worked example

It was required to find the approximate latitude of a place in the northern hemisphere whose longitude is $8^h \ 04^m \ 16^s$ West, using a sun observation on 10 October 1975. The instrument available was a glass-circle one reading to $01''$, with a vertical collimation error of $+06''$ to left-face altitude readings. The Standard Time of the place is 8^h West.

(i) Calculation of the approximate L.Std.T. of transit of the sun:
From *Star Almanac*:

E at 12^h U.T. on 10 October 1975 (see p. 100)	=	$12^h \ 12^m \ 50^s$
$(24^h - E)$ = approx. L.M.T. of transit (see **5.19**)	=	11 47 10
Standard Meridian East of local meridian	=	+04 16
Approx. L.Std.T. of transit	=	11 51 26

(ii) The observation was done on the left face of the instrument at the above time; the sun was observed south of the zenith, the altitude of its apparent lower limb being $45° \ 54' \ 20''$.

(iii) Calculation of the sun's declination:

Approx. L.Std.T. of observation, 10 October	=	$11^h \ 51^m \ 26^s$
Longitude of Standard Meridian West	=	8 00 00
Corresponding U.T., 10 October	=	19 51 26
Sun's dec. at 18^h U.T., 10 October	= S.	$6° \ 36.4'$
Change in $1^h \ 51^m \ 26^s$	=	+01.7
Sun's dec. when observed	= S.	6 38.1
	i.e. = S.	$6° \ 38' \ 06''$

(iv) Calculation of correct altitude of sun's centre:

Sun's observed altitude (app. lower limb)	=	$45° \ 54' \ 20''$
Correction for vertical collimation	=	+06
		45 54 26
Parallax (approx. = $0.1'$ for alts. up to 70°)	=	+06
		45 54 32
Refraction (take r_0 only, since no temp. or press. readings)	=	−56
		45 53 36
Semi-diameter (subtract because apparent *lower* limb, therefore actual *upper* limb observed)	=	−16 00
Corrected altitude of sun's centre	=	$45° \ 37' \ 36''$

236

Determination of latitude

(v) Draw diagram and deduce approximate latitude (see Fig. 11.2):

Altitude sun's centre	$= M_S OS$	=	45° 37′	36″
Declination	$= QOS$ (south)	=	6 38	06
$(90° - \phi)$	$= QOM_S$	=	52 15	42
Latitude (ϕ), N.	$= 90° - QOM_S$	=	37° 44′	18″

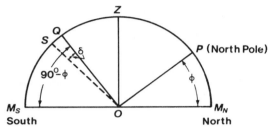

Fig. 11.2

Section 3

SECOND METHOD

LATITUDE BY ZENITH PAIR OBSERVATION OF STARS

11.05 Principle of method

A great improvement upon the accuracy of simple meridian observations may be effected by making observations upon two stars which culminate at approximately equal altitudes on opposite sides of the observer's zenith. The altitude of one star having been observed at culmination, the instrument is reversed in azimuth and the meridian altitude of the second star is then measured. The two stars must, of course, be chosen so that the second culminates at a convenient interval after the first. The method is commonly referred to as that of latitude determination by "zenith pair observations". No attempt is made to take two observations on the one star, and the combination of the two results largely eliminates errors of refraction as well as errors in vertical collimation and graduation of the vertical circle.

Referring to Fig. 11.1, let h_1 and h_3 be the altitudes of the stars at S_1 and S_3; let ψ_1 and ψ_3 be the celestial refractions at these altitudes, and let e be the vertical collimation error. Suppose further that h_1 is approximately equal to h_3; then, if each of the altitudes is measured within a relatively short period,

ψ_1 will be approximately equal to ψ_3. Suffixes to p and δ denote polar distances and declinations of the stars in those positions.

Then we have

$$(h_1 + e - \psi_1) - p_1 = \phi$$

and

$$90° - \{(h_3 + e - \psi_3) - \delta_3\} = \phi.$$

Adding these two equations together and dividing by 2, we get:

$$\frac{(h_1 - h_3)}{2} - \frac{(\psi_1 - \psi_3)}{2} + \frac{(\delta_1 + \delta_3)}{2} = \phi. \tag{11.1}$$

Similarly, if h_2 and h_4 are the altitudes of S_2 and S_4 in Fig. 11.1, and they are roughly the same, and if ψ_2 and ψ_4 are the appropriate refraction corrections, it can be proved that

$$\frac{(h_2 - h_4)}{2} - \frac{(\psi_2 - \psi_4)}{2} - \frac{(\delta_2 + \delta_4)}{2} + 90° = \phi. \tag{11.2}$$

Thus it is seen that in each case the collimation error disappears when the mean is taken, and furthermore it is the *difference* only of the two refractions which is involved. The two refractions are almost the same in each case, and any small error present in their assumed values will be practically identical for each pair; hence, as it is the difference of the refractions that we are concerned with, this small error will be eliminated.

In equation (11.1), δ_1 and δ_3 are both South and of the same "name" as the latitude; in equation (11.2), δ_2 is South and δ_4 is North. However, in both formulae, $\delta_1, \delta_2, \delta_3$ and δ_4 are treated as positive values. Instead of using (11.1) and (11.2), the latitude may be computed separately from each member of the pair concerned, and the mean taken; the same result will, of course, be achieved.

The face of the instrument is *not* reversed when reading the altitude of the second star, but the alidade bubble must be accurately centred before each altitude is read.

If the local time and longitude, and consequently the sidereal time, are known accurately, the best way is to intersect each star with the horizontal cross-hair at the instant when the sidereal time is equal to the star's right ascension. This is found from the *Star Almanac*, and the two stars will be selected for convenience, if possible, so that their right ascensions differ by a few minutes only. If the time is not known accurately, one of the other methods described in **11.03** should be used. The barometer and thermometer must also be read. If possible, the altitudes of the stars should be high (observed with a diagonal eyepiece if necessary) since the effect of refraction is least under these conditions.

The *Talcott method* of determining latitude (devised by Capt. Talcott of the U.S. Corps of Engineers) is based upon the same principle as that used above,

Determination of latitude

but instead of measuring the meridian altitude of each star (one north and the other south of the zenith), the *difference* between their meridian altitudes is measured very carefully. The two stars of each pair are selected so that their meridian altitudes differ by an amount small enough to be measured by an eyepiece micrometer without altering the inclination of the telescope. A very sensitive level is used to ensure that the inclination of the telescope is exactly the same for the two observations.

A special instrument called the Zenith Telescope was usually employed in the Talcott method, but modern geodetic theodolites adapted for accurate astronomical fieldwork, such as the Kern DKM3-A and the Wild T4, are usually supplied with the necessary (Horrebow) levels and eyepiece micrometers for precise latitude determination.

Talcott's method was for many years the most accurate for determining latitude, but its main disadvantage was the need for many pairs of stars with quite small zenith-distance differences. Star catalogues additional to the *Apparent Places of Fundamental Stars* were necessary, with consequent loss in declination and hence latitude accuracy. The precision of the new types of instrument mentioned immediately above has ousted the Talcott method in favour of meridian altitudes. (See section 27, *Field and Geodetic Astronomy*, A. R. Robbins, 1976, Ministry of Defence, United Kingdom.)

11.06 Worked example: *Zenith pair observation for latitude*

Date: 1975 July 18
Place: Arts Block, Pillar D,
    University of Otago,
    Dunedin, N.Z.
Longitude: $11^h\ 22^m\ 03.9^s$ E.
Chronometer: $11^h\ 59^m\ 53^s$ fast on U.T.C.
DUT1: $+0.2^s$

Instrument: Kern DKM2-A
    No. 182 907
Barometer: 1012 mb
Thermometer: 6 °C
Observer: J. B. M.
Booker: J. B. M.
Standard Meridian: 12^h E.

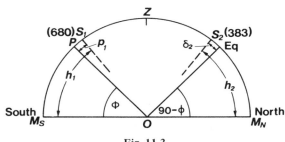

Fig. 11.3

Astronomy for surveyors

Face	Object	Vertical circle	Altitude	Chronometer
L	No. 680	322° 19′ 03″	52° 19′ 03″	$19^h\ 18^m\ 49^s$
L	No. 383	319 41 39	49 41 39	19 37 37

Chronometer times of upper transit
Star No. 680 δ Octantis

R.A. = L.S.T. of upper transit	=	$14^h\ 22^m\ 56.7^s$
Longitude, E.	=	11 22 03.9
Corresponding G.S.T.	=	3 00 52.8
Add 24^h	=	27 00 52.8
G.S.T. (= U.T. + R) at U.T. 6^h, 18 July 1975	=	25 41 43.2
Sidereal interval of transit after U.T. 6^h, 18 July	=	1 19 09.6
ΔR	=	-13.0
Solar interval of transit after U.T. 6^h, 18 July	=	1 18 56.6
i.e. U.T.1 of upper transit, 18 July	=	7 18 56.6
DUT1	=	0.2
U.T.C. of upper transit	=	7 18 56.4
Chronometer fast on U.T.C.	=	11 59 53.0
∴ Chronometer time of transit	=	19 18 49.4

Star No. 383 μ Virginis

R.A. No. 383	=	$14^h\ 41^m\ 47.5^s$
R.A. No. 680	=	14 22 56.7
Difference = sidereal interval between transits	=	0 18 50.8
ΔR	=	-3.1
Solar interval between transits	=	0 18 47.7
But chronometer time of transit of No. 680	=	19 18 49.4
∴ Chronometer time of transit of No. 383	=	19 37 37.1

Computation for latitude
Formula:
$$\frac{(h_1 - h_2)}{2} - \frac{(\psi_1 - \psi_2)}{2} + \frac{(\delta_1 + \delta_2)}{2} = \phi.$$

Determination of latitude

$$
\begin{aligned}
h_1 &= 52° 19' 03'' & \psi_1, 45'' \times 1.01 &= 45'' \\
h_2 &= 49\ 41\ 39 & \psi_2, 49'' \times 1.01 &= 49 \\
\hline
h_1 - h_2 &= 2\ 37\ 24 & \psi_1 - \psi_2 &= -04 \\
\tfrac{1}{2}(h_1 - h_2) &= 1° 18' 42'' & \tfrac{1}{2}(\psi_1 - \psi_2) &= -02''
\end{aligned}
$$

$$
\begin{aligned}
\delta_1 &= 83° 33' 49''\ S. \\
\delta_2 &= 5\ 33\ 10\ S. \\
\hline
\delta_1 + \delta_2 &= 89\ 06\ 59 \\
\tfrac{1}{2}(\delta_1 + \delta_2) &= 44\ 33\ 29.5 \\
\tfrac{1}{2}(h_1 - h_2) &= 1\ 18\ 42 \\
-\tfrac{1}{2}(\psi_1 - \psi_2) &= +02 \\
\hline
\phi &= 45\ 52\ 13.5\ S.
\end{aligned}
$$

The same result is obtained if we compute the latitude from each observation separately and take the mean (see Fig. 11.3):

	Star 680		Star 383
Observed altitude	52° 19' 03''		49° 41' 39''
Refraction correction	−45		−49
Corrected meridian altitude	52 18 18		49 40 50
p	−6 26 11	δ	−5 33 10
		$(90° - \phi)$	44 07 40
ϕ	45 52 07 S.		45 52 20 S.

Mean $\phi = 45° 52' 13.5''$ S.

Section 4

THIRD METHOD

LATITUDE BY CIRCUM-MERIDIAN OBSERVATIONS OF THE SUN OR A STAR

11.07 Principle of method

Observations of stars or the sun taken near to the meridian are commonly spoken of as circum-meridian observations. By taking a series of altitudes of a star or the sun, on both faces of the instrument, for some few minutes both

Astronomy for surveyors

before and after it crosses the meridian, instrumental errors may be largely eliminated, and by proper methods of reduction the results may be used to give an accurate determination of latitude. It is necessary to have the means of noting accurately the time of each observation, and then each altitude may be corrected for curvature or "reduced" so as to give us the corresponding altitude on the meridian itself. Thus a series of "circum-meridian" altitudes becomes equivalent to a series of measurements taken on the meridian itself, and in the taking of such a set of observations we will be able to eliminate instrumental errors in a way not possible with a single meridian observation. Still greater precision may be attained by taking such observations upon equal numbers of stars north and south of the zenith, at approximately equal altitudes.

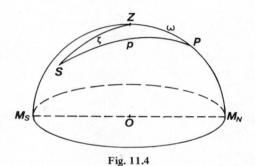

Fig. 11.4

In Fig. 11.4 let Z be the zenith, P the celestial pole and S the observed body. As this is to be near the meridian, the angle SPZ will be small.

Let

$\zeta = SZ$, the zenith distance,

$p = SP$, the polar distance of the body,

$\omega = PZ$, the co-latitude,

$t =$ the hour angle SPZ.

Then, from the triangle SPZ,

$$\cos \zeta = \cos \omega . \cos p + \sin \omega . \sin p . \cos t \qquad (11.3)$$

which may also be written

$$\sin h = \sin \phi . \sin \delta + \cos \phi . \cos \delta . \cos t. \qquad (11.3a)$$

We are interested in the change in altitude near the meridian in a relatively small change in time.

Let

Determination of latitude

h_z = altitude on meridian,
Δh = the change in altitude near the meridian,
T = the hour angle on the meridian (actually zero)
t = the change in the time corresponding to Δh (t, of course, is the east or west hour angle).

Thus (11.3a) may be re-written

$$\sin(h_z + \Delta h) = \sin\phi.\sin\delta + \cos\phi.\cos\delta.\cos(T+t). \tag{11.3b}$$

By Taylor's theorem,

$$\sin(h_z + \Delta h) = \sin h_z + \Delta h.\cos h_z - \tfrac{1}{2}(\Delta h)^2.\sin h_z + \ldots \tag{11.3c}$$

$$\cos(T+t) = \cos T - t.\sin T - \tfrac{1}{2}t^2.\cos T + \ldots$$
$$= 1 - 0 - \tfrac{1}{2}t^2 + \ldots \quad \text{(since } T = 0\text{)}. \tag{11.3d}$$

Using (11.3c) and (11.3d) in (11.3b), and neglecting powers higher than 2,

$$\sin h_z + \Delta h.\cos h_z - \tfrac{1}{2}(\Delta h)^2.\sin h_z$$
$$= \sin\phi.\sin\delta + \cos\phi.\cos\delta.(1-\tfrac{1}{2}t^2)$$
$$= \sin\phi.\sin\delta + \cos\phi.\cos\delta - \cos\phi.\cos\delta.\tfrac{1}{2}t^2.$$

But on the meridian, $\sin h_z = \sin\phi.\sin\delta + \cos\phi.\cos\delta$;

$$\therefore \Delta h.\cos h_z - \tfrac{1}{2}(\Delta h)^2.\sin h_z = -\tfrac{1}{2}t^2.\cos\phi.\cos\delta. \tag{11.3e}$$

However, near the meridian, $(\Delta h)^2$ is very small compared with t, so for ordinary purposes we may neglect the term containing it. Thus we have

$$\Delta h = -\frac{\cos\phi.\cos\delta}{\cos h_z}.\tfrac{1}{2}t^2. \tag{11.3f}$$

Since we are considering observations reasonably close to the meridian, t is small and we may assume that $t = \sin t$ or $t = 2.\sin\tfrac{1}{2}t$. Hence, if Δh is in seconds of arc (11.3f) may be written

$$(\Delta h)'' = -Bm, \tag{11.4}$$

where

$$B = \frac{\cos\phi.\cos\delta}{\cos h_z} \quad \text{and} \quad m = 2.\sin^2\tfrac{1}{2}t.\mathrm{cosec}\,1''.$$

Now $h_z + \Delta h = h$ at hour angle t, so if h is the observed altitude at H.A. = t, then

$$h_z = h - (\Delta h)'' = h + Bm. \tag{11.5}$$

243

Astronomy for surveyors

The value of m in seconds may be computed, knowing the value of t, or more conveniently it may be taken from the "Table for Circum-meridian Observations" on page 68 of the *Star Almanac* (see footnote, page 245).

Therefore, if we denote by h_{z0} the mean of the deduced meridian altitudes, by h_0 the mean of the actual observed altitudes, and by m_0 the mean of the computed factors m, we have

$$h_{z0} = h_0 + Bm_0. \tag{11.6}$$

With the aid of tables for m, the reduction of the observations thus becomes extremely simple. We take the mean of the values of m, multiply by B, and add the product to the mean of the observed altitudes (see Fig. 11.5).

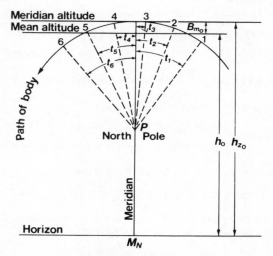

Fig. 11.5 Diagram to illustrate a circum-meridian observation
Hour angles $t_1, \ldots, t_6$ are exaggerated

The deduced mean meridian altitude is then corrected for refraction and the latitude is computed as an ordinary meridian altitude observation.

The value of B involves both the latitude and the meridian altitude, since

$$B = \frac{\cos\phi . \cos\delta}{\cos h_z}, \tag{11.7}$$

but the value of ϕ used in this is the approximate latitude as deduced either from the map or from a simple meridian observation, or as calculated from the maximum observed altitude. The value of h_z used is the meridian altitude computed from the approximate latitude and known declination of the body. The approximate value of B thus deduced is quite sufficiently accurate, when multiplied by m, to give the correction required.

Before starting the actual observations, it is necessary to calculate the time

Determination of latitude

of the body's meridian transit. The observations should then be made within about 10 minutes on each side of this. The t in the formula is the hour angle, and for stars is the interval of sidereal time between the instant of actual observation and the instant of meridian transit, expressed in angular measure; for the sun it is the corresponding interval of solar time similarly expressed in angular measure. In taking out the values of m from the table, we enter it with the hour angle in minutes and seconds of time. It should be noticed that, if the star observation is timed with a clock keeping mean solar time, the values of m corresponding to the hour angles will be in terms of mean solar time instead of sidereal time. Therefore, after m_0 has been found, it should be brought into terms of sidereal time by multiplying by 1.0055, a factor involving $\sin^2 \frac{1}{2} t$.

The method involves an accurate knowledge of the local time, and is then capable of giving very good results. For high-grade work the errors should be balanced by taking an equal number of observations on stars both north and south of the zenith. An equal number should be selected on each side at approximately equal altitudes. The errors are likely to be greatest for stars observed near the zenith, especially when the place of observation is near the equator. The range of observed altitudes should, if possible, lie between 40° and 75° above the horizon, and the closer the stars are observed to the meridian the better will be the results.

When balanced observations are being made to a north and a south star, the observing routine may be changed so that *all* the pointings (6 or 8 on each star) are done on the same face of the instrument, since vertical collimation error is eliminated in the mean (see **11.05**).

Formula (11.6), which uses mean values of h and m, assumes that all pointings are good, so any gross errors in reading altitudes and times may go undetected. For checking purposes therefore it is better, though more laborious, to reduce each pointing separately to its corresponding meridian altitude, when they should all lie within a range of a few arc-seconds. Any "wild" ones will be immediately obvious, and a reason should be sought for the mistake. One common blunder is for the observer to forget to level the alidade bubble before reading the altitude, and unfortunately this kind of error cannot be evaluated for correction purposes, whereas gross errors in the altitude and time readings have some chance of being rectified without re-observing. With separate reductions the 10-minute limit on either side of transit may be dispensed with, since any deterioration in quality due to the approximation used in the formula will soon be seen. The critical table of m in the *Star Almanac* is limited to about 10 minutes; for hour angles greater than this, interpolation is necessary in a noncritical table. An electronic calculator will obviate the need for any m tables, of course.*

---

* Mr L. P. Lee has kindly supplied the following simple formula for computing m to adequate accuracy:

$$m = (\pi/5760).t^2 = 0.0005454.t^2 \quad (t \text{ in seconds of time}).$$

Astronomy for surveyors

11.08 More exact methods of reduction of circum-meridian observations

The approximate formula (11.5) that we have given is the one usually adopted for the reduction of circum-meridian observations. A still closer approximation may be obtained by retaining another term of (11.3e) above, namely $-\frac{1}{2}(\Delta h)^2 . \sin h_z$, and substituting the right-hand side of (11.3f) in it. We then get an equation of the form

$$h_z = h + Bm + Cm',$$

where $C = B^2 . \tan h_z$ and $m' = 2.\sin^4 \frac{1}{2}t . \csc 1''$, h_z, B and m having the same significance as before.

The correction introduced by the third term in the formula is usually very small when the observations are made close to the meridian. If the value of t in minutes of time does not exceed two-fifths of the zenith distance of the star in degrees, then it can be shown that the correction introduced by the term Cm' is never more than $1''$, so that the more exact formula is only required where the highest precision is sought.

11.09 Circum-meridian observations of the sun

It is often more convenient for the surveyor to make observations of the sun than of the stars, and exactly the same method as we have described may be followed for circum-meridian observations of the sun. Obviously the sources of error cannot be balanced in the same way as with stars by taking observations both north and south of the zenith, although one observation to the sun in daytime may be partially balanced by another to a star at about the same altitude, on the other side of the zenith, at night. A difficulty arises from the fact that the sun's declination is not constant, and, if the observations extend over 20 minutes, it may vary as much as $20''$. If, however, a similar number of observations is made both before and after transit, the errors will very nearly balance in the mean, provided that in the computations the value of the declination used is the value at the time of transit.

11.10 Calculation of the time of transit of the celestial body observed

(a) *For the sun*
   See **5.19** and **10.25(A)(i)**.

(b) *For a star*

   EXAMPLE. Find the L.Std.T. of *lower* transit of γ Hydri at a place in Latitude $45° 52' 15'' $ S., Longitude $11^h 22^m 03.9^s$ E., on 13 July 1975. The Standard Meridian of the place is 12^h E. DUT1 = $+0.2^s$ From the *Star Almanac* the R.A.

Determination of latitude

of γ Hydri (No. 91) is $3^h\ 47^m\ 32.3^s$. The R.A. gives the L.S.T. when the star is at upper transit. To get the L.S.T. of lower transit, we must add 12^h, since it takes the star 12 sidereal hours to move from upper to lower transit.

Thus L.S.T. of lower transit, 13 July 1975	=	$15^h\ 47^m\ 32.3^s$
Longitude of place, E.	=	11 22 03.9
Corresponding G.S.T.	=	4 25 28.4
Add 24^h, G.S.T.	=	28 25 28.4
G.S.T. $(= $ U.T. $ + R)$ at U.T. 6^h, 13 July	=	25 22 00.4
Diff. = sidereal interval after U.T. 6^h, 13 July	=	3 03 28.0
ΔR for $3^h\ 03^m\ 28^s$	=	-30.1
Solar interval after U.T. 6^h, 13 July	=	3 02 57.9
i.e. U.T.1 of local lower transit, 13 July	=	9 02 57.9
DUT1 = U.T.1 $-$ U.T.C.	=	0.2
So U.T.C. of local lower transit, 13 July	=	9 02 57.7
Longitude of Standard Meridian, E.	=	12 00 00.0
$\therefore$ L.Std.T. of lower transit, 13 July	=	21 02 57.7

11.11 Observing procedure

We are not concerned here with azimuth, so intersection with the vertical cross-hair is not required. A star is therefore intersected with the horizontal cross-hair near the centre of the field of view of the telescope; similarly, tangency is made between this cross-hair and the sun's upper or lower limb near the centre of the field.

(a) *For a star*

The basic steps in the observation are:
  (i) At about 8 minutes before transit, intersect the star on L.F. with the horizontal cross-hair and note the time by one of the methods described in Chapter 6. Making sure the alidade bubble is exactly central,* read the vertical circle. Since the star near transit is moving very slowly in altitude, it is necessary to intersect it by using the vertical slow-motion screw rather than by waiting for it to make its own intersection. Indeed, if the cross-hair is preset slightly ahead of the star, intersection may not take place at all if the star has crossed the meridian in the meantime to proceed in the opposite direction in altitude.

---

* But see the end of **11.02**.

Astronomy for surveyors

(ii) Change to R.F. without delay and repeat the procedure of (i).
(iii) Keeping the instrument on R.F., repeat (ii).
(iv) Change to L.F. without delay and repeat procedure of (i).

As far as possible, the steps should be evenly spaced in time (the booker can assist greatly in this, and it is his duty to keep the observer informed on the timing).

The steps (i) to (iv) above should be repeated without interrupting the programme, to give a complete series of four L.F. and four R.F. observations, the whole period occupied being balanced about the calculated time of transit, i.e. four on each side.

Instead of using the above system of faces, *LRRLLRRL*, we may use *LRLRLRLR*, but although this has a very slight observational advantage, it has the practical disadvantage of taking longer and usually requiring more slight adjustments of the alidade level. The disadvantage of the *LRRLLRRL* sequence is that transit will probably occur between two pointings on the same face, and the non-cancellation of the vertical collimation error will result in a slight error in the mean of all the altitudes.

The barometer and thermometer must also be read by the booker, preferably in the middle of the series.

To increase the accuracy by reducing refraction error, another set of eight circum-meridian altitudes should be taken of a star on the other side of the zenith at approximately the same altitude and as soon after the first set as possible. Under good conditions, the mean of the two sets should yield a value within 3 seconds of the true value of the latitude, if a glass-circle theodolite reading to $01''$ is used. If the observer is reasonably proficient, one whole set of 8 observations should not take more than 16 minutes, or half this time if the observations are done on one face of the instrument only (see end of **11.07**). Accuracy may be increased slightly by doing two additional sets of 8 pointings each to another pair of stars, one on each side of the zenith, and taking the mean of the four sets.

(b) *For the sun*

The basic steps are the same as for a star, but they should be arranged so that alternate pointings are made to the sun's upper and lower limbs in order to eliminate semi-diameter in the mean. As has been mentioned in Chapter 10, the field book should show the position of the sun with respect to the cross-hair as seen through the telescope.

A typical set of eight pointings would be:

L.F., upper limb; R.F., lower limb; R.F., upper limb; L.F., lower limb; L.F., upper limb; R.F., lower limb; R.F., upper limb; L.F., lower limb.*

* The number of L.F.s must equal the number of R.F.s and the number of "uppers" must equal the number of "lowers", whatever the total number of pointings.

Determination of latitude

(But the observation may be done, with slightly more inconvenience, by changing face for each pointing (see (a) above).)

11.12 Worked examples: *Circum-meridian observations for latitude*

(a) *Using the sun*

Date: 1974 June 26
Place: Arts Block, Pillar F,
   University of Otago,
   Dunedin, N.Z.
Longitude: $11^h\ 22^m\ 03.9^s$ E.
Chronometer: $12^h\ 00^m\ 14.8^s$ fast on U.T.C.
DUT1: $+0.2^s$

Instrument: Wild T2 No. 76015
Barometer: 984 mb
Thermometer: 18 °C
Observer: M. D. Body
Booker: J. B. M.

Standard Meridian: 12^h E.

Face	Object	Vertical circle	Altitude	Chronometer
L	☉	69° 29′ 15″	20° 30′ 45″	$12^h\ 34^m\ 03^s$
R	☍	291 02 06	21 02 06	35 38
R	☉	290 31 07	20 31 07	37 35
L	☍	68 56 49	21 03 11	39 17
L	☉	69 28 18	20 31 42	41 08.5
R	☍	291 02 42	21 02 42	42 38.5
R	☉	290 31 03	20 31 03	44 12
L	☍	68 57 24	21 02 36	46 07

Mean altitude: 20° 46′ 54″

Chronometer time of transit:		
Approx. L.M.T. of transit $(24^h - E)$	=	$12^h\ 02^m\ 37^s$
Longitude, E.	=	11 22 04
Approximate U.T. of transit	=	0 40 33
E at U.T. 0^h 26 June 1974	=	$11^h\ 57^m\ 23.3^s$
ΔE for $0^h\ 40^m\ 33^s$	=	−0.4
E at transit	=	11 57 22.9
Subtract from 24^h to get L.M.T. of transit	=	12 02 37.1
Longitude, E.	=	11 22 03.9
U.T.1 of transit	=	$0^h\ 40^m\ 33.2^s$
DUT1	=	0.2
U.T.C. of transit	=	0 40 33.0

Astronomy for surveyors

Chronometer fast on U.T.C.		=	12 00	14.8
Chronometer time of transit,	$t_0 =$		12 40	47.8

Declination of sun at transit:
Sun's declination at U.T. 0^h 26 June = N. 23° 22.8′
Change for $0^h\ 40^m\ 33^s$ = 0.0

Sun's declination at transit, $\delta =$ N. 23° 22′ 48″

Sun's corrected altitude:
Mean observed altitude = 20° 46′ 54″
Refraction correction = 152″ × 0.94 = −02 23
Parallax correction = 9″ × $\cos h$ = +08

True mean altitude, $h_0 =$ 20 44 39

Approximate latitude:
Mean of two highest altitudes, one on each face = 20° 47′ 27″
Refraction and parallax corrections = −02 15

Approx. altitude on meridian, approx. $h_z =$ 20 45 12
(see Fig. 11.6)
∴ Add h_z to δ to get $(90° - \phi)$, $\delta =$ 23 22 48

$(90° - \phi)$ = 44 08 00
Approximate latitude, south, $\phi =$ 45 52 00

Factor B:

$$B = \frac{\cos\phi . \cos\delta}{\cos h_z} = 0.6835.$$

Mean value of m (m_0):

$t_0 - t$	m	
$6^m\ 44.8^s$	89″	The $t_0 - t$ values (hour angles)
5 09.8	52	are obtained by subtracting
3 12.8	20	the chronometer time of the
1 30.8	5	pointing from the chronometer
0 20.7	0	time of transit, or vice versa,
1 50.7	7	depending upon which is the
3 24.2	23	greater.
5 19.2	56	
	8)252	
	$m_0 = 31.5$	
	$m_0 B = 21.5″$	

250

Determination of latitude

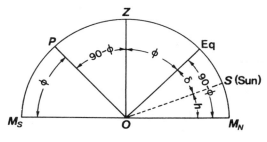

Fig. 11.6

True latitude:

True mean altitude,	$h_0 =$	20° 44′ 39″	
$m_0.B$	=	+21.5	
Meridian altitude,	$h_z =$	20 45 00.5	
Sun's declination at transit,	$\delta =$	23 22 48	
$(90° - \phi)$	=	44 07 48.5	
Latitude, south,	$\phi =$	45 52 11.5	

(b) *Using the stars*

Date: 1975 July 13
Place: Arts Block, Pillar F,
   University of Otago,
   Dunedin, N.Z.
Longitude: 11^h 22^m 03.9^s E.
Observer: J. B. M.
DUT1: $+0.2^s$

Instrument: Kern DKM2-A
   No. 182 907
Barometer: 1023 mb
Thermometer: 5 °C
Chronometer: 12^h 00^m 05.0^s
   fast on U.T.C.
Booker: S. G. M.
Standard Meridian: 12^h E.

(i) Star in North at upper transit:
   Star observed: *S.A.* No. 409, δ Serpentis

      R.A.: 15^h 33^m 39.5^s, Dec.: N. 10° 37′ 15″.

Face	Object	Vertical circle	Altitude	Chronometer	Stop-watch	Chronometer-stop-watch
L	Star	303° 31′ 37″	33° 31′ 37″	20^h 46^m 25^s	12.0^s	20^h 46^m 13^s
L	Star	303 31 47	33 31 47	20 48 00	17.0	20 47 43
L	Star	303 31 49	33 31 49	20 49 30	9.8	20 49 20.2
L	Star	303 31 49	33 31 49	20 50 45	8.6	20 50 36.4
L	Star	303 31 40	33 31 40	20 52 00	8.8	20 51 51.2
L	Star	303 31 27	33 31 27	20 53 10	14.0	20 52 56.0

Mean altitude: 33° 31′ 41.5″

Astronomy for surveyors

Chronometer time of transit:

L.S.T. of transit, = R.A.	=	15^h 33^m 39.5^s
Longitude, E.	=	11 22 03.9
G.S.T. of transit	=	4 11 35.6
Add 24^h	=	28 11 35.6
G.S.T. at U.T. 6^h (= U.T. + R), 13 July 1975	=	25 22 00.4
Difference = sidereal interval after U.T. 6^h	=	2 49 35.2
ΔR	=	−27.8
Solar interval after U.T. 6^h, 13 July	=	2 49 07.4
i.e. U.T.1 of transit	=	8 49 07.4
DUT1	=	0.2
U.T.C. of transit	=	8 49 07.2
Chronometer fast on U.T.C.	=	12 00 05.0
Chronometer time of transit, t_0 =		20 49 12.2

Star's corrected mean altitude:

Mean observed altitude	=	33° 31′ 41.5″
Refraction correction = 88″ × 1.02	=	−01 30
True mean altitude, h_0 =		33 30 11.5

Approximate latitude:

Highest observed altitude	=	33° 31′ 49″
Refraction correction	=	−01 30
Approximate meridian altitude, i.e. approx. h_z =		33 30 19
∴ Add h_z to δ to get $(90° − \phi)$ δ =		10 37 15
$(90° − \phi)$	=	44 07 34
Approximate latitude, south, i.e. approx. ϕ =		45 52 26
(see Fig. 11.7(a))		

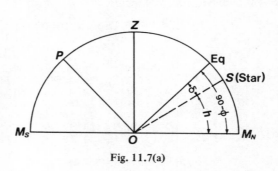

Fig. 11.7(a)

Determination of latitude

Factor B:

$$B = \frac{\cos\phi.\cos\delta}{\cos h_z} = 0.8207.$$

Mean value of m (m_0):

$t_0 - t$		m
2^m	59.2^s	$18''$
1	29.2	4
0	08.0	0
1	24.2	4
2	39.0	14
3	43.8	27
		6)67
	Mean =	11.17''

$m_0 = 11.17 \times 1.0055 = 11.23''$ (see **11.07**, near the end)
$m_0.B = 9.2''$

True latitude:

True mean altitude,	$h_0 =$	33°	30'	11.5''
$m_0.B$ (add for upper transit)	=			+09.2
Meridian altitude, upper transit,	$h_z =$	33	30	20.7
Star's declination,	$\delta =$	10	37	15
$(90° - \phi)$	=	44	07	35.7
Latitude, south,	$\phi =$	45	52	24

(ii) Star in South at lower transit
   Star observed: *S.A.* No. 91, γ Hydri

   R.A.: $3^h\ 47^m\ 32.3^s$, Dec.: S. 74° 18' 37''

Face	Object	Vertical circle	Altitude	Chronometer	Stop-watch	Chronometer–stop-watch
L	Star	300° 12' 35''	30° 12' 35''	$20^h\ 58^m\ 15^s$	10.7^s	$20^h\ 58^m\ 04.3^s$
L	Star	300 12 31	30 12 31	20 59 30	10.7	20 59 19.3
L	Star	300 12 29	30 12 29	21 01 20	11.0	21 01 09.0
L	Star	300 12 29	30 12 29	21 03 50	8.7	21 03 41.3
L	Star	300 12 28	30 12 28	21 04 50	5.4	21 04 44.6
L	Star	300 12 35	30 12 35	21 06 20	8.2	21 06 11.8

Mean altitude: 30° 12' 31.2''

Astronomy for surveyors

Chronometer time of transit:
This was worked out as an example in **11.10** to give:

L.Std.T. of lower transit, 13 July 1975		=	$21^h\ 02^m\ 57.7^s$
Chronometer fast on L.Std.T.		=	05.0
∴ Chronometer time of lower transit,	$t_0 =$		21 03 02.7

Star's corrected mean altitude:

Mean observed altitude		=	30° 12′ 31.2″
Refraction correction = $100'' \times 1.02$		=	−01 42
True mean altitude,	$h_0 =$		30 10 49.2

Approximate latitude:
Take the value computed in (b)(i) above, approx. $\phi = 45°\ 52′\ 24″$
(see Fig. 11.7(b))

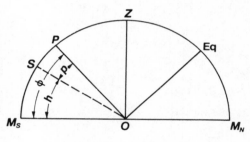

Fig. 11.7(b)

Factor B:

$$B = \frac{\cos\phi . \cos\delta}{\cos h_z}$$

Lowest observed altitude		=	30° 12′ 28″
Refraction correction		=	−01 42
Approx. altitude on meridian,	$h_z =$		30 10 46

$$\therefore B = 0.2178.$$

Mean value of m (m_0):

$t_0 - t$	m
$4^m\ 58.4^s$	49″
3 43.4	27

Determination of latitude

```
1   53.7        7
0   38.6        1
1   41.9        6
3   09.1       20
            ───────
            6)110
```

Mean = 18.33″

$m_0 = 18.33'' \times 1.0055 = 18.4''$ (see **11.07**, near the end)
$m_0 . B = 4.0''$

True latitude:

True mean observed altitude,	$h_0 =$	30° 10′ 49.2″
$m_0.B$ (subtract for lower transit)	=	−4.0
Meridian altitude, lower transit,	$h_z =$	30 10 45.2
Star's polar distance $p = 90° - \delta$,	$(90° - \delta) =$	15 41 23
Latitude, south,	$\phi =$	45 52 08

The observations for (b)(i) and (b)(ii) above were all done on Face Left, so the true latitude is now obtained by taking the mean, which eliminates vertical collimation error, as in **11.06**.

Latitude from north star (b)(i)	=	45° 52′ 24″
Latitude from south star, (b)(ii)	=	45 52 08
Mean latitude from the north-south pair	=	45 52 16

If only one star is used, an equal number of L.F. and R.F. pointings must, of course, be made.

Section 5

FOURTH METHOD

LATITUDE BY MEASURING THE ALTITUDE OF A CLOSE CIRCUMPOLAR STAR AT ANY HOUR ANGLE

11.13 Principle of the method

Brief reference was made to this method in **11.02**. The altitude of a star which is very close to the pole will never be very different from that of the pole. Hence, by measuring the altitude of such a star and applying a correction whose magnitude depends upon the hour angle and polar distance, we can arrive at the altitude of the pole, i.e. the latitude.

Astronomy for surveyors

The only two *Almanac* stars which are available for the method are Polaris in the northern hemisphere and σ Octantis in the southern; their polar distances are approximately 0° 48′ and 0° 58′ (1983) respectively. Polaris, of magnitude 2.1, is easily seen with the naked eye, but σ Octantis, of magnitude 5.5 is not, and its position has to be calculated before it can be observed (see **10.26**).

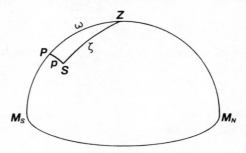

Fig. 11.8

In Fig. 11.8, let S be the circumpolar star, Z the zenith, and P the pole as before. Then, with the usual notation,

$\zeta = SZ$, the zenith distance,

$p = SP$, the polar distance of the star,

$\omega = PZ$, the co-latitude,

$t = $ the hour angle of the star, SPZ.

From the triangle SPZ we have

$$\cos\zeta = \cos\omega.\cos p + \sin\omega.\sin p.\cos t \quad \text{(see (1.2))}$$

or, if h is the observed altitude, and ϕ the latitude,

$$\sin h = \sin\phi.\cos p + \cos\phi.\sin p.\cos t.$$

Now h will differ from ϕ by a small quantity, which is always less than p (or equal to p at transit). In the case of Polaris or σ Octantis p is also small, being less than 1 degree.

Let

$$h = \phi + x,$$

where x is a small correction. Then

$$\sin\phi.\cos x + \cos\phi.\sin x = \sin\phi.\cos p + \cos\phi.\sin p.\cos t,$$

i.e.

$$\sin\phi(1 - \tfrac{1}{2}x^2 + \ldots) + \cos\phi(x - \tfrac{1}{6}x^3 + \ldots)$$
$$= \sin\phi(1 - \tfrac{1}{2}p^2 + \ldots) + \cos\phi.\cos t(p - \tfrac{1}{6}p^3 + \ldots).$$

Determination of latitude

Neglecting the square and higher powers of x and p in this equation, we get $x = p \cos t$, which is the value of x to a first approximation.

Next, retaining the squares of x and p but neglecting the higher powers, we get

$$x \cos \phi = p \cos \phi . \cos t - \tfrac{1}{2} p^2 \sin \phi + \tfrac{1}{2} x^2 \sin \phi.$$

Substituting for x^2 the value $p^2 \cos^2 t$, we obtain then as a second approximation

$$x = p \cos t - \tfrac{1}{2} p^2 \tan \phi . \sin^2 t.$$

The second term in this expression is very small, and as $\tan \phi$ differs from $\tan h$ by only a small quantity, the difference when multiplied by p^2 will be too small to take into account, so that we may write

$$x = p \cos t - \tfrac{1}{2} p^2 \tan h . \sin^2 t.$$

In this formula x and p are in circular measure, but if x and p are measured in seconds of arc we may write

$$x'' = p'' \cos t - \tfrac{1}{2} (p'')^2 \tan h . \sin^2 t . \sin 1'',$$

so that we have for the latitude

$$\phi = h - p \cos t + \tfrac{1}{2} p^2 \tan h . \sin^2 t . \sin 1'', \tag{11.8}$$

where p is in seconds of arc.

The formula is, of course, an approximation only, but it can be shown that it is sufficiently accurate to give the result within $1''$ of truth for any positions of the two stars concerned.

Formula (11.8) requires h, p and t for solution. h is measured, p is obtained from the *Star Almanac*, and t is derived from the formula

$$\text{L.S.T.} = \text{R.A.} + t.$$

The L.S.T. of the observation is therefore required; that is, if a sidereal clock is used, its error on L.S.T. must be known. If the clock keeps approximate L.Std.T., its error must be known and a knowledge of the longitude is also required for computing the L.S.T. The question then arises of how much error is introduced into the computed ϕ by an error in t.

The star is moving most rapidly in altitude at elongation, and it can be proved that $dh/dt = -\sin S . \sin p$. When $S = 90°$, at elongation, we find that, for $p = 0° 58'$, a change of 1 second of time in the hour angle produces a change of $0.25''$ (arc) in the altitude, and it takes a change of about 4 seconds of time to change the altitude by $1''$. This can be confirmed by computing $p \cos t$ for t at elongation, and then recomputing with t increased or decreased by $1'$ of arc (4 seconds of time). Thus, if L.Std.T. is known accurately, an error of one or two seconds of time in the longitude is not critical.

Astronomy for surveyors

A more serious source of error could arise in the refraction correction, hence it is advisable to balance an observation by this method with a circum-meridian observation to a star on the other side of the zenith at about the same altitude, and take the mean of the two. It is also feasible, of course, to use the zenith-pair technique by balancing a single-face observation to Polaris or σ Octantis at any hour angle with another on the same face of the instrument to a star on the meridian at about the same altitude on the other side of the zenith. This has the advantage that, if the meridian star is observed first, the second (close-circumpolar) star may be observed immediately afterwards, since its position does not matter.

11.14 Observing procedure

For the simplest observation, two altitudes may be taken in as quick succession as possible, one on L.F. and the other on R.F., the alidade level-bubble being carefully centred before reading the vertical circle, and the clock times noted. The mean of the altitudes and the mean of the corrected clock times are then taken as the basis for the reduction as a single observation. The procedure is simple, and several such L.F.-R.F. pairs may be observed one after the other quite quickly to provide a check and a good mean value of the latitude.

(*Note*. Tables for finding latitude from observations to Polaris are given on pages 56–59 of the *Star Almanac*. An explanation of the tables is also given on page xi, and a worked example on page xvi.)*

11.15 Worked example: *Latitude determination by observations to σ Octantis at any hour angle*

Date: 1974 August 19
Place: Arts Block, Pillar F,
    University of Otago,
    Dunedin, N.Z.
Longitude: $11^h \ 22^m \ 03.9^s$ E.
Observer: J. B. M.
Booker: M. D. Body
Instrument: Wild T2 No. 76015

Object observed: σ Octantis
R.A.: $20^h \ 45^m \ 35^s$
Dec.: S. $89°\ 03'\ 16''$
Polar distance $(p) = 3404''$
Barometer: 1020 mb
Thermometer: 5 °C
Chronometer: $12^h \ 00^m \ 12.6^s$ fast on U.T.C.

Face	Object	Vertical circle	Altitude	Chronometer
L	Star	43° 30′ 02″	46° 29′ 58″	$20^h \ 15^m \ 41^s$
R	Star	316 29 38	46 29 38	20 17 55

Mean observed altitude: 46° 29′ 48″
Mean chronometer time: $20^h \ 16^m \ 48^s$

* See Appendix I (**AI.04.02**) for examples of the use of Pole Star tables.

Determination of latitude

True altitude:
Mean observed altitude	=	46° 29′ 48″	
Refraction correction =55″ × 1.02	=	−56	
True observed altitude, h =		46 28 52	

Star's hour angle:
Mean observed chronometer time	=	20^h 16^m 48^s	
Chronometer fast on U.T.C.	=	12 00 12.6	
U.T.C. of observation	=	8 16 35.4	
R at U.T. 6^h, 19 August 1974	=	21 48 50.4	
ΔR in 2^h 16^m 35.4^s	=	22.4	
G.S.T. of observation	=	30 05 48.2	
Longitude observing station, E.	=	11 22 03.9	
L.S.T. of observation	=	41 27 52.1	
R.A. of σ Octantis	=	20 45 35	
Hour angle of σ Octantis, t =		20 42 17	
t, arc	=	310° 34′ 15″	

Computation for latitude
Formula:

$$\phi = h - p.\cos t + \tfrac{1}{2}p^2.\tan h.\sin^2 t.\sin 1''.$$

h	=	46° 28′ 52″
$p.\cos t$	=	−36 54
$\tfrac{1}{2}p^2.\tan h.\sin^2 t.\sin 1''$	=	+17
Latitude,	ϕ =	45 52 15 S.

Note. In the above example, no account was taken of DUT1 which was, in fact, $+0.1^s$—a quantity well able to be neglected since 2 seconds of time will only change the star's altitude by 0.5″ in the worst position for observing. For the same reason, the 0.1^s in the L.S.T. of observation was neglected.

It does not take long to do a series of F.L., F.R. pointings, and the mean of, say, three pairs will give a reliable value of the latitude.

Astronomy for surveyors

Section 6

FIFTH METHOD

LATITUDE BY OBSERVING TWO STARS AT THE SAME ALTITUDE

11.16 Principle of the method

In the triangle PZS_1 of Fig. 11.9, where P is the pole, Z the zenith and S_1 a star, and with the usual notation,

$$\cos \zeta = \cos \omega . \cos p_1 + \sin \omega . \sin p_1 . \cos t_1 \quad \text{(see (1.2))}$$

i.e.

$$\sin h_1 = \sin \phi . \sin \delta_1 + \cos \phi . \cos \delta_1 . \cos t_1.$$

If, now, we had another star S_2 at the *same* altitude, we would have

$$\sin h_1 = \sin \phi . \sin \delta_2 + \cos \phi . \cos \delta_2 . \cos t_2,$$

so that

$$\sin \phi . \sin \delta_1 + \cos \phi . \cos \delta_1 . \cos t_1$$
$$= \sin \phi . \sin \delta_2 + \cos \phi . \cos \delta_2 . \cos t_2.$$

Rearranging this equation, we get

$$\sin \phi (\sin \delta_1 - \sin \delta_2) = \cos \phi (\cos \delta_2 . \cos t_2 - \cos \delta_1 . \cos t_1)$$

or

$$\tan \phi = \frac{\cos \delta_2 . \cos t_2 - \cos \delta_1 . \cos t_1}{\sin \delta_1 - \sin \delta_2} \qquad (11.9)$$

Thus it will be seen that, if we observe two *Almanac* stars at the same altitude (which we do not need to know), noting the time when each was on the horizontal cross-hair of the telescope, we can compute the latitude. Since the altitude h is not required, we are not concerned with vertical collimation error, imperfection of circle graduation, or refraction, although the time between observations should be short to minimize the possible effect of changes in refraction. Air pressure and temperature are not required.

For good results, it will be noted that neither numerator nor denominator in formula (11.9) should be small. This means that δ_1 and δ_2 should differ considerably and indicates that one star should be on one side of the zenith and the other on the opposite side, and neither close to it. There is an advantage in observing first to the star on the same side as the pole and then to the second on the other side, where the stars are moving faster and are therefore more likely to reach the desired altitude more quickly.

The stars should be observed when they are not very far from the meridian,

Determination of latitude

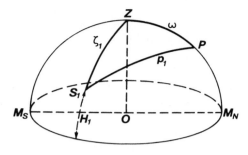

Fig. 11.9

if possible, although this may present difficulties in the case of the second star whose altitude will be changing more slowly the closer it approaches the meridian. A compromise is therefore necessary in order to cut down waiting time during which refraction may change.

11.17 Observing procedure

For accuracy, it is obvious that the altitude setting of the telescope must not change between the observations to the two stars. Thus great care must be paid to the stability and levelling of the theodolite. After it has been set up, the instrument should be given ten minutes or so to "settle down"; it should then be levelled by means of the sensitive alidade level-tube, until the bubble remains central in all positions. When making the actual observations, the observer should use a light touch on clamps and screws to reduce the chance of putting the instrument off level.

The observations can be done on one face only; a reversal of face and re-setting of the telescope elevation would change the altitude because of the inevitable presence of a small amount of vertical collimation error.

Once the instrument is stable and accurately levelled, the observation is a very simple one. The steps are:

(i) Point to the star on, say L.F. Set the horizontal cross-hair slightly in advance of the star so that it will transit the hair near the centre of the telescope field. As soon as it is on the cross-hair, note the time, using either a stop-watch or electronic clock. Enter the name of the star against the recorded time.

(ii) Point to the second star, again on L.F. Check the alidade level-bubble: if it is off-centre, the observation should be discarded.

(It is possible, of course, with modern instruments, to re-centre the bubble with the alidade screw and then re-set the altitude reading on the circle with the vertical slow-motion screw, but this is not altogether satisfactory since it is bound to introduce slight errors, mainly in the circle reading. The essence of the method is to keep the telescope inclination exactly the same for the two observations, and this can really be done only by having the instrument axis vertical so that the bubbles remain central in all positions.)

If the alidade bubble has not moved from centre, note the time when the star transits the horizontal hair and record the name of the star against this time.

The latest one-second theodolites are provided with automatic vertical-circle index levelling, thus obviating the need to centre an alidade bubble before reading the circle. This is particularly useful for latitude observations requiring accurate altitudes (also for some azimuth and time observations, of course). With such instruments it is easy to ensure that the inclination of the telescope is the same for pointings in different azimuths, since the micrometer setting remains unchanged while the index is quickly and accurately set, if necessary, with the vertical slow-motion screw.

Formula (11.9) requires only δ_1, δ_2, t_1 and t_2 for solution. The declinations are obtained from the *Star Almanac*, as are the R.A.'s which are used in the formula

$$\text{L.S.T.} - \text{R.A.} = t$$

for obtaining t_1 and t_2. Local time is therefore required.

By observing several pairs of stars and taking the mean of the computed latitudes, the accuracy of the final result will be improved. If the latitude is known approximately, it will be possible to compute in advance the approximate positions of stars suitable for observation.

The method is capable of good accuracy, although the observations are usually done with a first-order theodolite specially equipped for astronomical work, or with a prismatic astrolabe.

11.18 Worked example: *Latitude from observations to two stars at equal altitudes*

Date: 1975 July 18
Place: Arts Block, Pillar D,
   University of Otago,
   Dunedin, N.Z.
Longitude: $11^h\ 22^m\ 03.9^s$ E.
Observer: J. B. M.
Booker: J. B. M.
Instrument: Kern DKM2-A
   No. 182 907

Chronometer: $11^h\ 59^m\ 53.0^s$ fast on
   U.T.C.
DUT1: $+0.2^s$
Objects observed:
(1) *S.A.* No. 238, δ Velorum,
   Mag. 2.0
   R.A.: $8^h\ 44^m\ 00.4^s$
   Dec.: S. $54°\ 37'\ 16''$

Determination of latitude

(2) *Smithsonian Astrophysical Observatory Catalogue No. 141 586, Mag. 4.8*

R.A.: $17^h\ 15^m\ 23.3''$
Dec.: S. $0°\ 25'\ 00''$
Standard Meridian: 12^h E.

δ Velorum was observed at chronometer time $19^h\ 42^m\ 51.1^s$;
S.A.O. 141 586 was observed at chronometer time $19^h\ 53^m\ 40.4^s$.
(A stop-watch was used for the timing.)

To find the hour angles

	(1) δ Velorum	(2) S.A.O. 141 586
Chronometer time of observation	$19^h\ 42^m\ 51.1^s$	$19^h\ 53^m\ 40.4^s$
Chronometer fast on U.T.C.	11 59 53.0	11 59 53.0
U.T.C. of observation	7 42 58.1	7 53 47.4
DUT1	+ 0.2	+0.2
U.T.1 of observation	7 42 58.3	7 53 47.6
R at U.T. 6^h, 18 July 1975	19 41 43.2	19 41 43.2
ΔR	+16.9	+18.7
G.S.T. of observation	3 24 58.4	27 35 49.5
Longitude, E.	11 22 03.9	11 22 03.9
L.S.T. of observation	14 47 02.3	38 57 53.4
R.A. of star	8 44 00.4	17 15 23.3
Hour angle of star, in time	6 03 01.9	21 42 30.1
in arc,	$t_1 = 90°\ 45'\ 28''$	$t_2 = 325°\ 37'\ 31''$
	$\delta_1 = 54°\ 37'\ 16''$	$\delta_2 = 0°\ 25'\ 00''$

To find the Latitude

Formula:

$$\tan\phi = \frac{\cos\delta_2 \cdot \cos t_2 - \cos\delta_1 \cdot \cos t_1}{\sin\delta_1 - \sin\delta_2}$$

∴ Latitude $= \phi = \underline{45°\ 52'\ 13''\ \text{S}}$.

The above observation was done by selecting a "naked eye" star in the south-west at a suitable altitude, in this case $35°\ 27'\ 48''$, timing it with a stop-watch across the horizontal cross-hair, and noting its azimuth to within a few minutes of arc. The instrument was then swung round through about 180° of azimuth and a search made for a star which would cross the horizontal cross-hair. In the event, the observer had to wait for about 10 minutes before a suitable star was found. Maintaining the same altitude as for the first star

Astronomy for surveyors

(an easy matter with the Kern DKM2-A), the second one was timed across the horizontal hair and its azimuth noted. The first star was easily identified as *S.A.* No. 238 δ Velorum by the method of **9.04**, but the second one could not be found in either the *Star Almanac* or the *Apparent Places of Fundamental Stars* after its R.A. and Dec. had been calculated. However, it was located without trouble in the *Smithsonian Catalogue* because of its relative brightness. Its coordinates for 1950.0 were: R.A. 17^h 14^m 02.523^s, Dec. $-0°$ $23'$ $25.24''$. The reduction from this Mean Place to the Apparent Place at the time of observation was done as in **4.12**, except that the second-order terms were included. If a referring mark of known azimuth is not available, the stars' azimuths may be found as at the end of **9.04**, using the second star to avoid changing the altitude setting between pointings.

Section 7

CONCLUSION

11.19 Summary of methods described

The following methods of determining latitude have been described in this chapter:

(i) Latitude by observation of the sun or a star at meridian transit (approximate method).
(ii) By zenith pair observation of stars.
(iii) By circum-meridian observations of the sun or a star.
(iv) By measuring the altitude of a close circumpolar star at any hour angle.
(v) By observing two stars at the same altitude.

11.20 Choice of method

For the ordinary work of the surveyor who is using, say, a modern glass-circle instrument reading to $1''$, latitude may be obtained with a probable error of less than $3''$ by using a pair of circum-meridian observations, one to a star north of the zenith and the other to a south star. The same order of accuracy could probably be obtained by using method (iv), balanced by method (iii) of **11.19**, to a star on the opposite side of the zenith. A simple zenith pair observation of stars (method (ii) above) would yield comparable results with careful work and an accurate knowledge of local time.

For the highest order of accuracy, geodetic astronomers use the method of meridian altitudes described by Dr A. R. Robbins in section 27 of *Field and Geodetic Astronomy*, 1976, Ministry of Defence, United Kingdom.

A neat method of obtaining latitude and longitude simultaneously from position lines is described in Chapter 14.

Determination of latitude

References

Bhattacharji, J.C., A Method of Determination of Latitude and Longitude where only Time and Horizontal Angles are Observed. *Empire Survey Review,* **14**, No. 110, Oct. 1958, pp. 352–63.

Bhattacharji, J. C., A Method of Determining Astronomical Latitude from Observations of a Star-pair near their Times of Elongation. *Empire Survey Review,* **15**, No. 114, Oct. 1959, pp. 161–8.

Bhattacharji, J. C., Modification of Talcott's Method of Observation for Latitude Variation. *Bulletin Géodésique,* **77**, Sept. 1966, pp. 237–47.

Bowie, I. G., Latitude Determination without Vertical Circle Readings. *Empire Survey Review,* **13**, No. 102, Oct. 1956, pp. 380–3.

Capon, L. B., Latitude, Longitude and Azimuth. *Australian Surveyor,* **15**, No. 2, June 1954, pp. 87–90. (Also correspondence in **15**, No. 3, 4 and 5, and in **16**, No. 1, 2 and 6.)

Ezenwere, O. C., Latitude Determination from Calculated Azimuth and Observed Altitude. *Survey Review,* **XXI**, No. 166, Oct. 1972, pp. 348–54.

Ezenwere, O. C., Azimuth and Latitude from Observed Altitude: A Rigorous Solution. *Survey Review,* **XXIII**, No. 178, Oct. 1975, pp. 184–6.

Fitzgerald, R. D., Reduction to Meridian of Circum-meridian Observations. *Australian Surveyor,* **16**, No. 6 June 1957, pp. 343–7.

Ghosh, S. K., Determination of Azimuth and Latitude from Observations of a Single Unknown Star by a New Method. *Empire Survey Review,* **12**, No. 87, Jan. 1953, pp. 17–26. (See also correspondence in the same journal, **12**, No. 89, July 1953, p. 143; **12**, No. 91, Jan. 1954, p. 237; and in **13**, No. 96, April 1955, p. 95. Also an article by A. Gougenheim in **12**, No. 94, Oct. 1954, pp. 342–9, and another by D. A. Tarczy-Hornoch in **13**, No. 99, Jan. 1956, pp. 212–19.)

Husti, G. J., A Method of Determining Latitude and Azimuth Simultaneously by Star Altitudes. *Survey Review,* **XXIV**, No. 184, April 1977, pp. 66–70.

Lee, L. P., Note on the Reduction of Circum-meridian Altitudes to the Meridian. *Empire Survey Review,* **10**, No. 78, Oct. 1950, pp. 366–8.

Murthy, V. N. S., Simultaneous Determination of Latitude, Azimuth and Time by Observations to a Pair of Stars. *Empire Survey Review,* **15**, No. 111, Jan. 1959, pp. 15–22. (See also correspondence in the same publication, **15**, No. 114, Oct. 1959, p. 190; **15**, No. 115, Jan. 1960, p. 243; and **15**, No. 117, July 1960, p. 339.)

Pring, R. W., The Accuracy of Astro-observations for Position. *Empire Survey Review,* **13**, No. 96, April 1955, pp. 70–5.

Thom, A., Circum-meridian Altitudes. *Empire Survey Review,* **14**, No. 106, Oct. 1957, pp. 170–5. (See also correspondence in the same publication, **14**, No. 109, July 1958, p. 335.)

Thornton-Smith, G. J., Latitude, Longitude and Azimuth from Two Stars. *Empire Survey Review,* **13**, No. 97, July 1955, pp. 124–32.

Thornton-Smith, G. J., Azimuth-controlled Almucantar Observations for Position. *Empire Survey Review,* **14**, No. 109, July 1958, pp. 301–10.

Astronomy for surveyors

EXERCISES

1. Two stars S_1 and S_2 of southerly declination were observed, S_1 at eastern elongation and S_2 at western elongation. The difference between the azimuths of the two stars was $41° 55' 01''$. The declination of S_1 was S. $68°$ and that of S_2 was S. $78°$. Find the latitude (south) of the observing station.

(Hints:
$$\sin p_1 = \sin \omega . \sin Z_1 \qquad (1)$$
$$\sin p_2 = \sin \omega . \sin Z_2 \qquad (2)$$

Find $[(1) + (2)]/[(1) - (2)]$ which gives an expression containing only one unknown, $\cot \frac{1}{2}(Z_1 - Z_2)$. This gives $\frac{1}{2}(Z_1 - Z_2)$. $\frac{1}{2}(Z_1 + Z_2)$ is known from the observation. From $(Z_1 + Z_2)/2$ and $(Z_1 - Z_2)/2$, Z_1 and Z_2 are thus found, after which ω, and hence ϕ, may be found from (1) or (2).)

Ans. $35°$ S.

(*Note.* The above provides a method of determining latitude not described in the text; the observations are done on one face with an instrument in good adjustment. The altitudes of the two stars should not differ by too much, hence the difference between the declinations of the two stars selected should not be too great.)

2. At a place in north latitude, the meridian zenith distance of α Cephei (declination N. $61° 58' 21''$) is determined as $26° 54' 28''$ N. The meridian zenith distance of α Aquilae (declination N. $8° 29' 23''$) is found as $26° 34' 28''$ S. Find the latitude of the place.

Ans. N. $35° 03' 52''$.

3. An observation made in Antarctica on 19 November 1912 gave the altitude of the sun's centre as $42° 07.8'$, the temperature being $-8.3 °C$ and the barometer reading 921 mb. Correct for refraction and parallax, and compute the latitude of the place, given that the sun's declination was S. $19° 21.6'$. (The sun was at upper transit.)

Ans. $67° 14.7'$ S.

4. The declination of the sun being S. $20° 39.9'$, its meridian altitude is observed as $43° 17.0'$. The correction for refraction and parallax being $-00.9'$, determine the latitude of the place.

Ans. $67° 23.8'$ S.
or $26° 04.0'$ N.

5. The sun is observed on the P.V., morning and afternoon, the times by watch being $7^h 30^m$ and $4^h 14^m$. The sun's declination is $17° 21' 30''$. Compute the approximate latitude.

Ans. $37° 17'$.

6. The hour angle of Aldebaran (Dec. N. $16° 20' 15''$) when on the P.V. was found to be $4^h 35^m 19.5^s$. What was the latitude of the place of observation?

Ans. $39° 04' 03''$ N.

7. The altitudes of a star when it crossed the meridian and P.V. were

Determination of latitude

respectively 65° and 10° (corrected). Find the star's declination and latitude of the place.

Ans. Dec. 4° 58′ 39″ (S. in S. lat., or N. in N. lat.)
Lat. 29° 58′ 39″ N. or S.

8. At a place in south latitude the altitude of a star was observed at its upper and lower culminations, the altitude corrected for refraction at upper culmination being 60° 45′ 15″, and at lower culmination 10° 16′ 15″. Find the latitude of the place of observation and the declination of the star.

Ans. Lat. 35° 30′ 45″ S.
Dec. S. 64° 45′ 30″.

9. At a place in south latitude, the horizontal circle reading on β Hydri at its western elongation was 185° 47′ 35″, and that on θ Carinae at its eastern elongation was 137° 24′ 41″.

Declination of β Hydri: S. 77° 44′ 30″.
Declination of θ Carinae: S. 63° 56′ 36″.

Determine the latitude of the place.

Ans. Lat. 36° 25′ 02″ S.

10. The altitude of Regulus at 10^h 08^m 00^s sidereal time was 46° 52′ 32″ (fully corrected) in a place in S. latitude.

R.A. of Regulus: 10^h 03^m 17^s
Dec. of Regulus: N. 12° 26′ 00″.

What was the correct altitude when on the meridian?

Ans. 46° 53′ 26″.

11. On 9 March, at a place south of the equator in 140° 00′ 00″ E. longitude the following altitudes of α Virginis (Spica) were observed near its meridian passage and their times taken with a chronometer keeping L.M.T.:

Observed altitudes	L.M.T. (9 March)
57° 40′ 36″	2^h 02^m 18^s a.m.
44 34	05 54
48 40	10 50
50 10	15 58
49 30	22 10
46 40	27 00
42 35	31 02

The temperature was 12 °C and the barometer read 1023 mb.

R at U.T. 12^h on 8 March: 11^h 01^m 23.0^s
R.A. of Spica: 13^h 20^m 41.4^s
Dec. of Spica: S. 10° 43′ 00″

Find the latitude of the place.

Ans. 42° 52′ 54″ S.

Astronomy for surveyors

12. The declination of a star being S. 40°, what are the latitudes of the places where its meridian altitude will be 80°?

Ans. 50° or 30° S.

13. In south latitude two stars were observed on the meridian, one north and the other south of the zenith. The difference between their zenith distances was found to be 13′ 03.5″, and their declinations were S. 42° 44′ 04.6″ and S. 45° 38′ 37.5″ respectively. Find the latitude.

Ans. 44° 17′ 52.8″ S. or 44° 04′ 49.3″ S.

14. At $6^h 10^m 04^s$ p.m. L.Std.T. on 15 September, in longitude 151° 06′ 30″ E., the corrected altitude of σ Octantis was 34° 35′ 05″.

R.A. of σ Octantis: $20^h 32^m 20^s$
Dec. of σ Octantis: S. 89° 05′ 53″
R at U.T. 6^h on 15 September: $22^h 33^m 55.9^s$
Standard Meridian; 10^h E.

Compute the latitude of the observer.

Ans. Lat. 33° 54′ 18″ S.

15. On 6 March, the altitude of Polaris (fully corrected) was found to be 46° 27′ 27″ at L.Std.T. $7^h 43^m 35^s$ p.m. The longitude of the place is 37° 00′ 00″ W. (Standard Meridian 2^h W.). Find its latitude.

R.A. of Polaris: $1^h 57^m 06^s$
Dec. of Polaris: N. 89° 05′ 42″
R at U.T. 18^h on 6 March: $10^h 54^m 59.1^s$
$\sin 1″ = 0.00000485$.

Ans. 46° 03′ 36″ N.

16. The observatory at Stockholm is in Latitude 59° 20′ 33″ N., and that at the Cape of Good Hope in Latitude 33° 56′ 03.5″ S. The declination of Sirius is S. 16° 35′ 22″. Find the altitudes of Sirius when on the meridian at each of the two observatories.

Ans. At Stockholm, altitude = 14° 04′ 05″
At Cape of Good Hope, altitude = 72° 39′ 18.5″.

17. On 13 March, at a place south of the equator in longitude $9^h 15^m 00^s$ E., at 6 minutes before transit, the altitude of the sun's lower limb corrected for vertical collimation error was found to be 58° 04′ 20″, at which time clouds prevented further observation. The sun's declination at U.T. 0^h on 13 March is S. 3° 17.8′, decreasing 6.0′ in the next 6^h. Find the latitude of the place by reduction to the meridian, the sun's semi-diameter being 16′ 07″, its parallax 05″ and refraction 37″.

Ans. 34° 53′ 08″ S.

18. The altitudes of a star when it crosses the meridian and the prime vertical of a place are a and b. If ϕ is the latitude of the place, show that

$$\cot\phi = \tan a - \sec a . \sin b.$$

Determination of Latitude

19. The meridian altitude of Altair is 51° 55′ 45″, its declination being N. 8° 34′ 34″; and the meridian altitude of β Pavonis is 52° 54′ 32″, its north polar distance being 156° 36′ 18″. Find the latitude of the place of observation.

Ans.   29° 30′ 15.5″ S.

20. At a place south of the equator the meridian zenith distances of two stars A and B were observed, A to the south and B to the north. The observed difference of the zenith distances was found to be 19′ 21″. Find the latitude of the place of observation.

Declination of A: S. 49° 57′ 08″
Declination of B: S. 25° 23′ 31″.

Another observer, stationed some distance to the north, found the difference of the zenith distances of these stars to be exactly the same. Determine his latitude also.

Ans.   37° 50′ 00″ S. and 37° 30′ 39″ S.

21. The mean altitude reading from two observations of Polaris was 51° 39′ 34″, the mean readings of the alidade level, $E = 5.5$, $O = 6.5$, one division of level $= 15″$, mean chronometer time $19^h\ 09^m\ 54.8^s$, the chronometer being $3^m\ 24^s$ fast on L.M.T. The longitude of station was $0^h\ 02^m\ 09^s$ E. R at U.T. 18^h on the day of observation was $1^h\ 06^m\ 33.1^s$. Declination of Polaris N. 88° 45′ 50.8″; R.A. of Polaris, $1^h\ 22^m\ 26^s$. Barometer 1025 mb. Thermometer 5.5 °C. Compute the latitude of the place.*

Ans.   51° 23′ 34″ N.

22. At a place in south latitude, Longitude E. $11^h\ 34^m\ 20^s$, the following results were obtained from a circum-meridian observation of the sun on 26 January:

Sun	Face	L.Std.T.	Altitude
☌	L	$12^h\ 30^m\ 42.6^s$	67° 36′ 58″
☍	R	33 38.9	05 26
☌	R	35 53.6	38 33
☍	L	38 20.9	07 03
☌	L	40 11.3	39 10
☍	R	42 16.5	05 02
☌	R	43 55.2	36 19
☍	L	46 02.0	02 39

Barometer 948 mb; thermometer 29.4 °C.

Sun's semi-diameter 16′ 18″; sun's declination at U.T. 0^h on 26 January was S. 18° 57.0′, decreasing 03.7′ in the next 6^h. E at U.T. 0^h on 26 January was $11^h\ 47^m\ 36.2^s$, decreasing 3.4^s in the next 6^h. The Standard Meridian is 12^h E. Find the latitude.

Ans.   41° 33′ 59″ S.

* The actual observation in this example was done towards the end of the last century. Comparison of the R.A.s and decs. now and then shows how far Polaris has moved in nearly 100 years.

23. On 13 January, at a place in the northern hemisphere in longitude $5^h\ 59^m\ 38.0^s$ W., Betelgeuse was observed at a certain altitude east of the meridian at $20^h\ 00^m\ 36.8^s$ L.Std.T.; at $20^h\ 30^m\ 08.9^s$ on the same date δ Cassiopeiae was observed at the same altitude west of the meridian.

R.A. Betelgeuse:	$5^h\ 53^m\ 10.5^s$
Dec. Betelgeuse:	N. $7°\ 23'\ 59''$
R.A. δ Cassiopeiae:	$1^h\ 23^m\ 22.1^s$
Dec. δ Cassiopeiae:	N. $60°\ 02'\ 47''$

Standard Meridian is 6^h W.
R at U.T. 0^h on 14 January: $7^h\ 30^m\ 57.4^s$

Compute the latitude of the observer.

Ans. $29°\ 58'\ 44''$ N.

CHAPTER 12
The determination of local time by observation

Section 1
GENERAL

12.01 Introduction

In this chapter it is proposed to consider the principal methods available to the surveyor for the practical determination of the local mean or sidereal time by observation. Other methods have been devised, but those about to be described are the ones that have proved in practice to be the most convenient and satisfactory. Nearly all the ordinary time determinations of the surveyor are made by the second of the following methods, a convenient observation that may be carried out in daylight or at night and by which the time may be found quite accurately with ordinary instruments.

12.02 The need for astronomically determined time

In several of the methods for finding azimuth and latitude described in Chapters 10 and 11, we have seen that a knowledge of the time was necessary to enable us to calculate the hour angle of the body observed; this hour angle was required as one of the three elements needed to solve the astronomical triangle for Z or ϕ. We have usually started with the Local Standard Time of the observation as given by a clock or watch whose error and rate were found by comparison with radio time-signals. Application of the longitude of the standard meridian of the country to the correct L.Std.T. has then enabled us to find the U.T. of the observation. Alternatively, we could have used a clock, keeping U.T. checked by radio time-signals, as is done in ships. Our main object, then, has really been to find the U.T. of the observation so that we could enter the *Star Almanac* to find the quantities R, E and the sun's declination. U.T. + R has given us the G.H.A. Aries = the G.S.T., while U.T. + E has given us the G.H.A. sun; application of the longitude of the observing station to the G.S.T. has yielded L.S.T. and to the G.H.A. sun, the L.H.A. sun. From the equation L.S.T. − R.A. = L.H.A. we have been able to find the L.H.A. of a star, using the R.A. found from the *Star Almanac*.

In all these cases we have required a knowledge of the U.T. and the longitude

Astronomy for surveyors

of the observing station. Since we have been able to find our local time by means of clocks, radio time-signals and a knowledge of our longitude, why should we be concerned about finding it by observation? There are two good reasons:

(i) We might wish to find the longitude of the observing station; we have assumed so far that it has been known correctly. This matter will be dealt with in Chapter 13.
(ii) Our radio receiver may be put out of action, so depriving us of the means of checking clock error and rate by this means (cf. the pre-wireless days at sea).

Section 2

FIRST METHOD

LOCAL TIME BY MERIDIAN TRANSITS

12.03 Principle of method

We know that the local sidereal time at the instant when a star is on the meridian is measured by the R.A. of the star (L.S.T. = H.A. + R.A., but H.A. is zero for the body when on the meridian; ∴ L.S.T. = R.A. in these circumstances). Consequently, if we make the observation upon a star whose R.A. is known, by setting up a theodolite in the meridian and noting the time of transit across the vertical hair, we have clearly a very simple way of finding the sidereal time at that instant and thus of determining the error of a watch or clock. A knowledge of the direction of the meridian is, of course, necessary.

A similar observation may be made upon the sun, by noting the times of transit of the E. and W. limbs. The mean of these times will be the time of transit of the sun's centre, which takes place at *apparent* noon; the time of transit of the mean sun (i.e. 12 o'clock L.M.T.) may be found by application of the Equation of Time which is obtained from the quantity E in the *Star Almanac*. If only one limb be observed, then allowance must be made for the time taken by the sun's semi-diameter to cross the meridian.

12.04 Worked example: *Time determination by meridian transits*

On 1 December 1975, at a place in longitude $9^h\ 45^m$ E., the meridian times of transit of the E. and W. limbs of the sun across the vertical hair of a theodolite were taken with a watch supposed to keep the standard time of the meridian $9^h\ 30^m$ E. The observed times of transit being $11^h\ 32^m\ 27.5^s$ and $11^h\ 34^m\ 47.5^s$, determine the error of the watch.

Determination of local time by observation

At transit, Sun's L.H.A.	=	24^h	00^m	00^s
Longitude, E.	=	−9	45	00
Sun's G.H.A. at local transit	=	14	15	00
U.T. + E at 0^h U.T., 1 December 1975	=	12	11	19.0
Difference = time of local transit after 0^h U.T., 1 December	=	2	03	41.0
E at 0^h U.T. 1 December	=	12	11	19.0
Change in E in 2^h 03^m 45.7^s	=			−01.8
E at local transit	=	12	11	17.2
But U.T. + E (= G.H.A.) at local transit	=	14	15	00.0
∴ U.T. at local transit, 1 December	=	2	03	42.8
Standard meridian, E.	=	9	30	00
∴ L.Std.T. of local transit	=	11	33	42.8
But time of transit of Sun's centre, i.e. mean of observed times	=	11	33	37.5
∴ Watch is slow by				05.3

See **5.19**.

12.05 The effect of an error in the direction of the meridian

If the instrument be in accurate adjustment, but the direction of the meridian be in error, then the meridian set out will pass through the zenith of the observer, but not through the celestial pole. In Fig. 12.1, let ZC denote the erroneous meridian, making an angle that we will call e with the true meridian ZPM_S. Then a star will intersect the apparent meridian at S instead of the true meridian at S_1, and the time noted will be either too soon or too late, according as the meridian is wrongly marked out to the east or west of the true direction, the error being measured by the hour angle SPZ, which we will call t. In the diagram,

$$PZ = \omega = \text{co-latitude}$$

$$PS = p = \text{polar distance of star.}$$

Then, in the triangle PZS,

$$\cot p . \sin \omega = \cot e . \sin t + \cos \omega . \cos t \quad (vide\ (1.3a)).$$

Since e and t are both small, we may write, without appreciable error, t and e instead of $\sin t$ and $\sin e$ respectively, and may put $\cos t$ and $\cos e$ each $= 1$.

$$\therefore\ e(\cot p . \sin \omega - \cos \omega) = t$$

Astronomy for surveyors

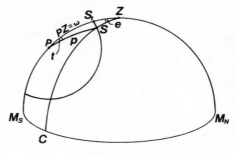

Fig. 12.1

$$\therefore\ t = e\frac{\sin(\omega - p)}{\sin p}.$$

Thus t will have its smallest value when p is nearly $= \omega$; that is to say, when the observed star makes its meridian transit near the zenith.

If in this equation $\omega = 60°$, or the latitude of the place is $30°$, and $p = 40°$, then if $e = 01'$ of arc, $t = 32''$ of arc or about 2 seconds of time. Thus, in this case, an error of 1 minute of arc in the direction of the meridian will make the time of transit wrong by two seconds.

It is clear, therefore, that the method requires the meridian to be very accurately set out, and the instrument must be in perfect adjustment if good results are to be obtained by this method.

In Fig. 12.1 we have illustrated the case where the star transits above the celestial pole. If the lower transit had been observed, then the angle t would be the supplement of the angle SPZ, and in this case the formula would become

$$t = e\frac{\sin(\omega + p)}{\sin p}.$$

Both are included in the general formula,

$$t = e.\frac{\sin(\text{zenith distance})}{\sin p} \quad \text{or} \quad e.\frac{\cos \text{alt.}}{\cos \text{dec.}} \tag{12.1}$$

which applies to all cases.

The error is thus very great if the polar distance of the star is small, and is least for those stars that transit near the zenith.

12.06 The effect of an error in the horizontality of the transverse axis

The direction of the meridian may be accurately set out with the telescope horizontal or nearly so, and yet, if the trunnion axis is not horizontal, the line of sight may depart considerably from the meridian at high altitudes. If

Determination of local time by observation

the angle made by the transverse axis with the horizontal be determined by means of the plate level, the necessary correction to the time of transit may be made as follows.

In Fig. 12.2, the meridian actually swept out by a telescope with the transverse axis slightly tilted is represented by $M_S S M_N$, M_N and M_S being the north and south points, and Z the zenith. The transit of the star is observed in consequence at a point S on this circle and the error in time is measured by the angle SPZ.

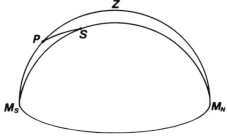

Fig. 12.2

In the triangle $M_N PS$,

$SP = p$ = polar distance of star

$M_N P = 180° - \phi$ = supplement of latitude

angle $PM_N S = e$ = error measured by plate level

angle $M_N PS = x$ = required error in time of transit

$\therefore \cot SP . \sin M_N P = \cot e . \sin x + \cos M_N P . \cos x.$ (vide (1.3a))

$\therefore$ treating x and e as small quantities,

$$\cot p . \sin \phi = \frac{x}{e} - \cos \phi$$

$$x = e \frac{\sin(\phi + p)}{\sin p} \quad \text{or} \quad e \frac{\text{sin altitude}}{\text{cos dec.}}.$$

This formula gives us the hour angle of the star at the moment of observation. Usually e, and therefore x, will be in seconds of arc, and x must then be divided by 15 to determine the error of the observed time of transit in seconds of time. Clearly the transit will be observed either too soon or too late according to the direction of tilt of the transverse axis.

If the star transits below the pole, x will be the supplement of the angle $M_N PS$, and we get

$$x = e\frac{\sin(\phi - p)}{\sin p}, \quad \text{which again} \quad = e\frac{\sin \text{alt.}}{\cos \text{dec.}}. \tag{12.2}$$

The error in time in this case increases with the altitude.

EXAMPLE. At a place in Latitude 30° S. the sidereal time of transit of a star across the meridian is observed to be $12^h\ 30^m\ 17.5^s$, the declination of the star being $-58°\ 30'$. The readings of the plate level, one division of which $= 20''$, are:

L	R
6.0	5.0
3.6	7.2
9.6	12.2
	9.6
	4) 2.6
	0.65

$$0.65 \times 20 = 13''$$

$$\therefore \text{ error in hour angle} = 13'' \times \frac{\sin 61°\ 30'}{\cos 58°\ 30'} = 21.8''.$$

This is equivalent to 1.5 seconds of time.

As the right-hand side of the axis is the higher, and the telescope is directed towards the south, the transit is therefore observed too soon by this amount and the corrected time of transit across the meridian is $12^h\ 30^m\ 19.0^s$.

12.07 Meridian transits on both sides of the zenith

A considerable improvement may be made in the accuracy of the method by taking observations of the times of transit of two stars, one on each side of the observer's zenith.

In Fig. 12.3 let Z denote the zenith, P the celestial pole, $M_N ZPM_S$ the direction of the true meridian, and CZD the direction of the meridian actually set out, the figure being drawn as though the celestial sphere were viewed from above. Suppose that the times of transit of two stars are observed, one at S_1 and the other on the opposite side of the zenith as at S_2. Then, since both stars move in the same direction, as shown by the arrows, if the observed time of transit of S_1 is later than it should be, owing to the faulty determination of the meridian, the time of transit of S_2 will be earlier. If the stars are well selected, it may be that the time errors of the two observations are equal and opposite, so that the mean of the two resultant clock errors will give

Determination of local time by observation

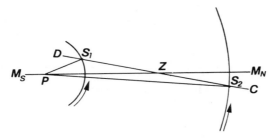

Fig. 12.3

the correct error in spite of the faulty setting out of the meridian. This will be the case if the hour angle S_1PZ is = the angle S_2PZ, for then one observation will be just as much too soon as the other is too late.

The conditions that this may be the case are readily obtained as follows:

Let angle $M_SZD = e$ = meridian error, and suppose that the hour angle $S_1PZ = S_2PZ = t$.

$$\omega = \text{co-latitude } PZ.$$

Then, from the triangles S_1PZ, S_2PZ,

$$\frac{\sin t}{\sin e} = \frac{\sin ZS_1}{\sin PS_1} = \frac{\sin ZS_2}{\sin PS_2}.$$

But, since the error e is small, we may write very approximately $PS_1 = \omega - ZS_1$ and $PS_2 = \omega + ZS_2$.

$$\therefore \frac{\sin(\omega - ZS_1)}{\sin ZS_1} = \frac{\sin(\omega + ZS_2)}{\sin ZS_2}$$

$$\therefore \sin\omega.\cot ZS_1 - \cos\omega = \sin\omega.\cot ZS_2 + \cos\omega$$

$$\therefore \cot ZS_1 - \cot ZS_2 = 2\cot\omega. \tag{12.3}$$

This, then, is the condition that has to be satisfied by the zenith distances of the two stars if the observations are to be so balanced that by taking the mean of the two we eliminate, or nearly so, the error due to a faulty setting-out of the meridian.

Table 12.1, based upon the above formula, gives the proper zenith distance of the star on the opposite side of the zenith to the pole, corresponding to different zenith distances of the other observed star, for different latitudes.

The advantage of selecting the two stars in this way may be illustrated by a computed example. Suppose that the place of observation is in latitude 30°, and that the polar distance of the star observed on the same side of the zenith as the pole is 40°, so that its zenith distance is about 20°. Suppose, further, that the marked meridian is as much as 1° in error.

Astronomy for surveyors

Table 12.1

Zenith distance of star on same side as pole	Zenith distance of star on opposite side of zenith to the pole							
	Lat. 10°	Lat. 20°	Lat. 30°	Lat. 40°	Lat. 50°	Lat. 60°	Lat. 70°	Lat. 80°
5°	5° 09′	5° 20′	5° 34′	5° 51′	6° 18′	7° 09′	9° 34′	85° 00′
10	10 39	11 26	12 29	14 30	16 55	24 22	80 00	–
20	22 40	26 21	32 08	43 05	70 00	–	–	–
30	35 56	44 53	60 00	86 55	–	–	–	–
40	50 00	65 07	87 53	–	–	–	–	–
50	64 04	83 39	–	–	–	–	–	–
60	77 20	–	–	–	–	–	–	–
70	89 21	–	–	–	–	–	–	–

Computing with these data the spherical triangle *SPZ* of Fig. 12.1, it may be shown that the hour angle *SPZ* is $2^m\ 04.8^s$. In other words, the observed transit will take place too soon by this amount.

Now, according to the table, the star observed on the opposite side of the zenith should have a zenith distance of 32° 08′. Suppose it actually has a zenith distance of 32°, equivalent to a polar distance of 92°. Then, computing in the same way the hour angle of this star when on the faulty meridian, we find that its observed transit will be too late by $2^m\ 04^s$.

Thus from one observation the clock would be set too fast by $2^m\ 04^s$, and from the other it would be set too slow by about the same amount, and the mean of the two observations would give the time correct to the nearest second, in spite of the fact that the direction of the meridian is 1° in error.

If, however, the zenith distances of the two stars are not balanced in the way indicated, the accuracy of the mean result is nothing like so great. If, for example, the two zenith distances were the same, the star observed on the opposite side of the zenith to the pole having a zenith distance of 20°, or a polar distance of 80°, then, on computing the spherical triangle, it would be found that the observed transit of this star is too late by $1^m\ 24^s$, so that the mean of the two observations is then in error to the extent of about 20 seconds.

It is obvious that

(a) the two pointings must be made on one face of the instrument;
(b) the instrument must be in good adjustment and be very carefully levelled.

Determination of local time by observation

Section 3

SECOND METHOD

LOCAL TIME BY EX-MERIDIAN ALTITUDES OF THE SUN OR STARS

12.08 Principle of the method

This is, as a rule, the most convenient and suitable method for the determination of time by the surveyor. It consists in the measurement of the altitude of sun or star when out of the meridian, at the same instant noting the clock time. Then, from a knowledge of the latitude of the place and the declination of the body observed we may compute the local hour angle of the body at the instant of observation, and hence the correct local time and the error of the clock.

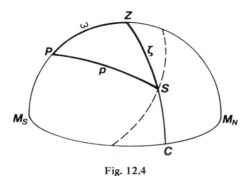

Fig. 12.4

The most favourable time for making such an observation will be when the altitude of the celestial body is changing most rapidly, and this will be the case when it is near the prime vertical. This position has also other advantages, as we shall see in the course of the discussion.

As an altitude has to be measured, refraction must be allowed for, and as there is considerable uncertainty about this at low altitudes, the star observed should have an altitude of at least $15°$.

The method involves the solution of the same spherical triangle as we have discussed in connection with ex-meridian altitude observations for azimuth.

From Fig. 12.4, with the usual notation and by reference to **1.06**, we can deduce the following formula for finding the hour angle, t:

$$\cos t = \frac{\cos \zeta - \cos \omega . \cos p}{\sin \omega . \sin p} \qquad (12.4)$$

Astronomy for surveyors

$$= \frac{\sin h - \sin\phi.\sin\delta}{\cos\phi.\cos\delta}. \qquad (12.4a)$$

In (12.4a) ϕ and δ may have opposite signs; if positive signs refer to the northern hemisphere in each case, then the north celestial pole is the reference pole, whether it be elevated or not. If the azimuth as well as the time is required from a single observation, the star must be observed on the intersection of the cross-hairs, or the sun in opposite quadrants of the graticule, as in **10.19**. The formula

$$\cos Z = \frac{\cos p - \cos\omega.\cos\zeta}{\sin\omega.\sin\zeta}$$

$$= \sin\delta.\sec h.\sec\phi - \tan h.\tan\phi \qquad (10.2)$$

may then be used to find Z, and hence the azimuth.

12.09 Having computed the hour angle, to find the local time of the observation

In the case of a star the calculated hour angle expressed in time measures the interval of sidereal time after or before the time of upper meridian transit, according as to whether the star is observed on the west or east of the meridian. But the R.A. of the star is equal to the sidereal time when the star is on the meridian at upper transit. Therefore, the sidereal time at the moment of observation is obtained by adding the value of t to the R.A. of the star if the latter was observed west of the meridian, or subtracting it from the R.A. if east. The sidereal time may be converted into mean time by the method of **5.14**; a knowledge of the longitude of the observing station is necessary for this.

If the sun has been observed, the computed value of the angle t is the Local Hour Angle of the apparent sun. We can obtain the L.M.T. from this by using the formula:

$$\text{L.H.A. sun} - E = \text{L.M.T.}$$

The quantity E varies, and if we want the L.M.T. correct to within one-tenth of a second we must know the U.T. (or L.Std.T.) of the observation to within 5 minutes, since the maximum rate of change in E amounts to 7.5 seconds in 6 hours at one particular time of the year (December).

Suppose, for example, that at a certain place on 4 December 1975 the L.H.A. of the sun was calculated from an afternoon observation to be $5^h\ 25^m\ 16.3^s$; the approximate U.T. of the observation (to within 5 minutes) was $8^h\ 27^m$. It is required to find the L.M.T. of the observation.

Approximate U.T., 4 December 1975 = $8^h\ 27^m\ 00^s$

Determination of local time by observation

E at U.T. 6^h, 4 December 1975 from *Star Almanac*	=	12 10	04.3
Change in E in $2^h\ 27^m$	=		−02.5
E at U.T. $8^h\ 27^m$, 4 December 1975	=	12 10	01.8
L.H.A. sun (from observation)	=	5 25	16.3
E	=	12 10	01.8
∴ L.H.A. sun $-E$ = L.M.T., 4 December 1975	=	17 15	14.5

Note. 24^h was added to L.H.A. sun because this was less than E.

12.10 Best position of celestial body for observation

In the case of azimuth observations, where we knew latitude and declination and measured the altitude, which gave us the three sides of the astronomical triangle, we considered the effect that small errors in these quantities had on the computed azimuth. We shall now use the differential calculus to see how small errors in the same quantities affect the computed hour angle.

(a) *The effect of an error in the assumed latitude*

In the spherical triangle, from equation (1.2), we can derive the formula

$$\cos \zeta = \cos \omega . \cos p + \sin \omega . \sin p . \cos t$$

or

$$\sin h = \sin \phi . \sin \delta + \cos \phi . \cos \delta . \cos t.$$

Differentiating with respect to ϕ, keeping h and δ constant, we get

$$0 = \cos \phi . \sin \delta + \cos \delta . \{(-\sin \phi) . \cos t + \cos \phi . (-\sin t . dt/d\phi)\}$$

or

$$\frac{dt}{d\phi} = \frac{\cos \phi . \sin \delta - \sin \phi . \cos \delta . \cos t}{\cos \phi . \cos \delta . \sin t}. \tag{12.5}$$

Simplification of the right-hand side of this equation leads to $dt/d\phi = \cot Z . \sec \phi$. This shows that $dt/d\phi$ will be large, i.e. a small change or error, $d\phi$, in the latitude will produce a large error, dt, in the hour angle, if Z is near zero or if ϕ is near $90°$. That is to say, a small error in the latitude will produce a very large error in the time if the body is observed near to the meridian, of if the observation is made in high latitudes near to either terrestial pole.

On the other hand, if Z is $90°$—i.e. if the body is observed on the prime vertical—$dt/d\phi$ is zero, and an error in latitude makes no difference. In this case the astronomical triangle is right-angled at Z, and we can get a relation between p, h and t that does not involve ϕ at all, so that a knowledge of the latitude is unnecessary. If the observation is made to the body when it is very

Astronomy for surveyors

near to the prime vertical, therefore, an error in latitude will produce very little effect on the time determination.

Table 12.2, based upon formula (12.5), gives the error in time corresponding to an error in 01' in the latitude for different zenith angles.

Table 12.2 Error in time correspondong to 01' error in latitude*

Zenith angle of observed body	Latitude of place				
	0°	30°	40°	50°	60°
	seconds	seconds	seconds	seconds	seconds
45°	4.0	4.6	5.2	6.2	8.0
60	2.3	2.6	3.0	3.5	4.5
80	0.7	0.8	0.9	1.1	1.4
90	0.0	0.0	0.0	0.0	0.0

* If the word *declination* be substituted for *latitude*, the same table will give the error in time due to an error of 01' in the declination, the first column representing not the zenith angle, but the angle ZSP in Fig. 12.4.

(b) *The effect of an error in the measured altitude*

By differentiating the second equation in **12.10**(a) with respect to h, keeping ϕ and δ constant, we get

$$\cos h = 0 + \cos\phi.\cos\delta\,(-\sin t.dt/dh)$$

$$\frac{dt}{dh} = -\frac{\cos h}{\cos\phi.\cos\delta.\sin t}.$$

But $\cos\delta = \sin p$, and from the sine rule for the astronomical triangle,

$$\sin p = \frac{\sin Z.\sin \zeta}{\sin t} = \frac{\sin Z.\cos h}{\sin t}$$

Substituting this, we get

$$\frac{dt}{dh} = -\frac{\cos h.\sin t}{\cos\phi.\sin Z.\cos h.\sin t}$$

$$= -\frac{1}{\cos\phi.\sin Z} \qquad (12.6)$$

$$= -\sec\phi.\text{cosec}\,Z.$$

From this we see that a small error in h will produce a large error in t if ϕ is large and Z is small. Thus, again, an error in observation of the altitude has the least effect when the body is observed on or near the prime vertical, and the most favourable place for making the observation is at the equator.

Table 12.3 deserves careful consideration, as it shows the degree of precision

Determination of local time by observation

Table 12.3 Error in the determination owing to an error of 01′ in the measured altitude, with different azimuths of the observed body

Zenith angle of observed body	Latitude of place				
	0°	30°	40°	50°	60°
	seconds	seconds	seconds	seconds	seconds
45°	5.6	6.4	7.3	8.7	11.2
60	4.6	5.3	6.0	7.1	9.1
80	4.1	4.7	5.3	6.3	8.1
90	4.0	4.6	5.2	6.2	8.0

with which altitudes must be measured if the time is to be determined accurately. Under the most favourable possible conditions, an error of $15''$ of arc will cause an error of one second in the time, and it may produce an error of two seconds or even more if the body is not observed in a very good position.

(c) *Effect of an error in the declination of the sun caused by a defective knowledge of longitude or local time*

With star observations the *Star Almanac* gives us the declination of the star with all the precision that is required, even if the date only of the observation is known, but with sun observations more than just the date is needed. The surveyor requires to know both his longitude and the approximate L.M.T., or else the approximate U.T. From the formula

$$\sin h = \sin \phi . \sin \delta + \cos \phi . \cos \delta . \cos t$$

it appears that the relation between an error in δ and an error in t will be of precisely the same nature as the relation between an error in ϕ and an error in t, and on differentiating this equation with respect to δ, keeping ϕ and h constant, we will get

$$dt/d\delta = \cot S . \sec \delta. \qquad (12.7)$$

Thus the table already given, showing the error in time caused by 01′ error in latitude, also gives the error in time caused by 01′ error in declination, provided that the first column is taken as representing the parallactic angle S instead of the zenith angle Z.

From (12.7) we see that a small error in the declination will cause a large error in the time if S is small and δ is large. Hence, to keep the error in t small if δ is doubtfully known, S should be kept large and bodies near the celestial equator should be observed. S is a maximum when the body is on the prime vertical, so observations near the prime vertical are indicated, which will also help to keep down errors due to uncertain knowledge of ϕ and imperfect measurement of h.

We have already seen that the maximum rate of variation of the declination

Astronomy for surveyors

of the sun is a little less than 1' per hour; therefore, to get the declination of the sun to the nearest minute it is sufficient to know the time to the nearest hour. But 1 hour of time corresponds to 15° of longitude, so that it is seldom that the surveyor will not know his longitude sufficiently well for this purpose.

It will be seen from the table that, in order to determine the time to the nearest second, it will be necessary to know the declination within only about one-fifth of a minute of arc under almost the worst conditions of observation considered in the table. For this it will usually be sufficient to know the local time within a quarter of an hour.

If the local time is not known with sufficient accuracy, its value must be assumed for the purpose of finding the approximate declination. This is then used in a preliminary calculation to determine the time. The calculation is then made over again, using the approximate local time so found in order to get a more accurate value of the sun's declination, which in turn is used in the computation to obtain a more accurate determination of the local mean time.

12.11   Observing procedure

The observing procedure is simple for this method. If the direction of the true meridian is not known it may be found fairly closely by a magnetic compass observation corrected for declination. The theodolite is pointed in this direction and the horizontal circle set to zero. Then, with the telescope azimuth near 90° or 270°, a search in altitude is made to locate a suitable star which will, of course, be near the prime vertical. Alternatively, if local time is known approximately, a programme may be made and the times when suitable stars are on the P.V., and their altitudes, may be calculated from the astronomical triangle which is right-angled at Z, with p and ω known.

When a suitable star has been got into the field of the telescope, say on L.F., the observation is done as follows:

(i) Preset the horizontal cross-hair a little ahead of the star, so that it will intersect it near the centre of the field; at the moment when the star makes the intersection, note the time by one of the methods described in Chapter 6, preferably by **6.03**(d). Making sure the vertical circle (alidade) bubble is central, read and note the altitude. (But see last paragraph of **11.02**.)

(ii) Without delay, change to R.F. and point again to the star, repeating the procedure of (i) above.

(iii) Note the barometer and thermometer readings.

The mean of the altitudes is taken as being the correct altitude at the mean of the recorded times, provided there has been minimum delay between faces (see **7.05**).

Determination of local time by observation

It is a great improvement in accuracy to take one set of observations upon a star in the east and another corresponding set, under conditions as similar as possible, upon a star in the west. The averaging of two such sets of observations tends to eliminate certain classes of errors, and this should always be done where the highest accuracy is sought. If, for example, the refraction assumed is too great, the corrected altitude will be too low, and the computed time will be too early for a star in the east, while it will be correspondingly too late for a star in the west. If the two errors are about equal, as will be the case if the east and west stars make about the same horizontal angle with the meridian, and are observed at about the same altitude, then the average of the two sets of results will be correct. Similarly, the effects of any systematic error in the measurement of altitude are eliminated by pairing sets of observations in this way. The same applies to ex-meridian observations for azimuth.

By making several pairs of observations on east and west stars and taking the mean of all, the accuracy will be improved still further. If balanced east–west stars are used, the observations may all be done on one face; four quick, timed altitudes can easily be measured in 2-3 minutes and a mean taken. As a check, the altitudes can be plotted against the times to confirm the straight-line path of a star, or discover a "wild" pointing.

If the sun is used for the observation, the procedure is similar to that for a star, except that the upper (lower) limb is observed on the first face and the lower (upper) on the second, the times of tangency being noted. If conditions are such that the sun does not cross the P.V., or crosses it at an altitude of less than $15°$, it should be observed as far from the meridian as possible, consistent with an altitude not less than $15°$.

12.12 Worked examples: *Determination of local time and clock error by observation of altitudes of east and west stars*

Date: 1974 August 19
Place: Arts Block, Pillar F,
   University of Otago,
   Dunedin, N.Z.
Latitude: $45° 52' 15''$ S.
Longitude: $11^h 22^m 03.9^s$ E.
Observer: J. B. M.
Booker: M. D. Body
Barometer: 1021 mb
Thermometer: $5 °C$

Instrument: Wild T2 No. 76015
Objects observed:
East star: *S.A.* No. 632 Fomalhaut
R.A.: $22^h 56^m 17.6^s$; Dec.: S. $29° 45' 09''$
West star: *S.A.* No. 366 π Hydrae
R.A.: $14^h 04^m 56.0^s$; Dec.: S. $26° 33' 46''$
Chronometer: $12^h 00^m 12.7^s$ fast on
   U.T.C.
Standard Meridian: 12^h E.
DUT1: $+0.1^s$

Astronomy for surveyors

(a) *East star*

Face	Object	Vertical circle	Altitude	Chronometer	Stop-watch	Chronometer time
L	Star	55° 10′ 55″	34° 49′ 05″	$20^h\ 50^m\ 00^s$	$18^m\ 09.3^s$	$21^h\ 08^m\ 09.3^s$
L	Star	54 56 27	35 03 33		19 34.2	21 09 34.2
L	Star	54 47 29	35 12 31		20 25.5	21 10 25.5
L	Star	54 37 34	35 22 26		21 23.2	21 11 23.2

Mean observed altitude	=	35° 06′ 53.8″
Refraction correction	=	−01 25.7
Corrected altitude, $h=$		35 05 28
Mean observed chronometer time	=	$21^h\ 09^m\ 53.1^s$

Formula:

$$\cos t = \frac{\sin h - \sin\phi.\sin\delta}{\cos\phi.\cos\delta}$$

where $h = 35°\ 05′\ 28″$, $\phi = 45°\ 52′\ 15″$, $\delta = 29°\ 45′\ 09″$ (ϕ and δ are both south, so both may be taken as positive).

On computing, $t =$		$-4^h\ 35^m\ 10.0^s$ (E. ∴ negative)
R.A. Fomalhaut	=	22 56 17.6
t + R.A. = L.S.T.	=	18 21 07.6
Longitude, E.	=	11 22 03.9
G.S.T.	=	6 59 03.7
G.S.T. + 24^h	=	30 59 03.7
G.S.T. at U.T. 6^h, 19 August 1974	=	27 48 50.4
Sidereal interval after U.T. 6^h	=	3 10 13.3
ΔR	=	−31.2
Solar interval after U.T. 6^h	=	3 09 42.1
i.e. U.T.1, 19 August	=	9 09 42.1
DUT1	=	0.1
U.T.C., 19 August	=	9 09 42.0
Standard Meridian, E.	=	12 00 00
L.Std.T. from observation	=	21 09 42.0
Mean chronometer time	=	21 09 53.1
∴ Chronometer fast on L.Std.T.	=	11.1^s

Determination of local time by observation

(b) *West star*

Face	Object	Vertical circle	Altitude	Chronometer	Stop-watch	Chronometer time
L	Star	54° 39′ 58″	35° 20′ 02″	$20^h\ 50^m\ 00^s$	$26^m\ 13.2^s$	$21^h\ 16^m\ 13.2^s$
L	Star	54 57 05	35 02 55		27 52.0	21 17 52.0
L	Star	55 12 02	34 47 58		29 17.8	21 19 17.8
L	Star	55 22 56	34 37 04		30 20.3	21 20 20.3

Mean observed altitude	=	34° 56′ 59.8″
Refraction correction	=	−01 25.7
Corrected altitude,	$h =$	34 55 34
Mean observed chronometer time	=	$21^h\ 18^m\ 25.8^s$

Using the formula in (a) above, with $h = 34°\ 55′\ 34″$, $\phi = 45°\ 52′\ 15″$, $\delta = 26°\ 33′\ 46″$ (ϕ and δ both south, or +):

On computing,	$t =$	$+4^h\ 24^m\ 42.5^s$ (west ∴ +)
R.A. π Hydrae	=	14 04 56.0
t + R.A. = L.S.T.	=	18 29 38.5
Longitude, E.	=	11 22 03.9
G.S.T.	=	7 07 34.6
G.S.T. + 24^h	=	31 07 34.6
G.S.T. at U.T. 6^h, 19 August	=	27 48 50.4
Sidereal interval after U.T. 6^h, 19 August	=	3 18 44.2
ΔR	=	−32.6
Solar interval after U.T. 6^h, 19 August	=	3 18 11.6
i.e. U.T.1, 19 August	=	9 18 11.6
DUT1	=	0.1
U.T.C., 19 August	=	9 18 11.5
Standard Meridian, E.	=	12 00 00
L.Std.T., from observation	=	21 18 11.5
Mean chronometer time of observation	=	21 18 25.8
∴ Chronometer fast on L.Std.T.	=	14.3
From (a), east star, chronometer fast	=	11.1
Mean value, chronometer fast	=	12.7^s

(From comparison with radio time-signals from VNG Australia, the chronometer was 12.7^s fast on L.Std.T.)

The two stars used in the above east-west balanced pair were both observed on left face only, but the mean value of the two results is free of instrumental error. Both stars were observed close to the prime vertical, at similar altitudes. A plot of the altitudes against the times shows a straight line for each star, indicating that there were no "wild" pointings.

If, for any reason, only one star is observed, an equal number of L.F. and R.F. pointings should be made to it to cancel instrumental errors. A balanced pair all done on one face, as in (a) and (b) above, should be observed as near together in time as possible, obviously on the same evening and with the same instrument.

Section 4

THIRD METHOD

LOCAL TIME BY NOTING THE INSTANTS WHEN STARS ARE AT THE SAME ALTITUDE

12.13 Using a theodolite

In 1884, S. C. Chandler, at the Harvard College Observatory, U.S.A., devised a form of instrument in which the telescope was fixed at a constant angle with the vertical, so that the line of sight traced out a horizontal circle on the celestial sphere, and observations for the determination of latitude and other purposes were made by noting the times of transit of stars across the fixed horizontal circle. The instrument was named an "almucantar", and it proved to be capable of very good work. The same principle may be readily applied with an ordinary theodolite, and experience has shown that quite accurate determinations of time are possible in this way.

Any horizontal circle may be used for the observations, but the most convenient is the one that passes through the pole of the observer. This has been named the "co-latitude circle", its zenith distance being everywhere equal to the co-latitude. The formulae for reduction then become very simple. The method consists in observing the times of transit of a series of east and west stars, somewhere near the prime vertical, across the horizontal wire of a telescope that is set to an altitude equal to that of the pole. Allowance must be made for refraction, and therefore the telescope is actually set so that its altitude as read off on the vertical circle is equal to the latitude of the place plus refraction.

Determination of local time by observation

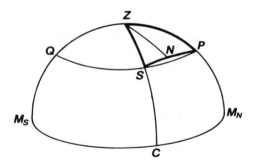

Fig. 12.5

In Fig. 12.5, Z denotes the zenith, P the celestial pole, M_N and M_S the north and south points, PSQ the co-latitude circle. Let S denote the position of a star, somewhere near the prime vertical, as it crosses the co-latitude circle.

Let $ZP = \omega =$ co-latitude.

$PS = p =$ star's polar distance, measured, of course, along the great circle arc PNS and not along the small circle PSQ.

angle $SPZ = t =$ hour angle of star

Angle $SZP = Z =$ zenith angle of star measured from elevated pole. Then, since $ZS = ZP$, ZSP is an isosceles triangle, and if ZN be drawn perpendicular to the great circle arc joining S and P, it will divide ZSP into two equal right-angled triangles.

From the triangle ZNP,

$$\cos NPZ = \tan PN . \cot ZP$$

$$\therefore \cos t = \tan \tfrac{1}{2} p . \cot \omega = \tan \tfrac{1}{2} p . \tan \phi, \qquad (12.8)$$

if ϕ is the latitude of the place.

To determine the azimuth at which a star will cross the co-latitude circle, from the same triangle:

$$\cos ZP = \cot NZP . \cot NPZ$$

$$\therefore \cos \omega = \cot t . \cot \tfrac{1}{2} Z,$$

or

$$\cot \tfrac{1}{2} Z = \sin \phi . \tan t. \qquad (12.9)$$

Formula (12.8) enables the time of transit to be computed, and formula (12.9) gives the azimuth if required.

Astronomy for surveyors

If an observation on one star in the east is balanced by a corresponding observation on a star in the west, of somewhere about the same declination, then the mean of the two time-observations will give a correct result, even if the co-latitude circle is considerably out. If, for instance, the co-latitude circle is set out too low, the observed time of transit in the east will be too soon, but that in the west will be too late, and if there is not much difference in the declinations of the stars the time of transit will be just as much too soon in the one case as it is too late in the other. Thus, by averaging the two results any small error in the setting out of the co-latitude circle is practically eliminated, and it is not necessary, therefore, in order to apply the method that the latitude of the place should be known with precision. An approximate latitude will suffice.

For precisely the same reasons as have been investigated when dealing with extra-meridian observations for time, slight errors in latitude, declination, and altitude will have least effect upon the result when the stars observed are near the prime vertical. The stars should be selected from a zone of about $20°$ on each side of the prime vertical.

EXAMPLE. On 13 July 1975, in Latitude $45°\ 52'\ 15''$ S., the transit of star No. 211 724 in the Smithsonian catalogue was observed in the east across the co-latitude circle at $21^h\ 27^m\ 57.0^s$ on a chronometer keeping approximate L.Std.T. The transit of γ Hydrae, No. 350 in the *Star Almanac*, was similarly observed in the west at $21^h\ 41^m\ 04.5^s$. Determine the error of the chronometer on L.Std.T. The Standard Meridian is 12^h E. and DUT1 $= +0.2^s$.

	S.A.O. No. 211 724	S.A. No. 350
R.A.	$19^h\ 58^m\ 48.5^s$	$13^h\ 17^m\ 36.3^s$
Declination	S. $33°\ 45'\ 59''$	S. $23°\ 02'\ 43''$
p	56 14 01	66 57 17
$\tfrac{1}{2}p$	28 07 00.5	33 28 38.5
ϕ	45 52 15	45 52 15
t, computed from $\tfrac{1}{2}p$ and ϕ	$-3^h\ 46^m\ 18.4^s$	$3^h\ 08^m\ 04.9^s$
R.A.	19 58 48.5	13 17 36.3
L.S.T. $= t +$ R.A.	16 12 30.1	16 25 41.2
Longitude, E.	11 22 03.9	11 22 03.9
G.S.T.	4 50 26.2	5 03 37.3
Add 24^h	28 50 26.2	29 03 37.3
G.S.T. at U.T. 6^h, 13 July 1975	25 22 00.4	25 22 00.4
Sidereal interval after U.T. 6^h	3 28 25.8	3 41 36.9
ΔR	-34.1	-36.3

Determination of local time by observation

Solar interval after U.T. 6^h	3 27 51.7	3 41 00.6
i.e. U.T.1, 13 July	9 27 51.7	9 41 00.6
DUT1	0.2	0.2
U.T.C., 13 July	9 27 51.5	9 41 00.4
Standard Meridian, E.	12 00 00	12 00 00
L.Std.T., 13 July	21 27 51.5	21 41 00.4
Chronometer time	21 27 57.0	21 41 04.5
Chronometer time is fast by	05.5^s	04.1^s
Mean amount, chronometer fast	04.8s	

(From comparison with time-signals from VNG, Australia, the chronometer was, in fact, 04.8^s fast at the time.)

The method is much the same as that described in Section 6 of Chapter 11 for determining latitude. However, for time determination the inclination of the telescope is pre-set to the altitude of the pole plus refraction, *and it must be maintained precisely at this inclination* during the observation of several pairs of east–west stars, preferably without making any intermediate slight adjustments to alidade bubble and vertical-circle slow motion screw (see precautions in **11.17**). The observations are done on one face only, of course. The instrument must be very solidly set up and levelled with great care (but see **11.17**, near the end).

An altitude different from that of the pole may be used, but the advantage of simplicity in the computations is then lost.

12.14 Using special equipment
(a) *The 60° prismatic astrolabe*

There is no little difficulty in maintaining the telescope of a theodolite precisely at a fixed inclination to the horizontal during rotation about the vertical axis, and another instrument, called the Prismatic Astrolabe, was devised by Claude and Driencourt of France at the beginning of the century. By incorporating a trough of mercury and a prism system in the instrument, they were able to bring two images of the same star into horizontal coincidence at the moment when the star was at altitude 60°.

Fig. 12.6 shows diagrammatically the construction of the 60° astrolabe. A 60° equilateral prism is mounted as shown in front of the objective of a horizontal telescope which has attached to it a mercury trough. The solid lines represent the paths of the parallel rays from a star at 60° altitude. The rays which are reflected down the telescope from the lower surface of the prism form an image of the star at c; the rays which are reflected from the surface of the mercury strike the upper surface of the prism where they suffer a further reflection down the telescope to form an image at c. The two images are there-

Astronomy for surveyors

fore coincident. Now consider the case when the star's altitude is slightly less than 60°. The parallel rays from it are represented by the dotted lines; those passing directly into the prism and reflected from its lower surface form an image at *b*, while those falling on the surface of the mercury form another image at *a* after the double reflection. It is easily seen that, as the star's altitude approaches closer and closer to 60°, the images *a* and *b* move closer together until they coincide when the star is at 60°; as the star's altitude increases above 60°, the images will separate, *a* moving upwards and *b* downwards.

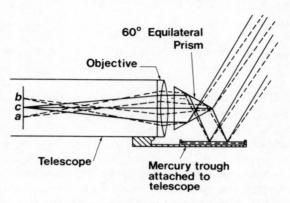

Fig. 12.6   Diagram of 60° prismatic astrolabe

Fig. 12.7 shows the paths of the images in the field of view of the telescope. Normally the two images would coincide where their paths intersect, but by rotating the prism very slightly about the optic axis of the telescope, the two images can be kept separate as they pass each other; this makes it easier to estimate the instant when they are at the same elevation. The clock time is noted when the two images coincide. No level bubbles are involved, since the

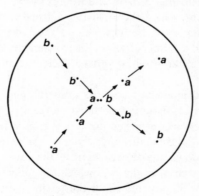

Fig. 12.7

surface of the mercury remains truly level in any position of the telescope in azimuth.

Wild Heerbrugg Ltd of Switzerland manufacture a prism-cum-mercury-trough attachment which may be fitted to their Universal Theodolite T2 and a larger model for their T3 Geodetic Theodolite; they also publish a brochure describing in detail the adjustment and use of the prismatic astrolabe attachment.

If the astrolabe is to be used, a star programme must be prepared beforehand; suitable tables to facilitate this have been published by the Institute of Geographical Exploration, Cambridge, Massachusetts. Wild Heerbrugg Ltd state in their brochure that "an experienced observer, using a series of 8 stars (2 in each quadrant), may determine the latitude and chronometer error to within $2-3''$ and 0.2^s respectively. By using a series of 30 stars fairly distributed over the horizon, an accuracy of within 1 second of arc for the latitude, and within 0.1 second of time for the chronometer error is obtainable." The timing should be done with the aid of a chronograph or other recording clock.

The firm of Carl Zeiss supply a prism attachment for their Ni2 automatic level which eliminates the necessity for using a trough of mercury. It is stated that, with this accessory, astronomic-geodetic positions and the time can be fixed with an accuracy of $\pm 1''$ mean square error by using a suitable programme of star observations. The instrument is simple to use and has considerable advantages over other instruments using an artificial mercury horizon (see Fig. 12.8). A suitable observing programme may be obtained by making use of a computer, and accuracy close to first order is obtainable from 16-24 quadrantal stars. An article entitled "Testing the Zeiss Ni2 Astrolabe for astronomic-geodetic determinations of latitude and longitude" by Albrecht Bartels was published in *Zeitschrift für Vermessungswesen*, 86th edition, 1961, vol. 10.

(b) *The 45° prismatic astrolabe*

Captain Baker of the British Royal Navy devised a more convenient 45° prismatic astrolabe to allow a wider choice of stars for observation. A number of improvements over the earlier 60° instrument were also incorporated, notably a duplicating prism which divides the direct rays into two pencils to give two side-by-side images between which passes the image reflected from the mercury surface; and a series of deflecting prisms which can be placed in the paths of the reflected rays and which allow observations to be made at $1\frac{1}{2}'$, $4\frac{1}{2}'$ and $7\frac{1}{2}'$ altitude greater and less than 45°. It is doubtful if even a second-hand Baker astrolabe could be obtained nowadays. (But note that the graticule in the Zeiss instrument permits up to 20 cross-hair transits of one star to be timed; the angular disposition of the cross-hairs is accurately known in relation to the centre of the field.)

(c) *Computations arising from astrolabe observations*

Observations with the prismatic astrolabe allow both clock error and latitude to be determined if the longitude is known. The computations, being repetitive,

Astronomy for surveyors

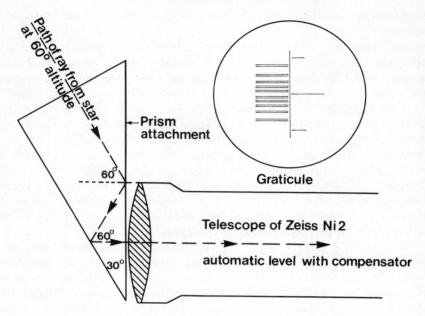

Fig. 12.8 Zeiss 60° prismatic astrolabe attachment on Ni2 level

are most conveniently done on a suitable electronic computer which can also be easily programmed to produce a least-squares solution from observations of any desired number of suitable stars. (See **14.08**.)

(d) *Further references*

Detailed descriptions of instruments and procedures used in modern geodetic astronomy are to be found in *Field and Geodetic Astronomy* (Robbins), 1976, Ministry of Defence, United Kingdom; and in *Geodesy* (Bomford), 1975, Clarendon Press, Oxford.

Section 5

SUNDIALS

12.15 Sundials

(a) *General*

Whilst the sundial does not provide the surveyor with a means of determining local time with anything like the precision obtainable by the methods that have been described, it enables the time to be fixed accurately enough for the ordinary purposes of life. Sundials were commonly used for checking clocks in country

Determination of local time by observation

places long before the advent of domestic radio, and although they are now somewhat rare, descriptions of the main types are included here as a matter of interest, and also as an aid to any surveyor who may be asked to design a dial for ornamental purposes.

When a sundial is illuminated by the direct light of the sun the shadow of a straight line or sharp straight edge is thrown upon a plane containing a graduated circle so marked that the apparent solar time is indicated by the reading at the place where the shadow intersects the circle. The plane containing the graduated circle may be either horizontal, vertical, or inclined. The straight edge, the shadow of which is thrown upon the circle, is always set up so as to be parallel to the earth's axis. It is called the *stile* or *gnomon* of the dial. When the graduated circle or "plane of the dial" is horizontal we have what is known as a *horizontal dial*, and as this is the most common form we will consider it first.

(b) *The horizontal dial*

In Fig. 12.9, let *MBLA* represent the plane of the dial, which we may suppose to be extended indefinitely so that *MBLA* is the circle in which it intersects the celestial sphere. *CP* is the direction of the gnomon, which again we may suppose to be produced to intersect the celestial sphere in the celestial pole *P*. *BPA* is the plane of the meridian.

If now *S* denotes the position of the sun, the line of intersection of the shadow of the gnomon *CP* with the plane of the dial will be the line of intersection of the plane containing *CP* and *S* with the plane *MBLA*. *MPL* represents in the figure the plane passing through *S* and *CP*, and *MCL* is the line of intersection of this plane with the plane of the dial, or *CL* is the direction of the shadow of the gnomon.

Neglecting the slight alteration in the declination of the sun during the hours

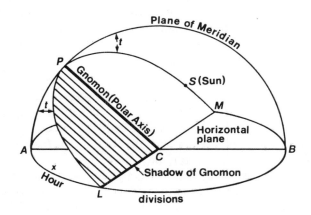

Fig. 12.9

of daylight, S will describe a circle uniformly on the celestial sphere about P as centre. The angle SPB is the hour angle of the sun, decreasing or increasing with the time according as the observation is made in the morning or in the afternoon.

Then, in the right-angled triangle LPA,

$AP = \phi =$ latitude of place;

Angle $APL = t =$ hour angle of sun;

$AL = x$, the required division along the dial corresponding to hour angle t.

$$\therefore \sin\phi = \cot t . \tan x, \quad \text{or} \quad \tan x = \sin\phi . \tan t. \tag{12.10}$$

Thus, to graduate the dial for the hourly intervals before and after noon, we must put $t = 15°, 30°, 45°$, etc., in succession and compute the corresponding values of x, knowing, of course, the value of ϕ.

Thus, if the latitude of the place is $30°$, the first hourly division on each side of noon will be marked out at an angle with CA given by

$$\tan x = \sin 30° . \tan 15°,$$

from which $x = 7° 38'$.

The next hourly division, indicating either 10 a.m. or 2 p.m., will make an angle with CA given by

$$\tan x = \sin 30° . \tan 30°,$$

from which $x = 16° 6'$, and so on.

The reading of the shadow of the gnomon gives the local solar (apparent) time which must be corrected by the Equation of Time, as given by the *Astronomical Ephemeris*, or by $(E - 12^{\text{h}})$ obtained from the *Star Almanac*, in order to obtain the mean time. A table of corrections may easily be drawn up for different times of the year to incorporate both Equation of Time and the difference for Standard Time; this information is usually engraved on a small plaque attached to the dial.

(c) *The prime vertical dial*

In this case the plane of the dial lies in the prime vertical. In Fig. 12.10 let $ALBM$ be the plane of the dial, which we will again suppose is continued on indefinitely, so as to cut the celestial sphere. CP, the direction of the stile or gnomon, is again parallel to the earth's axis, but this time P will be the celestial pole below the visible horizon. APB is the plane of the meridian.

Then if, as in the previous case, S denotes the position of the sun on the celestial sphere, the apparent movement of S is to describe a circle on the celestial sphere with P as centre, and the hour angle of S is the angle SPA.

Determination of local time by observation

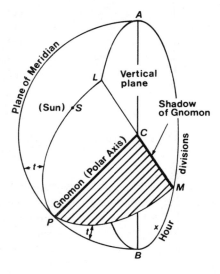

Fig. 12.10

The shadow of *PC* thrown by *S* upon the plane of the dial will be *CM*, the line of intersection of the plane passing through *S* and *PC* with the plane of the dial.

In the right-angled spherical triangle *PBM*,

$$PB = 90° - \phi = \text{co-latitude}$$

$$\text{angle } BPM = t = \text{hour angle of sun.}$$

$BM = x =$ required division along the dial corresponding to the hour angle t.
Therefore

$$\cos\phi = \cot t . \tan x$$

or

$$\tan x = \cos\phi . \tan t \qquad (12.11)$$

and by this formula the dial may be graduated in a similar manner to the horizontal dial.

Some good examples of vertical sundials are to be seen on old buildings in Europe. A particularly good one, possibly due to Sir Christopher Wren and dating from 1658, is on a wall of All Souls College, Oxford, England.

(d) *Oblique dials*

If the plane of the dial is inclined to the horizontal the dial is said to be "oblique". There is one case that is particularly simple, and has given rise to

Astronomy for surveyors

some of the simplest sundial constructions. This is the case in which the plane of the dial is tilted so as to be perpendicular to the stile, so that it coincides with the plane of the celestial equator. With this arrangement the shadow of the stile on the dial moves round uniformly with the revolution of the sun and the hour divisions on the dial are consequently uniformly spaced.

Section 6

CONCLUSION

12.16 Choice of method

If a prismatic astrolabe is not available, the best method for the surveyor of determining his local time is undoubtedly that of observing the altitudes of several pairs of east-west stars on or near the prime vertical, described in Section 3 of this chapter. The method is simple and effective, and the mean from four such pairs should be accurate to within 0.3 second of time if a modern 1" instrument in good adjustment and a stop-watch are used.

The method of position lines to be described in Chapter 14 can also yield accurate results in the determination of clock error and latitude simultaneously.

References

Kuznecov, A. N. and Baranov, V. N., On time determination at high latitudes (in Russian). *Izvest. Vysš. Učebn. Zaved., Geodez. i Aerofotos"emka Moskva*, **2**, 1980, 54–8.

Murthy, V. N. S., Simultaneous Determination of Latitude, Azimuth and Time by Observations to a Pair of Stars. *Empire Survey Review*, **15**, No. 111, Jan. 1959, pp. 15–22. (See also correspondence in the same publication, **15**, No. 114, Oct. 1959, p. 190; **15**, No. 115, Jan. 1960, p. 243; and **15**, No. 117, July 1960, p. 339.)

EXERCISES

1. On 17 May the following observations were made to stars near the P.V. to determine the error of a mean time chronometer on L.M.T.:

Star observed	Chronometer	Altitude
δ Serpentis (in east)	$18^h\ 55^m\ 48.9^s$	$17°\ 07'\ 38''$
Betelgeuse (in west)	19　13　23.8	17　57　50

Barometer 962 mb; thermometer: $24°$ C
Latitude of observing station: $22°\ 16'\ 25''$ N.
Longitude of observing station: $24°\ 38'\ 24''$ E.

Determination of local time by observation

Star observed	R.A.	Dec
δ Serpentis	$15^h\ 33^m\ 03.3^s$	N. $10°\ 39'\ 37''$
Betelgeuse	$5\ \ 53\ \ 09.0$	N. $7\ \ 24\ \ 00$

R at U.T. 12^h 17 May: $15^h\ 37^m\ 51.8^s$.

Find the chronometer error (neglecting rate).

       Ans. Chronometer fast $04^m\ 33.0^s$.

2. At a place in Latitude $30°$ S., the bearing of a wall is $110°$. Find the approximate L.M.T. at the equinox when it casts no shadow to either side. The value of E at the time concerned is $11^h\ 52^m\ 30^s$.

       Ans. $3^h\ 43^m$ p.m. L.M.T.

3. Find the approximate azimuth of the sun and the approximate L.M.T. at sunrise in Latitude $32°$ S. when the sun's declination is S. $20°$. (Take the sun's centre and neglect refraction and parallax.) The value of E at the time concerned is $12^h\ 14^m$.

       Ans. Bearing, $113°\ 47'$.
          L.M.T., $4^h\ 53^m$ a.m.

4. At a place in Latitude $32°$ S. a vertical rod 2 metres high casts a shadow 5 metres long in a direction bearing $75°\ 12'$. What is the time of the year and the approximate L.M.T.? The value of E at the time is $12^h\ 02^m$.

       Ans. $5^h\ 10^m$ p.m. L.M.T. The declination of the sun calculates as S.$23°\ 27'$, indicating that the time of the year is December (south latitude).

5. The following observations were made on 2 July to stars when they crossed the $60°$ altitude circle (assume corrected altitude = $60°$) at a place in Latitude S. $38°\ 03'\ 12''$, Longitude W. $64°\ 17'\ 33''$. The chronometer used was a mean time one.

Star observed	Chronometer time	Direction
ε Centauri	$21^h\ 12^m\ 20.8^s$	S.W.
σ Sagittarii	$21\ \ 44\ \ 50.9$	N.E.
σ Librae	$22\ \ 13\ \ 27.6$	N.W.
α Pavonis	$23\ \ 03\ \ 22.4$	S.E.

Right ascensions and declinations are:

Star	No.	R.A.	Dec.
ε Centauri	356	$13^h\ 37^m\ 3.33^s$	S. $53°\ 17'\ 01''$
σ Sagittarii	522	$18\ \ 53\ \ 00.4$	S. $26\ \ 20\ \ 34$
σ Librae	393	$15\ \ 01\ \ 55.5$	S. $25\ \ 08\ \ 22$
α Pavonis	564	$20\ \ 22\ \ 46.8$	S. $56\ \ 51\ \ 07$

R at U.T. 0^h 3 July: $18^h\ 41^m\ 11.7^s$
The Standard Meridian is 4^h W.

Find the error of the chronometer on L.Std.T. (the chronometer is known to gain $01^m\ 28.4^s$ in 24^h).

       Ans. $32^m\ 08.4^s$ slow (at time of observation to ε Centauri).

CHAPTER 13

The determination of longitude

13.01 Introduction

In previous chapters we have used the following relationships:

$$\text{L.Std.T.} \quad \begin{cases} -\text{East} \\ +\text{West} \end{cases} \text{Longitude of Std.} = \text{U.T.} \quad (13.1)$$
$$\text{Meridian}$$

$$\text{U.T.} \quad \begin{cases} +\text{East} \\ -\text{West} \end{cases} \text{Longitude of Std.} = \text{L.Std.T.} \quad (13.2)$$
$$\text{Meridian}$$

$$\text{L.M.T.} \quad \begin{cases} -\text{East} \\ +\text{West} \end{cases} \text{Longitude of observing} = \text{U.T.} \quad (13.3)$$
$$\text{stn.}$$

$$\text{U.T.} \quad \begin{cases} +\text{East} \\ -\text{West} \end{cases} \text{Longitude of observing} = \text{L.M.T.} \quad (13.4)$$
$$\text{stn.}$$

$$\text{L.S.T.} \quad \begin{cases} -\text{East} \\ +\text{West} \end{cases} \text{Longitude of observing} = \text{G.S.T.} \quad (13.5)$$
$$\text{stn.}$$

$$\text{G.S.T.} \quad \begin{cases} +\text{East} \\ -\text{West} \end{cases} \text{Longitude of observing} = \text{L.S.T.} \quad (13.6)$$
$$\text{stn.}$$

$$\text{L.H.A.} \quad \begin{cases} -\text{East} \\ +\text{West} \end{cases} \text{Longitude of observing} = \text{G.H.A.} \quad (13.7)$$
$$\text{star or sun} \qquad \text{stn.} \qquad \text{star or sun}$$

$$\text{G.H.A.} \quad \begin{cases} +\text{East} \\ -\text{West} \end{cases} \text{Longitude of observing} = \text{L.H.A.} \quad (13.8)$$
$$\text{star or sun} \qquad \text{stn} \qquad \text{star or sun}$$

The relationship in (13.1) and (13.2) concerns the time of a single meridian (the standard or zone meridian) in each country, and is useful to us in obtaining U.T., that is, in having available a time that is displaced from U.T. by a whole number of hours or half-hours.

It is a comparatively easy matter to find the error of a clock keeping approximate L.Std.T. by comparing it with radio time-signals (see Chapter 6). Instead of setting the clock to keep L.Std.T. we could, of course, set it to keep U.T., and compare it with radio time-signals to find its error. The important thing is that, by making use of accurate time-signals and a reliable clock (chronometer) we can always find the U.T.—i.e. the time of the meridian of 0° Longitude—at some instant when we make a *local* observation.

Determination of longitude

The relationships in (13.3) to (13.8) all involve the longitude of the observing station, and from them, therefore, if we know both U.T. and L.M.T. at the same instant, or G.S.T. and L.S.T. at the same instant, or G.H.A. and L.H.A. of a body at the same instant, we can find the longitude of the observing station by simple addition or subtraction. It will be obvious that, instead of using U.T., we could use the time of another meridian whose longitude is known accurately and whose time is broadcast in the form of time-signals.

Such a method of finding longitude is a *relative* one, in that we compare the time (found from radio signals) at a datum meridian with the time at the same instant (found by observation) at another place whose longitude is required. It is the method which will be described more fully later in this chapter.

There are other relative methods available to us, such as:

(a) Triangulation and EDM traverse,* where we determine the distance and azimuth between the known and unknown stations, and use them to calculate the difference of longitude.

(b) By finding the error on local time of a clock at the station whose longitude is known by making an astronomical observation there, and then carrying the clock to the station of unknown longitude and making a further observation there for clock error. The difference between the clock errors, corrected for rate, will yield the longitude difference. This was the method used at sea before the advent of radio; U.T., or the L.Std.T. of a home port, is still carried on rated chronometers by ships in case of radio breakdown.

(c) By using latitudes and azimuths (N. and S.). For this method the unknown and the known station should lie fairly close to the same meridian. The two are connected by a series of lines running as nearly north and south as possible whose azimuths are measured; latitude observations are made at all stations, and from these latitudes and azimuths the difference in longitude may be calculated.

Of all the relative methods, triangulation and EDM traverse are the most accurate, but they involve measurements which are not astronomical and are therefore outside the scope of this book.

Besides the relative methods of finding longitude which we have just touched upon, there are a number of *absolute* ones:

(i) By occultations of stars by the moon;
(ii) By moon photographs;
(iii) By moon culminations;
(iv) By moon altitudes;
(v) By lunar distances;
(vi) By eclipses of Jupiter's satellites.

---

* A traverse in which the distances are obtained by Electromagnetic Distance Measurement.

Astronomy for surveyors

The most accurate of these is (i) and, together with methods (iii) and (v), it was described in early editions of this book. Such absolute methods are not used by surveyors and navigators nowadays because of the ease with which accurate time may be obtained from radio signals.

13.02 Longitude by comparison of local time and radio time-signals

We have already seen in Chapter 6 how we can find the error of a clock on U.T. or L.Std.T. by comparing it with radio time-signals. This gives us accurate U.T. whenever we want it (or accurate time of another known meridian). If, now, we make an observation for local time by measuring the altitudes of a star near the P.V., as in Section 3 of Chapter 12, we shall be able to calculate the hour angle of the star and hence, knowing its R.A., the L.S.T., since

$$\text{H.A.} + \text{R.A.} = \text{L.S.T.} \tag{13.9}$$

If we noted the clock time of the observation, this would give us its U.T., and we could then, from the *Star Almanac*, find R at this U.T. But

$$\text{U.T.} + R = \text{G.S.T.}, \tag{13.10}$$

and the difference between the L.S.T. found in (13.9) and the G.S.T. in (13.10) would give us the longitude of the observing station. If the L.S.T. so found is greater than the G.S.T., the longitude of the observing station is east and vice versa. The sailor's mnemonic is:

Greenwich time *least*, longitude *east*;

Greenwich time *best*, longitude *west*.

It is more usual nowadays to use the difference between the Greenwich and local hour angles of a body at the same instant to compute longitude. For example, in the case of a star, our astronomical observation would give us its L.H.A. From the clock time of the observation, corrected by radio time-signal, we would find the U.T., and U.T. $+ R$ would then give us the G.S.T. But (G.S.T. $-$ R.A.) $=$ G.H.A. of the star. Comparison of the L.H.A. and G.H.A. star would yield the longitude, with the place east if the L.H.A. were greater than the G.H.A., and vice versa.

In the case of the sun, the observation would enable us to calculate its L.H.A. The clock time, corrected by radio time-signal, would give the U.T. of the observation and then, from the *Star Alamanc,*

$$\text{U.T.} + E = \text{G.H.A. sun.}$$

The difference between the L.H.A. and G.H.A. of the sun would, as for a star, be equal to the longitude.

Determination of longitude

13.03 Worked example: *Longitude by comparison of the local hour angle of a star with its Greenwich hour angle at the same instant found from time-signals*

For the observations given in **12.12 (a)** and **(b)** of Chapter 12, the chronometer used was known to be 12.7^s fast on L.Std.T. by comparison immediately afterwards with VNG time-signals. We shall now use the computed hour angles of the two stars concerned to find the longitude of the observing station.

(a) *From the observation to Fomalhaut* (see **12.12 (a)**)

Mean chronometer time of observation	21^h	09^m	53.1^s
Chronometer fast on L.Std.T. (by radio check)			12.7
Correct L.Std.T. of observation, 19 August 1974	21	09	40.4
Longitude of Standard Meridian E.	12	00	00
U.T.C. of observation, 19 August	9	09	40.4
R at U.T. 6^h, 19 August	21	48	50.4
ΔR in $3^h\ 09^m\ 40.4^s$			+31.2
G.S.T. = U.T. + R + ΔR	30	59	02.0
R.A. of Fomalhaut	22	56	17.6
G.H.A. of Fomalhaut = G.S.T. − R.A.	8	02	44.4
L.H.A. (W.) of Fomalhaut = $24^h - 4^h\ 35^m\ 10.0^s$	19	24	50.0
Subtract G.H.A. from L.H.A. for east longitude	11	22	05.6

(b) *From the observation to π Hydrae* (see **12.12 (b)**)

Mean chronometer time of observation	21^h	18^m	25.8^s
Chronometer fast on L.Std.T. (by radio check)			12.7
Correct L.Std.T. of observation, 19 August 1974	21	18	13.1
Longitude of Standard Meridian, E.	12	00	00
U.T.C. of observation, 19 August	9	18	13.1
R at U.T. 6^h, 19 August	21	48	50.4
ΔR in $3^h\ 18^m\ 13.1^s$			+32.6
G.S.T = U.T. + R + ΔR	31	07	36.1
R.A. of π Hydrae	14	04	56.0
G.H.A. of π Hydrae = G.S.T. − R.A.	17	02	40.1
L.H.A. (W.) of π Hydrae	4	24	42.5
Subtract G.H.A. from L.H.A. for east longitude	11	22	02.4
Mean longitude from (a) and (b), east	11	22	04.0

Astronomy for surveyors

The reader has probably noticed that DUT1 has not been mentioned in the above reductions. Its value (see **12.12**) is $+0.1^s$ and it will have an effect on the longitude. Reference to page 61 of the *Star Almanac* shows that we may make the correction either to the time or to the position (longitude). The correction in longitude in seconds of time, measured positively to the east, is equal to *minus* the difference DUT1.

Thus in the example given above, the correct longitude from the observation is $11^h\ 22^m\ 04.0^s - (+0.1^s) = 11^h\ 22^m\ 03.9^s$ E.

References

Bhattacharji, J. C., A Method of Determination of Latitude and Longitude where only Time and Horizontal Angles are Observed. *Empire Survey Review*, **14**, No. 110, Oct. 1958, pp. 352–63.

Bomford, A. G., Astronomic Longitudes in Australia. *Australian Surveyor*, **20**, 1964–5, pp. 273–7.

Fallon, N. R., Two Short Notes on Field Astronomy. *Empire Survey Review*, **14**, No. 104, April 1957, pp. 89–90.

Freislich, J. G., A Comparison of Three Methods used for Calculating Longitude from a Pair of well-balanced Stars. *Empire Survey Review*, **13**, No. 96, April 1955, pp. 75–81.

Gregerson, L. F., Weighted Solution of Longitude Equations. *Canadian Surveyor*, **XXI(5)**, 1967, pp. 370–9.

Krasnorylov, I. I., Accuracy of Longitude Determinations at Laplace Stations. *Geodesy and Aerophotography* (English edn), 1967, pp. 318–23.

McCarthy, D. D., The Effect of Systematic Errors in UT1 on the Determination of Astronomic Longitude. *Proc. 2nd Internat. Symp. on Problems related to the Redefinition of North American Networks 1978*, pp. 143–7. U.S. Govt. Printing Office, Washington, D.C.

Thornton-Smith, G. J., Latitude, Longitude and Azimuth from Two Stars. *Empire Survey Review*, **13**, No. 97, July 1955, pp. 124–32.

Thornton-Smith, G. J., Azimuth-controlled Almucantar Observations for Position. *Empire Survey Review*, **14**, No. 109, July 1958, pp. 301–10.

EXERCISES

1. At a place in Latitude 44° 36′ 25″ S. and in east longitude, the following observation was made to Spica near the P.V. in the east for longitude on 11 February:

L.Std.T. of observation: $11^h\ 36^m\ 12.3^s$ p.m. on 11 February
Mean observed altitude of Spica: 17° 50′ 00″.
Barometer: 982 mb; thermometer: 11° C.

Additional information:
R.A. Spica: $13^h\ 23^m\ 03.3^s$. Dec. Spica: S. 10° 56′ 57″.
Standard Meridian: 12^h E.
R at U.T. 6^h 11 February: $9^h\ 22^m\ 13.9^s$

Determination of longitude

Find the longitude of the observing station and the azimuth of the star.
$$\text{Ans.} \quad \text{Longitude: } 11^h\ 20^m\ 42.2^s\ \text{E.}$$
$$\text{Azimuth of Spica: } 87°\ 55'\ 24''.$$

2. At the same place as that in Problem 1 above, a further observation was made to α Leporis near the P.V. in the west for longitude. The results were:

L.Std.T. of observation: $1^h\ 19^m\ 00.4^s$ a.m. on 12 February.
Mean observed altitude of α Leporis: $27°\ 58'\ 24''$,
Barometer: 982 mb; thermometer, $9°$ C.

Additional information:
 R.A. α Leporis: $5^h\ 30^m\ 56.5^s$.
 Dec. α Leporis: S. $17°\ 51'\ 20''$.
 R at U.T. 12^h 11 February: $9^h\ 23^m\ 13.0^s$.

Find the longitude of the observing station and the azimuth of the star.
$$\text{Ans.} \quad 11^h\ 20^m\ 42.5^s\ \text{E.}$$
$$\text{Azimuth of α Leporis: } 272°\ 02'\ 46'.$$

3. At a place in Latitude S. $40°\ 55'\ 48''$ in New Zealand (east longitude), the sun was observed in the west near the P.V. for longitude, on 26 February, with the following results:

L.Std.T. of observation: $4^h\ 46^m\ 06^s$ p.m. on 26 February.
Mean observed altitude of sun's centre: $27°\ 30'\ 51''$.
Sun's parallax: $08''$. Barometer 897 mb. Thermometer $15.5°$ C.

Additional information:
 Standard Meridian: 12^h E.
 Sun's declination at 0^h U.T. 26 February; S. $8°\ 59.0'$, decreasing
  $05.5'$ in the next 6^h.
 E at 0^h U.T. 26 February: $11^h\ 46^m\ 52.4^s$, increasing 02.5^s in the next 6^h.

Find the longitude of the observing station.
$$\text{Ans.} \quad 11^h\ 31^m\ 35.4^s\ \text{E.}$$

4. At a place in Latitude $18°\ 58'\ 25''$ N. in west longitude, the sun was observed in the west near the P.V. for longitude, on 19 April, with the following results:

L.Std.T. of observation: $4^h\ 02^m\ 32.0^s$ p.m. on 19 April.
Mean observed altitude of sun's centre: $35°\ 52'\ 45''$.
Sun's parallax: $07''$. Barometer: 989 mb. Thermometer: $31.5°$ C.

Additional information:
 Standard Meridian: 10^h W.
 Sun's declination at 0^h U.T. 20 April: N. $11°\ 11.5'$, increasing
  $05\ 2'$ in the next 6^h.
 E at 0^h U.T. 20 April: $12^h\ 00^m\ 50.8^s$, increasing 03.3^s in the next 6^h.

Find the longitude of the observing station.
$$\text{Ans.} \quad 155°\ 08'\ 15''\ \text{W.}$$

CHAPTER 14

Simultaneous determination of latitude and longitude from position lines

14.01 Introduction

Although this method is commonly used by navigators, both on sea and in the air, it has only come generally into use among surveyors in comparatively recent times. This is rather surprising, since it is very neat in conception and consists of the blend of computation and plotting so common in many surveying operations. Not only is the method neat, but it has the merit of simplicity in its observations, flexibility in its application, and accuracy in its results. The requirements are:

(a) An approximate knowledge of the position of the observing station.
(b) The altitudes of at least two known stars whose azimuths differ by an angle which will allow a good "cut" of the position lines.
(c) An accurate knowledge of the times of observations.

The method is applicable to sun observations, but from the surveyor's point of view, observations to stars are preferable because of the greater accuracy obtainable, greater flexibility, and shorter time required for the "astrofix", as determination of position by this method is usually called.

14.02 Principle of the method

This is based on the fact that, at any instant, there is an infinite number of points on the earth's surface from which the zenith distance of a particular star, or the sun, is the same. These points, if plotted on the earth, would form a *small* circle with centre the geographic position of the star (i.e. the substellar point, where the line joining the star to the centre of the earth intersects the earth's surface), and angular radius equal to the zenith distance of the star. If two suitable stars were observed, and their zenith distance circles plotted on the earth's surface, these circles would, in general, intersect at two points, at either of which the observing station could lie. The selection of the correct point would be simple if the approximate latitude and longitude of the station were known. In practice, of course, it is not possible to plot the entire zenith-distance circles, but the portions in the vicinity of their intersection may be regarded as straight lines (called the "position lines") which are plotted as such.

Latitude and longitude from position lines

The directions of these lines will be at right angles to the respective radii of the zenith-distance circles joining the observing point to the geographic positions of the stars. The radii are arcs of great circles and give the azimuths of the stars in terms of the observer's meridian. Hence, the directions of the two position lines can be found if these azimuths are calculated. The method is not confined to two stars and two position lines, of course. It is normal practice to observe at least four stars, one in each quadrant. In Fig. 14.1, A and B are two places on the earth's surface; Z_A and Z_B are the zeniths of A and B, S_1A, S_1O and S_1B are parallel rays from a star falling at A, O (the centre of the earth) and B respectively. S is the point on the earth's surface where S_1O cuts it (geographic position or substellar point).

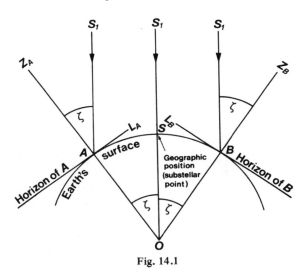

Fig. 14.1

The altitude of S_1 at A is S_1AL_A, and its zenith distance $= Z_AAS_1 = \zeta = AOS$; if at the same instant, S_1 had the same altitude at B as at A, Z_BBS_1 would be equal to $\zeta = BOS$. So the arc AS would be equal to the arc BS.

We have considered only one plane in the discussion, but if we rotate that plane about the axis S_1O, A and B will trace out a circle on the earth's surface with centre S, and we can see that there will be an infinite number of points on its circumference at which, for the same instant, the zenith distance of the star will be the same.

Consider now the diagram Fig. 14.2. Here POP' is the axis of the earth, and EQ the equator. S_1 and S_2 are two stars whose geographic positions are, at the instants under consideration, S_1' and S_2'. The two small circles with centres S_1' and S_2' represent the zenith distance circles of the two stars; they intersect at two points A and K. The zenith distances of S_1 would be equal to ζ_1, say, at the same instant at all points on the right-hand circle; similarly, at all points

Astronomy for surveyors

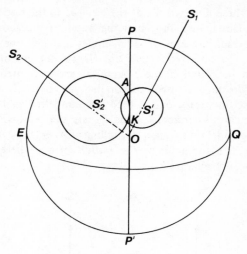

Fig. 14.2

on the left-hand circle, the zenith distances of S_2 would be equal to ζ_2 at one particular instant. However, there would be only two points, A and K, where S_1 would have zenith distance ζ_1 at the one instant, and S_2 would have zenith distance ζ_2 at the other instant. Thus, if at the instants concerned, star S_1 were observed at ζ_1 and star S_2 at ζ_2 the observer must have been either at A or K. Since A and K would normally be separated by hundreds if not thousands of miles, and the observer must know his approximate position anyway, there will usually be no difficulty in deciding whether the observing station is at A or K.

The next diagram, Fig. 14.3, carries the discussion a stage further. AS'_1 and AS'_2 are the arcs of two great circles joining A to the substellar points; they therefore define the azimuths of the stars from A at the instants of observation. The zenith distance circles are of large diameter in miles, and therefore the portions of their circumferences where they intersect at A may be regarded,

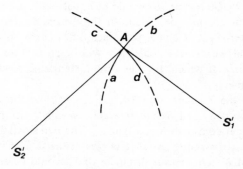

Fig. 14.3

308

Latitude and longitude from position lines

for large-scale plotting purposes, as straight lines ab and cd perpendicular to AS'_1 and AS'_2 respectively; ab and cd are, in fact, the position lines referred to in the title of this chapter.

14.03 Application of the principle

(a) *Observations and computations*

As noted in **14.01**(a), the observer must know his position approximately, preferably to within a minute or two of arc; he may obtain this assumed position from a map or from rough observations, expressing its latitude and longitude to the nearest minute of arc. Assuming for the moment that he will observe only two stars, he selects these so that their azimuths differ by 60° to 120° and so that their altitudes are not less than 15°; he then observes these stars, noting the time by stop-watch and clock when they cross the horizontal crosshair. The observed altitudes are corrected for refraction.

Using the assumed latitude, the known declinations of the stars, and their hour angles computed from the observed times and known R.A.'s, the surveyor may solve the astronomical triangles for the azimuths and altitudes. For each star he now has an observed altitude, as well as the (calculated) value of altitude and azimuth it should have from the assumed position. If, by some extraordinary chance, the calculated altitudes agreed exactly with the observed ones, the assumed position would be the correct one, but this is highly unlikely.

The next point for consideration is the relationship between the observed and calculated altitude of each star. If the observed altitude is greater than the computed one, we shall see that the true position of the observing station is nearer to the substellar point than the assumed one; if less, it is further away from this point. For each star the difference between observed and computed altitudes measures the difference between the radius of the true zenith distance circle (obtained by observation) and the radius of another (computed) zenith distance circle passing through the assumed position.

In Fig. 14.4:
A is assumed position with zenith Z_A.
T is true position with zenith Z_T.
$S_1 A$, $S_2 T$ and $S_3 G$ are parallel rays from a star.
G is the geographical position or substellar point of the star, and O is the earth's centre.
AN is the direction of the horizon at A (at right angles to $Z_A A$); TL is the direction of the horizon at T (at right angles to $Z_T T$).
AN and TL intersect at Q.
NAS_1 is the altitude of the star at A, and LTS_2 its altitude at T.
AN intersects TS_2 at N.
The quadrilateral $AQTO$ is cyclic, therefore the supplement of angle AQT,

Astronomy for surveyors

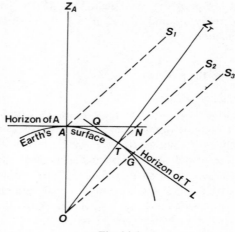

Fig. 14.4

i.e. angle NQT, is equal to angle AOT, which is a measure of the arc AT.

But in the triangle NQT, the exterior angle LTS_2 = angle QNT + angle NQT. Also, angle QNT = angle NAS_1.

Therefore angle LTS_2 (altitude at T) = angle NAS_1 (altitude at A) + angle AOT.

That is, the difference between the altitudes at A and T is a measure of the arc AT.

If, in the diagram, T were on the left of the line $Z_A AO$, it is seen quite readily that the star's altitude there would be less than $NAS1$, the difference again being the angle AOT.

The arc AT is known as the "intercept", and there are two mnemonics relating this intercept and the observed altitude:

GOAT. Greater **O**bserved **A**ltitude, intercept **T**owards star's geographic position.
LOAF. Lesser **O**bserved **A**ltitude, intercept **F**rom star's geographic position.

(b) *Plotting*

The intercepts and azimuths are now used in a simple graphical plot to obtain the position lines and, from them, the correct position of the observing station (see Fig. 14.5).

A line NM is drawn to represent the meridian, N being in the direction of north, and any point Z is selected on NM to represent the assumed position. Lines ZS_1 and ZS_2 are then laid off with a protractor (or by co-ordinates if greater accuracy is required) in the directions given by the computed azimuths of the stars S_1 and S_2.

Now, assuming that the observed altitude of S_1 was greater than the computed one, the difference Za is measured off along ZS_1 in the direction of the

Latitude and longitude from position lines

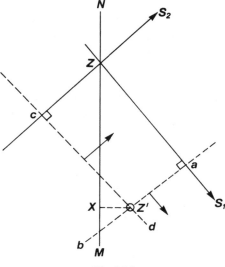

Fig. 14.5

star, to a suitable scale (so many seconds of arc to the centimetre); the position line *ab* is then drawn through *a* at right angles to ZS_1.

Assuming that the observed altitude of S_2 was less than the computed one, the difference Zc is scaled off from Z away from the star and the position line *cd* is drawn through *c* at right angles to ZS_2.

The two position lines *ab* and *cd* intersect at Z' the true position of the observing station (assuming no errors in instrument, observation, or time). It should be remembered that $aZ'b$ and $cZ'd$, although plotted as straight lines, are in reality arcs of the zenith-distance circles, but they may be treated as straight lines, since the radii of the circles are very large at the scale of the plot.

If now a perpendicular be dropped from Z' to meet *NZM* in X, ZX may be scaled off directly to yield the difference of latitude between assumed and true positions. XZ', the difference of longitude, obviously cannot be scaled in the same way because degrees of latitude and longitude are approximately equal only at the equator. It can easily be proved, however, that $(XZ' \times \sec \text{latitude})$ is equal to the true difference of longitude. Thus if XZ' is scaled from the plot and multiplied by the secant of the latitude of Z', the difference of longitude is obtained. The differences of latitude and longitude found in this way are now applied to the assumed latitude and longitude (in the sense indicated by the plot) to give the observer's true position.

(*Note*. The intercepts are scaled directly along ZS_1 and S_2Z because these are, in reality, portions of great circles through Z; similarly the latitude difference is scaled directly along *NM*, the meridian which is, of course, also a great circle.)

Astronomy for surveyors

In practice at least three, or more usually four, stars are observed to give three, or four, position lines, as shown in Fig. 14.6. Short arrows are attached to the position lines to indicate the directions in which the stars were observed. If no errors of any kind were present in instrument, observation, or time, the position lines should all pass through one point, the true position of the observing station. However, there is usually some vertical-circle index error present in the instrument, and if the observations are all made on one face (as they often are) this will displace the position lines a constant amount towards or away from the star observed. For example, if there was a vertical-circle index error correction of +05″ to all altitudes read on the left face, *all* observed altitudes would be too *small*, and the position lines would all have to be moved 05″ *towards* their respective stars (keeping them parallel to their originally plotted positions, of course). A constant error in the assumed refraction correction would have a similar effect.

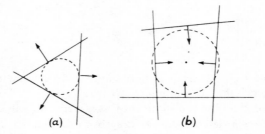

Fig. 14.6

In the case of three position lines forming a triangle as in Fig. 14.6(a), the centre of the inscribed circle is taken as the true position, since, by moving them all inwards by the same amount (the radius of the inscribed circle) they would all pass through this point. If the azimuths of the three stars all lay within 180°, the position would be as in Fig. 14.7, and the centre of one of the escribed circles would be accepted. Note the directions of the arrows.

In the case of four or more position lines, e.g. Fig. 14.6(b), it is unlikely that one circle can be inscribed in the resultant figure to touch each line because of slight observational errors in intersecting the stars, reading the circle and recording the time. However, the circle is drawn in such a way that it "just misses" each line by as nearly as possible the same amount (+ or −), and its centre is accepted as the most probable position of the observing station. Where there are four or more lines, the attached arrows should *all* point either inwards or outwards. If this is not so, there is a gross error present, since translation of all the lines towards or away from their stars by a constant amount will not make them concurrent or nearly so. For this reason, the arrows should always be drawn as a check.

With careful observations the method gives good results. A left- and right-face

Latitude and longitude from position lines

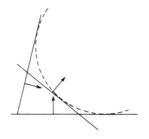

Fig. 14.7

pointing on each of eight stars, two in each quadrant, should yield latitude to ±5" and longitude to ±5" sec ϕ, assuming that a modern 1" instrument and a good stop-watch are used. Since longitude is generally more weakly determined than latitude, it is good practice to expand the programme to include additional pairs of stars east and west near the prime vertical. Forgetting to centre the alidade level bubble before reading altitudes is probably the greatest source of error with inexperienced observers.

14.04 Prismatic astrolabe observations

Observations with the prismatic astrolabe consist of timing the passage of stars across a constant altitude (see **12.14**) and they can therefore be used in the same way as has been described above for determining latitude and longitude by means of position lines. The scope of the observations is slightly more limited than with a theodolite, however, since an astrolabe cannot be used for observing stars near the meridian.

It is customary to make the plot in a somewhat simpler form than that of **14.03**(b).

An east–west line, EW in Fig. 14.8, is drawn on a sheet of paper, and a convenient scale of longitudes is marked off along it. The assumed position Z is marked on the line. Longitudes and azimuths are calculated from the star observations by using altitudes, declinations and assumed latitude to solve for the hour and zenith angles, and using formulae

$$\text{H.A.} + \text{R.A.} = \text{L.S.T.}$$

$$\text{U.T.} + R = \text{G.S.T.} \ (\text{U.T. from times of observations})$$

$$\text{L.S.T.} - \text{G.S.T.} = \text{longitude}.$$

Assuming, for simplicity, that only two stars are observed, the calculated longitudes are then marked on the EW line as at A and B in Fig. 14.8. Position lines AZ' and BZ' are now drawn through A and B in directions at right angles to the calculated azimuths; these meet at Z' to give the true position of the

Astronomy for surveyors

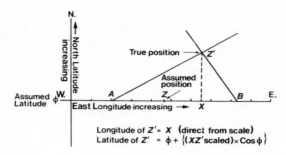

Fig. 14.8

observing station (assuming no errors). A perpendicular is dropped from Z' to meet EW at X to give the true longitude directly from the scale. The scaled length of XZ' is multiplied by the cosine of the assumed latitude to give the true latitude difference. It is customary, of course, to observe more than two stars, and the true position is found as in **14.03**(b) by taking the centre of the circle which most nearly touches all the position lines.

The accuracy obtainable from astrolabe observations has already been indicated in **12.14**(a) (end).

14.05 Observing procedure for theodolite work

The actual observations for altitude, and the recording of the time are done as in **12.11**; this entails a L.F. and R.F. pointing to each star, and acceptance of the mean of the altitudes as correct at the mean of the times. Since vertical collimation error and any (constant) error in the assumed value of refraction are eliminated by accepting the centre of the "average circle" of the plot as the correct station position, the observations may all be done on one face without apparent loss of accuracy. However, provided the L.F. and R.F. readings are taken close together, it is probably better to use both faces, since the means will tend to "even out" some errors of star intersection, bubble setting, and circle graduation. If the vertical collimation error is comparatively large and there is no time to adjust the instrument to remove it, two-face observations are best, otherwise the "average circle" of the plot will have a large radius.

It is good practice in ordinary work to select a minimum of four stars for observation, one in each quadrant, and all at roughly equal altitudes not less than, say, $30°$ if this is feasible, and to observe them within as short a period as possible to avoid errors arising from changes in refraction. Often a previously prepared programme is not necessary, as the surveyor may be able to select suitable stars in the right positions, after setting up the instrument, by visual observation. If eight stars are selected, two may be observed near the meridian north and south of the zenith, two near the prime vertical, and the remaining

Latitude and longitude from position lines

four distributed so that one is approximately in the middle of each of the four quadrants. The greater the number of stars used, the lengthier are the calculations involved. For ordinary work, four stars should be sufficient.

14.06   Use of tables to reduce calculations

Various sets of tables are available for position-line work; they are designed to help air and marine navigators, and most of them have explanations describing how they are to be used. A list is given below:

> *Tables of Computed Altitude and Azimuth* published by the Hydrographic Dept. of the British Admiralty (Hydrographic Publication H.D. 486). 6 vol., each covering a 15° latitude belt.
> *Hughes' Tables for Sea and Air Navigation* published by Kelvin Hughes Ltd., Hainault, Ilford, Essex.
> *Navigation Tables for Mariners and Aviators* (Dreisonstok) published by the Hydrographic Office, U.S. Navy Dept. (publication H.O. 208).
> *Sight Reduction Tables for Air Navigation* published by H.M.S.O., London, as Air Publication A.P. 3270.

For the surveyor these tables are rendered more or less redundant by electronic computers, but they could have some value in field situations where computed predictions were not available.

14.07   Worked example: *Position line method (theodolite observations)*

Date: 1974 August 19
Place: Arts Block, Pillar F,
   University of Otago,
   Dunedin, N.Z.
Assumed Latitude: 45° 52′ 00″ S.
Assumed Longitude: 170° 31′ 15″ E.
Standard Meridian: 12^h E.
            DUT1: $+0.1^s$

Observer: J. B. M.
Booker: M. D. Body
Instrument: Wild T2,
   No. 76 015
Barometer: 1020 mb.
Temperature: 5° C
Timing: Chronometer and
   split-second-hand stop-watch

(i) *Stars observed*

Object	S.A. No.	Direction	R.A.	Dec.
ε Centauri	356	S.W.	$13^h\ 38^m\ 16.6^s$	S. 53° 20′ 32″
γ Ophiuchi	490	N.W.	17 46 38.9	N. 2 43 05
γ Aquarii	613	N.E.	22 20 23.0	S. 1 30 42
χ Eridani	40	S.E.	1 54 59.8	S. 51 43 43

(ii) Observations

Face	Object	Approx. L. Std. T.	Corrn to L. Std. T.	Altitude	Approx. observed azimuth
R	356	22^h 24^m 59.4^s	−12.4^s	35° 24′ 35″	226° 53′
R	356	26 11.0	−12.4	35 15 28	
R	356	27 20.2	−12.4	35 06 40	
R	356	28 19.6	−12.4	34 59 09	
R	490	22 30 32.2	−12.5	35 08 10	323 25
R	490	32 06.8	−12.5	34 58 25	
R	490	33 17.2	−12.5	34 51 04	
R	490	34 04.3	−12.5	34 46 08	
R	613	22 38 54.0	−12.6	34 50 38	47 29
R	613	39 49.6	−12.6	34 57 50	
R	613	40 42.9	−12.6	35 04 40	
R	613	41 30.8	−12.6	35 10 46	
R	40	22 46 57.0	−12.6	34 46 07	130 52
R	40	48 09.4	−12.6	34 55 35	
R	40	49 16.2	−12.6	35 04 20	
R	40	49 58.9	−12.6	35 10 01	

Comments on the observations. It will be noted that all observations were done on right face, which makes the derivation of the altitudes slightly simpler with the Wild T2 instrument, since subtraction of 270° from the readings gives the required values, as against subtracting the circle readings from 90°, for left face.

The times were measured by starting the split-second-hand stop-watch at a known minute on the chronometer and using the No. 2 second hand for timing the events; the watch was stopped, after completion of the observations, on a known minute on the chronometer, thus enabling the watch rate relative to the chronometer rate to be determined. The chronometer rate and offset were found from radio time-signals, and this allowed the correction in the column headed "Corrn to L.Std.T." to be calculated.

The stars were selected from a computer printout giving objects crossing the 35° altitude circle at azimuths within ±10° of quadrant centres between given time-limits. This enabled all four stars to be observed within a total period of 25 minutes. For each star a plot of altitude against time was made, and straight lines were obtained in each case, indicating that all pointings were good. These plots are not reproduced here in the interests of space, but they are done quite simply on ordinary graph paper to a suitable scale.

Formulae used in the computations:

For altitude, h_c: $\sin h = \cos \omega . \cos p + \sin \omega . \sin p . \cos t$.

Latitude and longitude from position lines

(iii) Computations

	356	490	613	40
Mean observed approx. L.Std.T.	22^h 26^m 42.6^s	22^h 32^m 30.1^s	22^h 40^m 14.3^s	22^h 48^m 35.4^s
Correction to L.Std.T.	12.4	−12.5	−12.6	−12.6
L.Std.T.	22 26 30.2	22 32 17.6	22 40 01.7	22 48 22.8
Standard Meridian, E.	12 00 00	12 00 00	12 00 00	12 00 00
U.T.C.	10 26 30.2	10 32 17.6	10 40 01.7	10 48 22.8
DUT1	0.1	0.1	0.1	0.1
U.T.1	10 26 30.3	10 32 17.7	10 40 01.8	10 48 22.9
R at U.T. 6^h, 19 August 1974	21 48 50.4	21 48 50.4	21 48 50.4	21 48 50.4
ΔR	43.8	44.7	46.0	47.4
G.S.T.	8 16 04.5	8 21 52.8	32 29 38.2	8 38 00.7
Longitude, E. (assumed)	11 22 05	11 22 05	11 22 05	11 22 05
L.S.T.	19 38 09.5	19 43 57.8	43 51 43.2	20 00 05.7
R.A.	13 38 16.6	17 46 38.9	22 20 23.0	1 54 59.8
L.H.A.	5 59 52.9	1 57 18.9	21 31 20.2	18 05 05.9
t, arc	89° 58′ 14″	29° 19′ 44″	322° 50′ 03″	271° 16′ 29″
ω (= 90° − assumed ϕ)	44 08 00	44 08 00	44 08 00	44 08 00
p (= 90° − δ)	36 39 28	92 43 05	88 29 18	38 16 17
h_c (computed altitude)	35 10 06	34 54 53	35 00 16	34 57 52
A_c (computed azimuth)	226° 55′	323° 22′	47° 30′	130° 55′
Mean observed altitude	35° 11′ 28″	34° 55′ 57″	35° 00′ 59″	34° 59′ 01″
Refraction correction	−01 24	−01 25	−01 25	−01 25
Altitude corrected	35 10 04	34 54 32	34 59 34	34 57 36
Curvature correction	−01	+02	−02	−01
Corrected observed altitude = h_o	35 10 03	34 54 34	34 59 32	34 57 35
h_c from above	35 10 06	34 54 53	35 00 16	34 57 52
Intercept = $h_o - h_c$	−03″	−19″	−44″	−17″

Astronomy for surveyors

For azimuth, A: $\cot Z = \cot p . \sin \omega . \csc t - \cos \omega . \cot t$.

For curvature: $(\bar{h} - h)'' = f''\{\Sigma(T_i^s - \bar{T})^2 .225/(2.n.\rho)\}$ (see **7.05**).

(iv) *Plotting*: see Fig. 14.9.

The perpendicular dropped from Z' on to the north line through the assumed position has its foot at a point which scales 15″ south of Z, so that the true position is 15″ further south than the assumed position, i.e. S. 45° 52′ 15″. The length of the perpendicular from Z' scales 13.5″ to the west, so that the true position is (13.5″ × sec 45° 52′ 15″) west of the assumed position.

Assumed Longitude	= E. 170° 31′ 15″
$Z'Z = 13.5''.\sec\phi$	= −19.4
True Longitude	= E. 170 30 55.6
	or = E. 11ʰ 22ᵐ 03.7ˢ
True Latitude	= S. 45° 52′ 15″

The radius of the circle in the plot scales 19″, and since all the pointings were on right face, this indicates that the vertical collimation error of this theodolite was of the order of 19″, and that all the altitudes read on R.F. were too small by this amount (observed altitudes all less than the computed values).

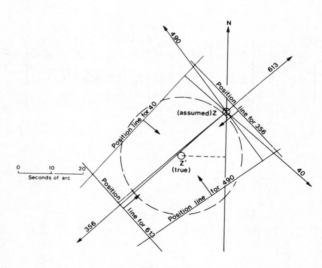

Fig. 14.9

Z is the assumed position, Latitude 45° 52′ 00″ S, Longitude 170° 31′ 15″ E.
Z' is the true position, i.e. it is the centre of the circle most nearly touching all the position lines.

Latitude and longitude from position lines

14.08 Use of computer to calculate results

Since the routine for computing the intercept and the azimuth of each star is the same, it is clear that the reductions are ideally made by an electronic computer which will deal with a large number of stars in a matter of a few minutes once the observation data have been converted into suitable input form. As each intercept and azimuth are calculated, they can be stored at the end of the first phase of the computation and used to determine the most probable value of the latitude and longitude by least squares, instead of making a complicated graphical plot. See **AII.03.02**, page 386.

In Fig. 14.10, T is the most probable position of the observing station, v is the displacement of the position line from the true position, TP is perpendicular to OS, TQ is perpendicular to SQU, TR is perpendicular to OR. $OR = x$, $TR = y$.

Angle $TOR = \theta$; angle A = azimuth of star from north;

angle $POT = 90° - (A + \theta)$

$$OP = OT \cos\{90° - (A + \theta)\} = OT \sin(A + \theta)$$
$$= OT(\sin A . \cos\theta + \cos A . \sin\theta)$$
$$= x \sin A + y \cos A.$$

But
$$OS = \Delta t = OP + PS = OP + TQ = OP + v.$$
$$\Delta t = x \sin A + y \cos A + v$$

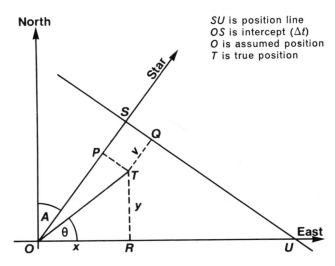

Fig. 14.10

Astronomy for surveyors

Thus
$$v + x \sin A + y \cos A - \Delta t = 0. \tag{14.1}$$

For each star, A and Δt are known. If three stars are observed, each yields an equation of the same form as (14.1), so that v, x and y can be found. With more than three stars observed, the method of least squares can be used to yield the most probable values of v, x and y.

The difference of latitude from the assumed position is equal to y, while the difference of longitude is equal to $x \sec \phi$. If all the observations have been carefully done, on one face of the theodolite, the value of v should be close to the vertical collimation error of the instrument (see **14.03** (b)).

EXAMPLE. *Least-squares solution of position line observation*
We shall apply the method to the worked example in **14.07**.
The four equations (one for each star in the example) are:

Star	Equation
356	$v + x.\sin 226° 55' + y.\cos 226° 55' + 3 = 0$
490	$v + x.\sin 323° 22' + y.\cos 323° 22' + 19 = 0$
613	$v + x.\sin 47° 30' + y.\cos 47° 30' + 44 = 0$
40	$v + x.\sin 130° 55' + y.\cos 130° 55' + 17 = 0$

Three normal equations are formed from these four observation equations by the usual method to be found in textbooks on survey adjustments, such as H. F. Rainsford's *Survey Adjustments and Least Squares*. They are:

$$4v + 0.1659x + 0.1400y + 83.0000 = 0$$
$$+ 2.0041x + 0.0232y + 31.7584 = 0$$
$$+ 1.9959y + 31.7894 = 0$$

Solution of these normals gives:

$$v = -19.66''; \quad x = -14.05''; \quad y = -14.38''.$$

v gives us the radius of the best-fit circle, $-19.7''$.
$x.\sec \phi$ gives us the difference between the assumed and true longitudes $= -20.2''$.
y gives us the difference between the assumed and true latitudes $= -14.4''$.

Latitude and longitude from position lines

Thus

Assumed longitude	=	E. 170° 31′ 15″
$x.\sec\phi$	=	−20.2
True longitude	=	E. 170 30 54.8
	or =	E. $11^h\ 22^m\ 03.7^s$
Assumed latitude	=	−45° 52′ 00″ (south)
y	=	−14.4
True latitude	=	−45° 52′ 14.4″

These values agree well with those from the plot.

EXERCISE

The following stars were observed on 6 Feburary 1977 with a 60° prismatic astrolabe at a place whose approximate position was Latitude −41° 18′ 45″, Longitude 173° 14′ 34″ E. Barometer: 1016 mb; thermometer: 16 °C; DUT1: +0.6^s.

Star FK4	R.A.	Dec.	U.T.C.	Aspect
271	$7^h\ 02^m\ 45.0^s$	−15° 36′ 12″	$9^h\ 12^m\ 35.6^s$	NE
1198	7 35 08.3	−52 29 15	8 10 36.5	SE
84	2 24 14.6	−60 25 12	8 18 06.4	SW
204	5 27 17.1	−20 46 55	10 32 05.0	NW

R at U.T. 6^h, 6 February 1977 = $9^h\ 05^m\ 02.9^s$.
Find the true latitude and longitude of the place of observation.

Ans. Latitude: −41° 18′ 32″
Longitude: 173 14 31.

References

Angus-Leppan, P. V., A Note on the Calculation of Position Lines. *Empire Survey Review,* **13,** No. 98, Oct. 1955, pp. 184–6.

Bennett, G. G., Position and Azimuth from the Time Rate of Change of Altitude and Azimuth of the Sun. *Survey Review,* **25,** No. 198, Oct. 1980.

Bowring, B. R., Astrofixes by the Method of Equal Zenith Distances, *Survey Review,* **25,** No. 196, April 1980.

Carroll, J. E., A New Instrument for the Determination of Astronomic Position. *Surveying and Mapping,* **XXIX,** No. 3, Sept. 1969, pp. 447–61.

Chiat, B., Notes on the Position Line. *Empire Survey Review,* **13,** No. 97, July 1955, pp. 137–40.

Ezenwere, O. C., A Rigorous Solution for Azimuth, Latitude and Longitude from Direction and Time. *Survey Review,* **26,** No. 201, July 1981.

Fallon, N. R., Effect of Tilt on Position Lines. *Empire Survey Review*, **14**, No. 103, Jan. 1957, pp. 40–42. (See also correspondence in the same publication, **14**, No. 105, July 1957, pp. 139–41.)

Kazanskiy, K. V., Evaluating the Accuracy of a Point Determined by Position Lines. *Geodesy and Aerophotography* (English edn), No. 5, 1969, pp. 318–23.

Rish, R. F., Position Finding by Rate of Change of Altitude of the Sun. *Survey Review*, **24**, No. 189, July 1978.

Stepec, W. A. Simultaneous Position Lines from Altitude and Azimuth Observations. *Survey Review*, **XXII**, No. 174, Oct. 1974, pp. 347–58.

Tait, G. B., Proposed Method of Solution of Position Line Problem. *Empire Survey Review*, **14**, No. 107, Jan. 1958, pp. 220–6. (See also a further paper on the same subject in **14**, No. 109, July 1958, pp. 310–25, and correspondence in **14**, No. 110, Oct. 1958, p. 383; **15**, No. 111, Jan. 1959, p. 46; and **15**, No. 112, April 1959, p. 92.)

Thornton-Smith, G. J., Almucantar Position Lines. *Empire Survey Review*, **12**, No. 88, April 1953, pp. 77–84.

Thornton-Smith, G. J., The Straight Position Line. *Empire Survey Review*, **13**, No. 100, April 1956, pp. 269–72.

A short general bibliography

Adams, L. P., Astronomical Position and Azimuth by Horizontal Directions. *Survey Review*, **19**, No. 148, April 1968, pp. 242–51.

Anon., The Astronomer's Drinking Song. *Australian Surveyor*, **20**, 1964–5, pp. 384–5.

Anon., Azimuth Variation from Observations of Close Stars. *Survey Review*, **XVIII**, No. 139, Jan. 1966, pp. 208–13.

Bennett, G. G., Yearly Sets of Fourier Coefficients for the Evaluation of Astronomical Data. *Survey Review*, **24**, No. 189, July 1978, pp. 326–8.

Bennett, G. G., Position and Azimuth from the Time Rate of Change of Altitude and Azimuth of the Sun. *Survey Review*, **25**, No. 198, Oct. 1980, pp. 373–80.

Bennett, G. G. and Freislich, J. G., *Field Astronomy for Surveyors*. New South Wales University Press, Sydney, 1980.

Biddle, C. A., Standard Practice in Field Astronomy. *Jour. R.I.C.S.*, **33**, Part v, Nov. 1953, pp. 340–50.

Black, A. N., Laplace Points in Moderate and High Latitudes. *Empire Survey Review*, **11**, No. 82, Oct. 1951, pp. 177–84.

Böhme, D., Contribution on Geodetic and Astronomical Utilization of Occultations of a Star by the Moon (in German). *Geodät. Geophys. Veröff. R.*, **3**, Berlin, **42**, 1980, 84.

Bomford, A. G., Surveying in Northern Australia. *The Chartered Surveyor*, **93**, No. 6, Dec. 1960, pp. 321–4. (See also discussion in **93**, No. 9, March 1961, pp. 509–10.)

Bomford, G. *Geodesy*. Clarendon Press, Oxford, 1975. (Contains considerable material on, and references to, geodetic astronomy. Also 4th edition, 1980.)

Brenchley, D. R., First-order Astronomical Observations. *Jour. N.Z. Inst. Surveyors*, **20**, No. 201, Dec. 1951, pp. 5–15.

Butkevič, A. V. and Lavnikevič, A. S., On Calculation of Azimuth and Chronometer Correction from Solar Observations with Mean Elements (in Russian). *Geodez., Kartogr. i. Aerofotos"emka, Lvov*, **32**, 1980, 3–8.

Chauvenet, W., *Manual of Spherical and Practical Astronomy.* Philadelphia, 1891.
Clark, D., *Plane and Geodetic Surveying,* **2**, 5th edn. Constable, London, 1963.
Clark, D., *Plane and Geodetic Surveying,* **2**. 6th edn (as revised by J. E. Jackson). Constable, London, 1973.
Close, Sir C. F. and Winterbotham, H. St. J. L., *Text Book of Topographical and Geographical Surveying.* London, 1926.
Ezenwere, O. C., A Rigorous Solution for Azimuth, Latitude and Longitude from Direction and Time. *Survey Review,* **26**, No. 201, July 1981, pp. 141-51.
Fominas, D., The Significanace of Laplace Stations in the Geodetic Network. *Australian Surveyor,* **16**, No. 3, Sept. 1956, pp. 161-5.
Freislich, J. G., A System of Conventions for use in Field Astronomy. *Empire Survey Review,* **12**, No. 90, Oct. 1953, pp. 174-83. (See also correspondence in **12**, No. 92, April 1954, pp. 284-7.)
Freislich, J. G. Identification of Stars by Learners. *Empire Survey Review,* **16**, No. 120, April 1961, pp. 50-54.
Fricke, W., On the Determination of the Equinox and Equator of the New Fundamental Reference Co-ordinate System, the FK5. *Sdr. Celestial Mech.,* **22**, 1980, 113-25.
Gadd, B., A Graphical Method of Selecting Stars for Astronomical Observation. *Empire Survey Review,* **15**, No. 118, Oct. 1960, pp. 377-83.
Kivioja, L. A., The Vertical Mirror, its Potential Applications to Theodolites and Two Star Image Stopping Micrometers. *Bulletin Géodésique,* **93**, Sept. 1969, pp. 263-75.
Köchle, R., Simultaneous Determination of Latitude and Azimuth from Observation of the Sun. *Survey Review,* **19**, No. 145, July 1967, pp. 116-32.
Lambert, B. P., the Role of Laplace Observations in Geodetic Survey. *Australian Surveyor,* **20**, 1964-5, pp. 81-96.
Mathisen, O., Determination of Deflection of Vertical with a Small Instrument. *Bulletin Géodésique,* **93**, Sept. 1969, pp. 283-6.
Mueller, I. I., *Spherical and Practical Astronomy as Applied to Geodesy.* New York, 1968.
O'Keefe, J. A. Satellite Methods in Geodesy. *Surveying and Mapping,* **18**, No. 4, Oct.-Dec. 1958, pp. 418-22.
Papworth, K. M., Field Astronomy. *Jour. R.I.C.S.* **30**, Part V, Nov. 1950, pp. 425-41.
Pettey, J. E., Uncertainties of Astronomical Positions and Azimuths. *Proc. 2nd Internat. Symp. on Problems related to the Redefinition of North American Networks 1978,* pp. 135-41. U.S. Govt. Printing Office, Washington, D.C.
Prestcott, N. J. D., A Photographic Method of Determining Azimuth, Latitude and Longitude. *Survey Review,* **17**, No. 134, Oct. 1964, pp. 355-63.
Rainsford, H. F., *Survey Adjustments and Least Squares.* Constable, London, 1968.
Reeves, E. A., *Hints to Travellers.* 11th edn, London, 1935 and later.
Rish, R. F., Position Finding by Rate of Change of Altitude of the Sun. *Survey Review,* **24**, No. 189, July 1978, pp. 323-5.
Rish, R. F., Rate of Change Methods in Astronomical Surveying, *J. Survey. & Mapp. Div. ASCE, New York, 106, SU1, Proc. Papers 15837,* Nov. 1980, pp. 137-41.

Rish, R. F., A Three Star Resection Method. *Australian Surveyor,* **Sydney**, **30**, 1981, pp. 313-16.

Robb, A. G. and Lawson, W. F., Practical Astronomy for the Cadastral Surveyor. *Jour. N.Z. Inst. Surveyors,* **XXVI(2)**, No. 236, Sept. 1969, pp. 149-97.

Robbins, A. R., *Field and Geodetic Astronomy.* Ministry of Defence, United Kingdom, 1976.

Robbins, A. R., Geodetic Astronomy in the Next Decade. *Survey Review,* **XXIV**, No. 185, July 1977, pp. 99-108.

Roelofs, R., *Astronomy Applied to Land Surveying.* Ahrend & Zoon, Amsterdam, 1950.

Roy, S. N., Correspondence in *Empire Survey Review*, **16**, No. 124, April 1962, pp. 285-6, on adaptation of Napier's rule for a right-angled spherical triangle in which one side = $90°$.

Sadler, D. H., The Reduction of Astronomical Sights. *Roy. Inst. Navigation,* London, **33**, No. 1, 1980, pp. 97-113.

Smart, W. M., *Spherical Astronomy.* Cambridge, 1960.

Stoch, L. and Papo, H., Pointing Accuracy on Stars. *Survey Review,* **XXI**, No. 164, April 1972, pp. 242-53.

Tait, G. B., Identification of Stars by Learners. *Empire Survey Review,* **15**, No. 116, April 1960, pp. 277-81.

Vanderaa, M. H., The Solar Pointing Error. *Surveying and Mapping,* **XXIV**, No. 2, June 1964, pp. 277-82.

War Office, *Text Book of Field Astronomy* (as revised by Lt.-Col. C. A. Biddle). H.M.S.O., London, 1958.

White, L. A., General Theory for Horizontal Angle Observations in Astronomy. *Survey Review,* **18**, No. 141, July 1966, pp. 316-27; also **18**, No. 142, Oct. 1966, pp. 346-56.

APPENDIX I

EXPLANATIONS AND EXAMPLES OF THE USE OF
The Ephemeris of the Sun, Polaris and other Selected Stars with Companion Data and Tables PUBLISHED BY THE UNITED STATES DEPARTMENT OF THE INTERIOR, BUREAU OF LAND MANAGEMENT

Supplementary notation to that on pages xi-xii.

L.A.T. local apparent time; see **12.15**; equal to L.H.A. sun $+ 12^h$ (-24^h if L.H.A. $+ 12^h > 24^h$)
G.A.T. L.A.T. at Greenwich
L.A.N. local apparent noon (sun at upper transit)
G.A.N. Greenwich apparent noon (sun at upper transit at Greenwich)
Eq.T. Equation of Time
Ephemeris *The Ephemeris of the Sun, Polaris and other Selected Stars with Companion Data and Tables.* Published annually. U.S.A.
SALS *The Star Almanac for Land Surveyors.* Published annually. England.

AI.01 Introduction

In order to avoid confusion within the main text of this book, the explanation and examples of the use of the *Ephemeris* are kept separate in this Appendix. In the examples the working is done firstly by using the *Ephemeris* and secondly by using *SALS* where this is useful, so that the surveyor in the U.S.A. may see the relationship between the two sets of tables. *SALS* is available in the U.S.A. from Pendragon House Inc., 2595 East Bayshore Road, Palo Alto, California 94303 for those who might prefer to use it. *The appropriate chapters in the main body of "Astronomy for Surveyors" should, of course, be read in conjunction with the material in this appendix.* In the examples which follow, watch times are given in terms of standard times, e.g. Pacific Standard Time (P.S.T.), Eastern Standard Time (E.S.T.), etc. In many areas clocks are advanced in the summer months to show Daylight or Summer Time, and readers in those areas should remember to make the appropriate adjustments when reducing their own observations.

Facsimiles of the pages of the *Ephemeris* and *SALS* referred to in this Appendix are included; see **AI.08**.

AI.02 Identification of stars and prediction of star positions (see Chapter 9 also)

Information is given on 28 stars only in the "Selected Stars" section of the *Ephemeris*, plus Polaris in the special tables relating to that star. Using one of the many star charts which may be bought, or those on pages 60 and 61 of this

Astronomy for surveyors

book, the surveyor may easily learn to recognize each one of these stars since they are all "naked eye" objects of magnitude 2.6 or brighter. Selection of stars for observation should thus be easy after a little practice, and the only problem might be identification of a star observed through a gap in the clouds, when its position in relation to the others in the constellation cannot be determined.

AI.02.01   Identification of an observed star

EXAMPLE (refer to **AI.08** for the appropriate tables)

At a place in California whose latitude and longitude are 38° 05′ N. and 121° 40′ W. ($8^h\ 06^m\ 40^s$), a star of magnitude about 1 was observed through a gap in the clouds at an azimuth of 243° 50′ and altitude 41° 54′ at Pacific Standard Time 6.30 p.m. on 24 November 1980. What was the star?

Referring to Fig. 2.6a, Chapter 2, page 26, we see that the information given enables us to deduce two sides and the included angle of the astronomical triangle. With the notation on page xi herein, these are:

$$90° - \phi = \omega;\quad 90° - h = \zeta;\quad 360° - A = Z.$$

These data, with the appropriate formulae, will allow us to solve for p, the polar distance which will give us the declination, δ from $90° - p$; and for the hour angle, t from which we can derive the right ascension, R.A. The coordinates R.A. and δ identify the star. The formulae are:

from (1.2). $\cos p = \cos\omega.\cos\zeta + \sin\omega.\sin\zeta.\cos Z$; and

from (1.15), $\cot t = (\cot\zeta.\sin\omega - \cos\omega.\cos Z)/\sin Z$.

In this case,

$$\omega = 90° - 38° 05′ = 51° 55′$$
$$\zeta = 90° - 41° 54′ = 48° 06′$$
$$Z = 360° - 243° 50′ = 116° 10′ \text{ (see } \mathbf{10.02}).$$

Since the problem is one of identification only, we may ignore the effect of atmospheric refraction in deriving ζ.

From the first formula above, p calculates as 81° 10′, hence δ = N. 8° 50′. The polar distance p is referred to the elevated, or north celestial pole, and since it is less than 90°, the declination must be north.

From the second formula above, t calculates as 42° 32′ 09″, or $2^h\ 50^m\ 09^s$, and since the star was observed at azimuth 243° 50′, in the south-west quadrant, t in this case must be the west hour angle.

In order to find the star's R.A. we shall need to use the formula:

$$\text{L.S.T.} = t + \text{R.A.}\quad (\text{see } \mathbf{2.09})$$

which entails finding the local sidereal time, L.S.T.

There is no specific table in the *Ephemeris* from which can find the L.S.T. knowing the local standard time (P.S.T.), but there is a way of getting it. We

Appendix I

note that, if a star is on the meridian at upper transit, its hour angle t is zero, and at that instant the formula last given reduces to,

$$\text{L.S.T.} = \text{R.A.}$$

and at Greenwich,

$$\text{G.S.T.} = \text{R.A.}$$

If, then, we select any star from the table on page 15 of the *Ephemeris*, on 24 November, we can find its G.M.T. of transit; from the table on page 9 we can also find its R.A. Thus, if we select star no. 1 whose R.A. is $0^h\ 07^m$, we find by interpolation that its G.M.T. of transit (page 15) is $7^h\ 51.7^m$ p.m. Thus at $7^h\ 51.7^m$ p.m. G.M.T. on 24 November 1980, the G.S.T. is $0^h\ 07^m$.

P.S.T. of the observation = 6.30 p.m., 24 Nov.	=	18^h	30^m	00^s
Standard meridian, W.	=	8	00	00
Corresponding G.M.T. of observation	=	26	30	00
G.M.T. of transit of star no. 1 = $7^h\ 51.7^m$ p.m.	=	19	51	42
Thus, the G.M.T. of obsn. was later than G.M.T. of transit of no. 1 by:		6	38	18

This $6^h\ 38^m\ 18^s$ is the difference between two G.M.T.'s and is therefore a mean time-interval. Converting it to a sidereal time-interval by using the table on page 27 of the *Ephemeris*, we add

			01	06
Thus, at Greenwich, the corresponding instant of our observation was later than the instant of the transit of star no.1 by the sidereal interval	=	6	39	24
But the G.S.T. of transit of no.1	=	0	07	
Thus, the G.S.T. of the instant of our observation	=	6	46	24
Longitude of the place of observation, W.	=	8	06	40
∴ the corresponding L.S.T.	=	22	39	44
From our earlier astro triangle calculation, hour angle t	=	2	50	09
∴ R.A. of unknown star = L.S.T. $-t$	=	19^h	49^m	35^s

Thus we have found the R.A. and Dec. of the observed star to be $19^h\ 49^m\ 35^s$ and N. $8°\ 50'$ respectively. The only star in the table on page 9 of the *Ephemeris* which fits these coordinates is no.25, Altair, with R.A. $19^h\ 50^m$ and Dec. N. $9°$.

The above is a fairly roundabout way of finding the R.A. compared with that using *SALS*, which is also more accurate:

P.S.T. of observation, 24 Nov. 1980	=	18^h	30^m	00^s
Standard meridian, W.	=	8	00	00
G.M.T., or U.T.C. of observation	=	26	30	00
i.e., U.T.C. of observation, 25 Nov.	=	2	30	00

R at U.T. 0^h, 25 Nov. 1980	=	4	16	21.1
ΔR in 2^h 30^m	=			24.6
G.S.T. of obsn. = U.T. of obsn. + R at U.T. 0^h + ΔR	=	6	46	45.7
Longitude of observing station, W.	=	8	06	40
L.S.T. of observation	=	22	40	05.7
Hour angle of star, from above,	=	2	50	09
∴ R.A. of star	=	19^h	49^m	57^s

It should be noted that, strictly, we should have applied the DUT1 correction (see **6.07** herein) to the U.T.C. of observation above, i.e. to 2^h 30^m 00^s, 25 Nov., in order to get the U.T.1 of the observation. However, it was not necessary to do so in this (approximate) calculation.

AI.03.02 Observation of the sun for azimuth by finding its hour angle

Since it should not be too difficult to identify any one of the 28 stars in the *Ephemeris* at night, it is perhaps more appropriate to relate this method to a daylight star in the late afternoon or early morning.

EXAMPLE (refer to **AI.08** for the appropriate tables)

Find the azimuth and altitude of Regulus (α Leonis) at 4 p.m. Eastern Standard Time on 1 May 1980 at a place in Georgia whose latitude and longitude are 33° 43' N. and 84° 21' W. (5^h 37^m 24^s W.) respectively. Regulus is no. 15 in the list of "Selected Stars" on page 9 of the *Ephemeris*.

From p. 12 of the *Ephemeris*, G.M.T. of transit of no.15 =		7^h	27.8^m	p.m.
	=	19^h	27^m	48^s
Sidereal correction for longitude W. (5^h 37^m 24^s); see page 27 of the *Ephemeris*	=		–0	55
∴ L.M.T. of local transit of Regulus	=	19	26	53
Longitude correction for standard meridian which is fast on the meridian of the observing station	=	0	37	24
∴ Eastern Standard Time of local transit	=	20	04	17
But Eastern Standard Time when observation is to be made	=	16	00	00
Difference	=	4	04	17
This difference is the mean time-interval between the time of observation and the time of local transit, i.e. it is the east hour angle of Regulus in mean time-units. Convert to sidereal time-units (table on p.27, *Ephemeris*)			+00	40
East hour angle of Regulus at observation time, t	=		4 04	57
i.e. t =		61°	14'	15"

Appendix I

From (1.2), $\cos \zeta = \cos \omega . \cos p + \sin \omega . \sin p . \cos t$;

From (1.15), $\cot Z = (\sin \omega . \cot p - \cos \omega . \cos t)/\sin t$.

$\omega = 90° - \phi = 90° - 33° 43' = 56° 17'$

$p = 90° - \delta = 90° - 12° 03.8' = 77° 56' 12''$ (*Ephemeris* p.12 for δ)

$t = 61° 14' 15''$ (east)

Computing, we get:

$\zeta = 59° 30' 30''$, $\therefore h = 30° 29' 30''$;

$Z = 95° 49' 07''$, $\therefore A = 95° 49' 07''$ (see **10.02**)

So, with azimuth set on 95° 49' and altitude on 30° 31' (allowing 01' 30" for refraction), the star should be visible near the centre of the telescope field (but see **10.04**). At magnitude 1.3, a good theodolite telescope properly focused at infinity, and keen eyesight would be needed to see this star in daylight.

If we use *SALS* to find the L.H.A. the procedure is:

E.S.T. when observation is to be made, May 1	=	16ʰ 00ᵐ 00ˢ
Longitude of standard meridian. W.	=	5 00 00
Corresponding U.T.C., May 1	=	21 00 00
R at U.T. 18ʰ, May 1, 1980	=	14 39 15.2
ΔR in 3ʰ 00ᵐ	=	29.6
G.S.T. of observation time = U.T. + R at U.T. 18ʰ + ΔR	=	35 39 44.8
Longitude of observing station, W.	=	5 37 24
Corresponding L.S.T.	=	30 02 20.8
R.A. of Regulus (star no. 273, *SALS*)	=	10 07 19.1
L.H.A. of Regulus at 4 p.m. E.S.T., May 1 (West H.A.), t	=	19 55 01.7
$\therefore$ east H.A.	=	4 04 58.3
	t =	61° 14' 35"
Declination of Regulus, from *SALS*,	δ =	*N*. 12° 03' 48"

Using the two formulae in the earlier part of this example, calculation gives:

$h = 30° 29' 13''$

$A = 95° 48' 55''$,

which are much the same, but slightly more accurate than the values calculated using t obtained by using *Ephemeris* tables.

AI.03 Determination of azimuth (see Chapter 10 also)

AI.03.01 Observation of the sun for azimuth by measuring its altitude

In this type of determination, we must know the latitude ϕ, the declination δ of the sun (from tables) and the approximate L.Std.T. (in order to find the sun's declination). We measure the altitude h of the sun carefully, and also the horizontal directions to the sun and some fixed ground-mark or reference object (R.O.). We also need to know the atmospheric temperature and pressure to enable us to find the refraction correction to the altitude. Essentially, we must know the three sides ω, p and ζ of the astronomical triangle to allow us to compute the zenith angle Z and hence the azimuth of the sun, and that of the R.O. by application of the angle between it and the sun.

EXAMPLE (refer to **AI.08** for the appropriate tables)

Place: California, Stn. 22
Latitude: 36° 52′ 06″ N.
Longitude: 118° 48′ 18″ (7^h 55^m 13.2^s) W.
Date: 1980 September 8

R.O: Stn. 25
Plate level: 30″ per division
Standard meridian: 8^h W.
Watch: 16^s fast on P.S.T.

Object observed: sun in the east

Face	Object	Horizontal circle	Altitude	Plate bubble L	Plate bubble R	Watch time	Baro. Temp.
L	R.O.	25° 42′ 20″					
L	☉	112 19 20	30° 23′ 20″	2	6	8^h 06^m 38.5^s a.m.	1016 mb (30.0 in.)
R	☉	292 18 00	30 07 00	5	3	8 08 58.5	22.2 °C (72 °F)
R	R.O.	205 42 00					

Plate bubble correction:
$\{(\Sigma L - \Sigma R)/n\} \times v \times \tan h = \{(7 - 9)/4\} \times 30'' \times 0.583 = -09''$

Mean horizontal circle reading on sun	=	112° 18′ 40″
Plate bubble correction	=	−09
Corrected horizontal circle reading on sun	=	112 18 31
Mean horizontal circle reading on R.O.	=	25 42 10
Included angle, sun to R.O., anticlockwise	=	86° 36′ 21″

Sun's altitude:

Mean observed altitude of sun	=	30° 15′ 10″
Refraction correction = 99″ × 1.01 × 0.96 (*Ephemeris*, p. 25)	=	−01 36
Parallax correction (*Ephemeris*, p. 25)	=	+08
Corrected observed altitude, $h =$		30 13 42
$\zeta = 90° - h =$		59° 46′ 18″

Sun's declination:

Mean of observed watch times, Sept. 8	=	8^h 07^m 48.5^s
Watch fast on P.S.T.	=	−16.0

Appendix I

P.S.T. of observation, Sept. 8	=	8 07	32.5
Standard meridian, W.	=	8 00	00
Corresponding G.M.T. of observation, Sept. 8	=	16 07	32.5
Equation of Time, by interpolation (*Ephem.*, p. 6)	=	+02	29.5
G.A.T. of observation, Sept. 8	=	16 10	02.0
Sun's declination at G.A.N., Sept. 8 (*Ephem.*, p. 6)	= N.	5° 31'	56.7"
Change in 4^h 10^m 02^s after G.A.N. = $4.1672^h \times (-56.5")$	=	−03	55.4
Sun's declination when observed, δ = N.		5° 28	01.3
$90° - \delta = p =$		84° 31'	59"

The three sides of the astronomical triangle are now known:

$$\omega = 90° - \phi = 90° - 35° \, 52' \, 06" = 54° \, 07' \, 54"$$
$$p = 84 \ 31 \ 59$$
$$\zeta = 59 \ 46 \ 18$$

A suitable formula is:

$$\cos Z = (\cos p - \cos \omega . \cos \zeta)/\sin \omega . \sin \zeta$$

From this,

$$Z = 106° \, 34' \, 21"$$

The sun was in the east, $\therefore Z = A$, sun (see **10.02**)

Thus A, sun = 106° 34' 21"
From above, included angle = 86 36 21
∴ Azimuth of R.O. = 19° 58' 00"

To check the determination, another, similar observation should be made in the afternoon when the sun is near the prime vertical in the west. With careful work, the mean of the two results should be within 01' of the true value.

AI.03.02 Observation of the sun for azimuth by finding its hour angle

In this type of determination we must know the latitude ϕ, the declination δ (from tables), the L.Std.T. of the observation as accurately as we can get it (by checks against time-signals), the longitude λ, and the approximate altitude h (to calculate the plate bubble correction). Essentially, we must know two sides ω, p, of the astronomical triangle and the included angle t, to compute Z and A.

EXAMPLE (refer to **AI.08** for the appropriate tables)
Place: California, Stn. 22 Longitude: 118° 48' 18" (7^h 55^m 13.2^s) W.
Latitude: 35° 52' 06" N. Date: 1980 September 8

Astronomy for surveyors

R.O. Stn. 25
Plate level: 30″ per division
Standard meridian: 8^h W.
Watch: 15^s fast on P.S.T.
Object observed: sun in the west

Face	Object	Horizontal circle	Approx. altitude	Watch time	Plate bubble L	Plate bubble R
L	R.O.	19° 57′ 48″				
L	☼	269 34 20	10° 00′	$5^h\ 17^m\ 17.6^s$ p.m.	3	5
R	☼	88 51 30		5 18 29.3	4	4
R	R.O.	199 57 20				

Plate bubble correction:
$\{(7-9)/4\} \times 30″ \times 0.176 = -03″$

Mean horizontal circle reading on sun	=	269° 12′ 55″
Plate bubble correction	=	−03
Corrected horizontal circle reading on sun	=	269 12 52
Mean horizontal circle reading on R.O.	=	19 57 34
Included angle, sun to R.O., clockwise	=	110° 44′ 42″

Sun's hour angle:

Mean of watch times, Sept. 8	=	$17^h\ 17^m\ 53.5^s$
Watch fast on P.S.T.	=	−15.0
P.S.T. of observation, Sept. 8	=	17 17 38.5
Standard meridian, W.	=	8 00 00
Corresponding G.M.T. of obsn (= U.T.C), Sept. 9	=	1 17 38.5
DUT1 (see **6.07**)	=	0.0
U.T.1 of observation, Sept. 9	=	1 17 38.5
Eq. T., interpolated from *Ephemeris*, p. 6	=	+02 37.4
G.A.T. of observation, Sept. 9	=	1 20 15.9
G.A.T. of sun's transit, Sept. 8 (G.A.N.)	=	12 00 00
G.H.A. sun at time of observation	=	13 20 15.9
Sun's declination at G.A.N., Sept. 8	= N.	5° 31′ 56.7″
Change in $13^h\ 20^m\ 15.9^s = 13.3378 \times (56.50″)$ (*Ephemeris*, p. 6)	=	−12 33.6
Sun's declination at time of observation	δ = N.	5° 19′ 23.1″
	$p = 90° - \delta =$	84° 40′ 36.9″

Sun's local hour angle (L.H.A.):

G.H.A. sun, from above	=	$13^h\ 20^m\ 15.9$
Longitude of observing station, W.	=	7 55 13.2
L.H.A. sun when observed	=	$5^h\ 25^m\ 02.7^s$
	i.e. $t =$	81° 15′ 41″

Appendix I

A suitable formula is:

$$\cot Z = (\cot p . \sin \omega - \cos \omega . \cos t)/\sin t$$

where
$\omega = 90° - \phi =$ 54° 07′ 54″
$p =$ 84 40 37 from above
$t =$ 81 15 41 from above

From this, $Z =$ 90 47 00
so that A, sun $= 269$ 13 00
Included angle $= 110$ 44 42

and A, R.O. 19 57 42

As a check, and to balance the observation properly, another similar observation should be made in the morning, when the sun is near the prime vertical in the east.

If we use *SALS* to find the hour angle of the sun, the procedure would be as follows:

From above, G.M.T. (= U.T.C.) of obsn, Sept. 9	=	1^h 17^m 38.5^s
DUT1	=	0.0
U.T.1 of observation, Sept 9	=	1 17 38.5
E at U.T. 0^h, Sept. 9	=	12 02 36.3
ΔE in 1^h 17^m 38.5^s	=	+01.1
G.H.A. sun = U.T.1 + E at U.T. 0^h + ΔE	=	13 20 15.9
Longitude of observing station, W.	=	7 55 13.2
L.H.A. sun,	$t =$	5 25 02.7

This value agrees with that obtained by using the *Ephemeris*.

AI.03.03 Observation of stars for azimuth by measuring their altitudes
See preamble to **AI.03.01** for the kind of information that is required.

EXAMPLE (refer to **AI.08** for the appropriate *Ephemeris* tables)
Place: California, Stn. 22   R.O.: Stn. 25
Latitude: 35° 52′ 06″ N.   Plate level: 30″ per division
Longitude: 118° 48′ 18″ (7^h 55^m 13.2^s) W.   Standard meridian: 8^h W.
Date: 1980 May 1   Watch: 28^s fast on P.S.T.

Object observed: Procyon (no. 12 in *Ephemeris*) in the west

Face	Object	Horizontal circle	Altitude	Plate bubble L	R	Watch time	Baro. Temp.
L	R.O.	10° 58′ 30″					1009 mb
L	*	241 21 47	33° 08′ 01″	4	3	8^h 22^m 34.5 p.m.	(29.8 in.)
R	*	61 34 53	32 54 20	4	3	8 23 44.5	17.8 °C
R	R.O.	190 58 00					(64 °F)

333

Plate bubble correction:
$\{(8-6)/4\} \times 30'' \times \tan 33° = +10''$

Mean horizontal circle reading on star =	241° 28′	20″
Plate bubble correction =		+10
Corrected horizontal circle reading on star =	241 28	30
Mean horizontal circle reading on R.O. =	10 58	15
Included angle, star to R.O., clockwise =	129 29	45

For stars' altitude:

Mean observed altitude of star = 33° 01′ 11″
Refraction correction = $88'' \times 1.01 \times 0.97$
(*Ephemeris*, p. 25) = −01 26

Corrected observed altitude, $h =$ 32° 59′ 45″
$\zeta = 90° - h =$ 57 00 15

Declination of Procyon (*Ephemeris*, p. 12), $\delta = $ +5° 16.4′
$p = 90° - \delta =$ 84 43 36
Co-latitude of observing station, $\omega = 90° - \phi =$ 54 07 54

We now have the three sides of the astronomical triangle, ω, p, ζ.
Using formula:
$$\cos Z = (\cos p - \cos \omega . \cos \zeta)/\sin \omega . \sin \zeta$$
we get $Z = 109° 31′ 36''$, star in the west
$\therefore A$, star $= 250\ 28\ 24$
Included angle $= 129\ 29\ 45$
and Azimuth R.O.$= 19° 58′ 09''$

This observation should be checked and balanced by another, similar one to a star in the east near the prime vertical, and the mean of the two resulting azimuths accepted (they should agree to within a minute or so of arc).

AI.03.04 Observation of stars for azimuth by finding their hour angles

See preamble to **AI.03.02** for the kind of information that is required.

EXAMPLE (refer to **AI.08** for the appropriate *Ephemeris* tables)

We shall use the observation in **AI.03.03** above, but assume that no altitudes were measured.

Mean of watch times =	20ʰ 23ᵐ	09.5ˢ
Watch fast on P.S.T. =		−28.0
P.S.T. of observation, May 1 =	20 22	41.5
Longitude of standard meridian =	8 00	00
Corresponding G.M.T. (= U.T.C.), May 2 =	4 22	41.5
DUT1 =		+0.3
U.T.1 of observation =	4 22	41.8

Appendix I

Longitude of observing station, W.	=	7	55	13.2
L.M.T. of observation, May 1	=	20	27	28.6 P
G.M.T. of transit of Procyon, May 1 *(Ephem.,* p.12)	=	16	59	06
Sidereal correction for longitude W. (*Ephem.,* p. 27)	=		−01	18
L.M.T. of local transit of Procyon	=	16	57	48 Q
Hour angle, west, in mean time-units (= P − Q)	=	3	29	40.6
Convert to sidereal units, since we are dealing with a star	=		+0	34
Hour angle, west, in sidereal units, t =		3	30	14.6
i.e. t =		52°	33′	39″
From **AI.03.03**, we know that ω =		54°	07′	54″
and p =		84	43	36

Using formula:
$$\cot Z = (\cot p.\sin\omega - \cos\omega.\cos t)/\sin t,$$
we get $Z = 109°\ 30'\ 53''$, star in west
$\therefore A$, star $= 250\ \ 29\ \ 07$
Included angle $= 129\ \ 29\ \ 45$ (see **AI.03.03** above)
and Azimuth R.O. $= \overline{19°\ 58'\ 52''}$

This value of the azimuth differs from that obtained in **AI.03.03** (19° 58′ 09″ by using measured altitudes) by 43″ which is a significant amount. It is due mainly, however, to the fact that the G.M.T.'s of transits of stars are given in the tables of the *Ephemeris* to 0.1ᵐ or 06ˢ of time, and so could be in error by nearly 03ˢ of time (or 45″ of arc). A more accurate answer in this method may be found by using *SALS*:

U.T.1 of observation, from above, May 2	=	4ʰ	22ᵐ	41.8ˢ
R at U.T. 0ʰ, May 2	=	14	40	14.3
ΔR in 4ʰ 22ᵐ 41.8ˢ	=			43.2
U.T.1 of obsn + R at U.T. 0ʰ + ΔR = G.S.T. of obsn	=	19	03	39.3
R.A. of Procyon (No. 212 in *SALS*)	=	7	38	15.2
∴ G.H.A. Procyon = G.S.T. − R.A.	=	11	25	24.1
Longitude of observing station, W.	=	7	55	13.2
∴ L.H.A. Procyon, = t =		3	30	10.9
i.e. t =		52°	32′	44″
From **AI.03.03**, we know that ω =		54	07	54
From *SALS*, declination of Procyon = N. 5° 16′ 24″ ∴ p =		84	43	36
From formula given above, Z =		109	31	35
∴ A, star =		250	28	25
Included angle =		129	29	45
and Azimuth R.O. =		19°	58′	10″

Astronomy for surveyors

This agrees closely with the azimuth found from the altitudes of the star. It would thus appear that using the *Ephemeris* for ex-meridian azimuths by hour angles of stars could produce approximations to the order of 01'. If more accuracy is required with this method, *SALS* should be used.

The observation should be checked and balanced by another, similar one on the other side of the meridian to a star near the prime vertical, and the mean of the two accepted.

AI.03.05   Observation of the Pole Star, Polaris, for azimuth

A convenient method of determining azimuth is to use the Pole Star (α Ursae Minoris), and a special set of tables is provided in both the *Ephemeris* and *SALS* for this purpose. In contrast with the pole star of the southern hemisphere (σ Octantis) which, at magnitude 5.5, is hard to see with the naked eye, Polaris at 2.1 is almost 23 times as bright and is easily identified in the constellation of the Little Bear.

AI.03.05.1   Use of the table "Azimuth of Polaris at all Hour Angles" in the *Ephemeris*

EXAMPLE (refer to **AI.08** for the appropriate *Ephemeris* tables)
Place: California, Stn. 22   R.O. Stn. 25
Latitude: 35° 52' 06" N.   Plate level: 30" per division
Longitude: 118° 48' 18" (7^h 55^m 13.2^s) W.   Standard meridian: 8^h W.
Date: 1980 May 1   Watch: 28^s fast on P.S.T.
                   Object observed: Polaris

Face	Object	Horizontal circle	Approx. altitude	Watch time	Plate bubble L	R
L	R.O.	20° 42' 20"				
L	*	0 00 50	35° 17'	8^h 22^m 34.5^s p.m.	3	4
R	*	180 00 30		8 23 44.5	4	4
R	R.O.	20 42 00				

Plate bubble correction:
$\{(7 - 8)/4\} \times 30" \times 0.708 = -05"$

Mean horizontal circle reading on Polaris	=	0° 00' 40"
Plate bubble correction	=	—05
Corrected horizontal circle reading on Polaris	=	0 00 35
Mean horizontal circle reading on R.O.	=	20 42 10
Included angle, star to R.O., clockwise	=	20° 41' 35"

The watch times and the watch offset from P.S.T. are the same as those in the example of **AI.03.04**, so:

L.M.T. of observation, from **AI.03.04**, May 1	=	20^h 27^m 28.6^sP
G.M.T. of upper transit of Polaris (*Ephem.*, p. 12, no. 4) =		11 32 42
Sidereal correction for west longitude (*Ephem.*, p. 27)	=	—01 18

Appendix I

L.M.T. of transit of Polaris	=	11 31 24 Q
Mean time hour angle = P − Q	=	8^h 56^m 04.6^s

Entering the table on p. 19 on the *Ephemeris* with
latitude 35° 52.1′ N. and mean time hour angle
8^h 56.1^m, and interpolating between rows and
columns, Azimuth of Polaris when observed   =   0° 43.3′

The declination of Polaris (no. 4) for May 1 at transit
is given as +89° 10.3′ on p. 12 of the *Ephemeris*. It is
changing so slowly that there is no need to correct this
value for the short interval since the star was at
transit, but there *is* a need to correct the azimuth we
have just found, since the table we entered is based
upon a mean declination of +89° 10′ 30″ (see top of
p. 19). On the right-hand side of page 19, under
"10′ 20″" (the nearest value to the actual declination
of +89° 10′ 18″), and for the mean time hour angle
8^h 56.1^m, the correction to the azimuth for difference
of declination   =   +0° 00.2′

Thus, corrected azimuth of Polaris when observed   =   0° 43.5′

This value, 0° 43.5′, is actually the zenith angle (*Z*)
of the star, and since the hour angle is between 0^h
and 12^h, the star must have been west of the meridian,
hence its true azimuth is 360° − *Z* (see **10.02**).

∴ azimuth of Polaris on 0° − 360° system	=	359° 16.5′
But included angle, star to R.O. (found earlier)	=	20 41.6
Thus, azimuth of R.O.	=	19° 58.1′

AI.03.05.2   Use of the formula given in the *Ephemeris*

At the foot of pages 18 and 19 of the *Ephemeris* a formula is provided for use "when values in seconds are required. If A = azimuth angle; t = sidereal hour angle; ϕ = latitude of station; δ = declination of Polaris;

$$\tan A = \sin t / (\cos\phi . \tan\delta - \sin\phi . \cos t).$$

The product, $\sin\phi . \cos t$, is subtracted for hour angles 0° to 90° and added for hours angles from 90° to 180°." In the formula given, A is equal to Z, the zenith angle as used elsewhere in this book (see **10.01**).

Applying the formula to this example:

Mean time hour angle (from **AI.03.05.1** above)	=	8^h 56^m 04.3^s
Convert to sidereal units (*Ephemeris*, p. 27)	=	+01 28
Sidereal hour angle of Polaris,	t =	8 57 32.3
	i.e. t =	134° 23′ 05″
From *Ephemeris*, p. 3, declination of Polaris, May 1,	δ =	+89 10 18
Latitude of observing station, N.,	ϕ =	35 52 06

Astronomy for surveyors

Applying the formula of the *Ephemeris*,	$A(=Z) =$	0°	43'	31"
Polaris was west of the meridian ∴ its azimuth				
	$A = 360° - Z =$	359	16	29
Included angle, star to R.O., clockwise as above	$=$	20	41	35
∴ Azimuth of R.O.,	A, R.O. $=$	19°	58'	04"

AI.03.05.3 Use of *SALS* and the method of Chapter 10, Section 6

Mean watch time of observation, May 1	$=$	8^h	23^m	0.95^s p.m.
Watch fast on P.S.T.	$=$			-28.0
∴ P.S.T. of observation, May 1	$=$	20	22	41.5
Standard meridian, W.	$=$	8	00	00
∴ U.T.C. of observation, May 2	$=$	4	22	41.5
DUT1	$=$			$+0.3$
U.T.1 of observation, May 2	$=$	4	22	41.8
R at U.T.1 0^h, May 2	$=$	14	40	14.3
ΔR in $4^h\ 22^m\ 41.8^s$	$=$			43.2
G.S.T. of obsn. = U.T.1 + R at U.T. 0^h + ΔR	$=$	19	03	39.3
Longitude of observing station, W.	$=$	7	55	13.2
∴ L.S.T. of observation	$=$	11	08	26.1
R.A. of Polaris, May 1 (*SALS*, p. 54)	$=$	2	10	54.4
Hour angle of Polaris at observation = L.S.T. − R.A. = $t =$		8	57	31.7
	i.e. $t =$	134°	22'	56"
Declination of Polaris, May 1 (*SALS*, p. 54),	$\delta =$ N.89°	10'	19"	
Thus polar distance $(90° - \delta)$,	$p =$	0	49	41
	$=$			2981"

Formula:
$$Z'' = \{p.\sin t + (p'')^2.\sin t.\cos t.\tan\phi.\sin 1''\}\sec\phi$$
$$= \{2130.5'' - 15.6''\}\sec\phi = 2609.8''$$
i.e. $Z = 0° 43' 30''$

Thus, A Polaris $= 360° - Z$		$=$	359° 16' 30"	
Included angle, as before		$=$	20 41 35	
Azimuth of R.O.	$= A$, R.O. $=$		19° 58' 05"	

AI.03.05.4 Use of the "Pole Star Table" in *SALS*

Yet another method can be found in the Pole Star Table of *SALS* (pp. 56-9) where the azimuth of Polaris is given by $(b_0 + b_1 + b_2) \sec\phi$. The table (see **AI.08**) is entered with the local sidereal time (L.S.T.) of the observation, the latitude, and the month. In this observation (recorded in **AI.03.05.1**), the L.S.T. (from **AI.03.05.3** above) $= 11^h\ 08^m\ 26.1^s$

Appendix I

the latitude,	$\phi =$	N. 35° 52′ 06″
the month	=	May

In column headed "11^h" (*SALS*, p. 57), by interpolation,	$b_0 =$	−35.1′
In the same column, "Lat" table, by interpolation,	$b_1 =$	−0.15
In the same column, "Month" table, for May,	$b_2 =$	0.0
	$b_0 + b_1 + b_2 =$	−35.25′
Thus, $Z = (-35.25')\sec\phi$	=	−0° 43′ 30″
∴ A, Polaris	=	359 16 30
Included angle, as before	=	20 41 35
Azimuth of R.O.	$= A$, R.O. $=$	19° 58′ 05″

AI.03.05.5 Comment

Before leaving Polaris azimuths, we should note that pages 1 through 8 of the *Ephemeris* give the G.M.T.'s of Eastern and Western Elongation of Polaris for latitude 40 °N., as well as that of Upper Culmination. When Polaris is at E. or W. elongation it is moving vertically in the field of view (upwards for W. and downwards for E. in an inverting telescope)—very slowly—and these are, therefore, very good times to observe it for azimuth since a face-left and a face-right observation may easily be made before it moves more than a couple of seconds of arc in azimuth. The observation is thus a simple one which should be made quite comfortably within a period of 4 minutes of time starting 2 minutes before and ending 2 minutes after elongation, reading the plate bubble ends in the usual way on each face, and the approximate altitude, to correct for any dislevelment of the trunnion axis. The mean of the two horizontal circle readings of the pointings to Polaris should be taken, corrected for plate bubble, and compared with the mean of the pointings to the R.O. to get the included angle between star and R.O. The azimuth of Polaris (really the Z angle) is readily obtained from the table on page 22 of the *Ephemeris*.

EXAMPLE (refer to **AI.08** for the appropriate *Ephemeris* tables)

Suppose an observation had been made at the station in the example in **AI.03.05.1** above at E. elongation on the same date, 1980 May 1.

For latitude 40 °N. the table on the right-hand side of page 3 of the *Ephemeris* gives the G.M.T. of E. elongation of Polaris as 5^h 36.5^m a.m. The latitude of the observing station is 35° 52′ 06″ N., and the declination of Polaris is +89° 10′ 18″. Since the latitude of the station is not 40 °N., a small correction must be made to the tabular G.M.T. of E. elongation:

Tabular G.M.T. of E. elongation of Polaris for $\phi = 40°$N. =		5^h 36.5^m a.m.
Interpolating in the second column in table on p. 22, the correction for the latitude difference	=	−0.4
Hence, G.M.T. of E. elongation for station latitude	=	5^h 36.1^m

Astronomy for surveyors

Sidereal correction for W. longitude 7^h 55^m 13.2^s (p. 27) =	−1.3
L.M.T. of E. elongation at station	5^h 34.8^m
Standard meridian W. of local meridian (slow) ($\lambda - 8^h$) =	−04.8
Thus, P.S.T. of E. elongation, May 1 =	5^h 30.0^m a.m.

From the table on page 22 of the *Ephemeris*, the
   azimuth (*Z* angle) of Polaris at E. elongation may
   now be found by interpolation using the appropriate
   latitude and declination values   =   1° 01′ 20″
Since Polaris was at E. elongation, this *Z* value
   is the true astronomical azimuth (see **10.02**), *A*, star   =   1° 01′ 20″
Application of the included angle between the star
   and R.O. will give the azimuth of the R.O.

AI.04   **Determination of latitude** (see Chapter 11 also)

AI.04.01   **Observation of the sun for latitude by circum-meridian pointings**

EXAMPLES (refer to **AI.08** for the appropriate *Ephemeris* tables)

Place: Florida, Stn. 105	Standard meridian: 5^h W.
Approximate latitude: 28 °N.	Watch: 16.9^s fast on E.S.T.
Longitude: 82° 13′ 40″ (5^h 28^m 54.7^s) W.	Object obsd: sun near transit
Date: 1980 September 24	Barometer: 1009 mb (29.8 in.)
DUT1: 0.0^s	Temperature: 25.6 °C (78 °F)

     Remarks: eyepiece prism needed for high altitude of sun

			Watch
Face	Object	Altitude	(approx. E.S.T.)
L	☉	60° 58′ 55″	12^h 15^m 44^s
R	☌	61 31 44	17 24
R	☉	61 00 18	18 44
L	☌	61 32 32	20 38
L	☉	61 00 34	21 38
R	☌	61 32 15	23 12
R	☉	60 59 46	24 46
L	☌	61 30 45	26 27

Mean altitude: 61° 15′ 51′

For declination:

Sun's dec. at transit at Greenwich (*Ephemeris*, p.6) =	S.	0° 37′ 41.6″
Change in Dec. because of longitude diff. (interpolated) =		+05 20.2
Sun's Dec. at transit at place of observation =	S.	0° 43′ 01.8″

For watch time of transit:

L.A.T. of transit at place of observation =	12^h 00^m 00^s
Equation of Time, transit, Greenwich = 08^m 03.99^s (p.6)	
Change in longitude difference (interpolated) = +04.78	
Equation of Time at transit at station = 08^m 08.77^s =	−08 08.8

Appendix I

L.M.T. of transit of sun at place of obsn	=		11ʰ 51ᵐ 51.2ˢ
Standard meridian east of (fast on) local meridian (= $\lambda - 5^h$)	=		+28 54.7
∴ E.S.T. of transit at Stn. 105	=		12 20 45.9
Watch fast on E.S.T.	=		16.9
∴ Watch time of transit at Stn. 105	=		12ʰ 21ᵐ 02.8ˢ

For mean altitude:
Mean of all observed altitudes = 61° 15′ 51″
Refraction correction = $31'' \times 1.01 \times 0.94$ (*Ephem.*, p. 25) = −29
Parallax correction = (*Ephem.*, p. 25) = +04
Corrected mean observed altitude = 61° 15′ 26″

Two highest altitudes, L and R, $\bar{\odot}$ and $\underline{\odot}$ = 61° 00′ 18″
 and = 61 32 32
Mean of these two highest altitudes = 61 16 25
Refraction correction = $31'' \times 1.01 \times 0.94$ = −29
Parallax correction = +04

Corrected mean of highest two altitudes, h_z = 61° 16′ 00″
Sun's declination, from above, δ = S. 0° 43′ 02″

Approximate co-latitude, ω_{ap} = 61° 59′ 02″
∴ Approximate latitude, ϕ_{ap} = 28° 00′ 58″
Factor B (see **11.07**) = $(\cos\phi_{ap}.\cos\delta)/\cos h_z$ = 1.8362

Sun's hour angles at individual pointings (watch time of transit minus watch time of pointing before transit, and vice versa after transit), and corresponding "m" value, where $m = 0.0005454 t^2$ (t in seconds)

H.A.	m
5ᵐ 18.8ˢ	55″
3 38.8	26
2 18.8	11
0 24.8	0
0 35.2	1
2 09.2	9
3 43.2	27
5 24.2	57
sum =	186

Mean = 23.3″ = m_0
Bm_0 = 42.8″

Astronomy for surveyors

For latitude:

Corrected mean altitude, from above	=	61° 15′ 26″	
Bm_0 (positive because at upper transit)	=	+43	
Mean altitude corrected to meridian altitude	=	61 16 09	
Declination of sun at transit	= S.	0 43 02	
Co-latitude, ω (see Fig. 11.1) = $(90° - \phi)$ =		61 59 11	
∴ Latitude of Stn. 105, ϕ =		28° 00′ 49″ N.	

We could use *SALS* to find the watch time of local transit:

Sun's L.H.A. at local transit	=	$24^h\ 00^m\ 00^s$
Longitude of observing station 105, W.	=	5 28 54.7
Sun's G.H.A. at local transit (= U.T. + E, both as yet unknown)	=	29 28 54.7 P
U.T. + E at U.T. 12^h, Sept. 24	=	24 08 04.1
Interval after U.T. 12^h until local transit	=	5 20 50.6
E at U.T. 12^h, Sept. 24 = $12^h\ 08^m\ 04.1^s$		
Change in E in $5^h\ 20^m\ 50.6^s$ = +04.6		
E at local transit = 12 08 08.7		12 08 08.7 Q
U.T.1 of local transit (= P – Q)	=	17 20 46.0
DUT1	=	0.0
U.T.C. of local transit (= U.T.1 – DUT1)	=	17 20 46.0
Standard meridian, W.	=	5 00 00
∴ E.S.T. of local transit	=	12 20 46.0
Watch fast on E.S.T.	=	+16.9
∴ Watch time of local transit	=	$12^h\ 21^m\ 02.9^s$

(This is 0.1^s different from that using the *Ephemeris*.)

When using *SALS*, find the m values from the table on page 68, which may be quicker and more convenient than computing them individually from the formula.

AI.04.02 Observation of the Pole Star for latitude by measuring its altitude

The approximate latitude (±1°) of a place in the northern hemisphere may be found immediately by just measuring the altitude of Polaris. If the time of observation is noted, the hour angle and polar distance may be used to compute a correction which will convert the measured altitude to the altitude of the celestial pole, i.e. the latitude of the place of observation. Instead of computing this small correction, it may be found from tables in the *Ephemeris* or *SALS*.

Appendix I

AI.04.02.1 Using *Ephemeris* tables

EXAMPLE (refer to **AI.08** for appropriate tables)

The mean of two altitudes (left and right faces) of Polaris measured at a place in Nebraska, U.S.A., whose approximate latitude is 41 °N. and longitude 100° 46' 30" (6^h 43^m 06^s) W., on 1980 May 16, was 40° 20' 18"; the mean of the two times was 11^h 00^m 47^s p.m. taken with a watch which was 51.0^s fast on Central Standard Time. The standard meridian of the region is 6^h W. The temperature was 23.9 °C (75 °F) and the air pressure 914 mb (27.0 in.). What is the latitude of the place?

Watch time of observation, May 16	=	23^h 00^m 47^s
Watch fast on C.S.T.	=	51
C.S.T. of observation, May 16	=	22 59 56
Standard meridian east of (fast on) local meridian ($6^h - \lambda$)	=	−43 06
∴ L.M.T. of observation, May 16	=	22 16 50 P
G.M.T. of transit of Polaris, May 16 (*Ephem.*, p. 12)	=	10 33 54 a.m.
Sidereal correction for W. longitude (*Ephem.*, p. 27)	=	−01 06
∴ L.M.T. of transit of Polaris, May 16	=	10 32 48 a.m. Q
Mean time hour angle of Polaris (= P − Q)	=	11^h 44^m 02^s
(this H.A. is close to 12^h, ∴ we infer that Polaris was near lower transit and ∴ moving only very slowly in altitude)		
Declination of Polaris, by interpolation (*Ephem.*, p. 4)	=	+89° 10' 14"
Observed altitude of Polaris	=	40 20 18
Refraction correction = 68" × 0.91 × 0.95	=	−59
Corrected observed altitude of Polaris	=	40 19 19 R
Entering the table on p. 23 of the *Ephemeris* with the mean time hour angle and declination found above, and interpolating, Polaris is below the pole by the amount	=	0 49 41 S
Supplemental correction to vertical angle reading (p. 23)	=	00 T
Latitude (= R + S + T), ϕ =		41° 09' 00" N.

AI.04.02.2 Using *SALS* tables (refer to **AI.08**)

We may also use *SALS* tables to reduce this observation:

C.S.T. of observation, 1980 May 16, from above	=	22^h 59^m 56^s
Standard meridian, W.	=	6 00 00
U.T.C. of observation, May 17	=	4 59 56
DUT1	=	+0.3
U.T.1 of observation, May 17 (= UTC + DUT1)	=	4 59 56.3
R at U.T. 0^h, May 17	=	15 39 22.6
ΔR for 4^h 59^m 56.3^s	=	49.3

Astronomy for surveyors

G.S.T. of observation = U.T.1 + R @ U.T. 0^h + ΔR	=	20 40	08.2
Longitude, W.	=	6 43	06
∴ L.S.T. of observation	=	$13^h\ 57^m$	02.2^s

Entering the Pole Star Table on p. 58 of *SALS* (1980) with L.S.T. = $13^h\ 57^m\ 02.2^s$, Latitude 40 °N., and Month = May, we find

a_0 =		+49.4'
a_1 =		0.0
a_2 =		+0.3
Sum =		+49.7
	=	+0° 49' 42"
Corrected observed altitude, from above	=	40 19 19
Latitude = Corrd obsd alt. + $a_0 + a_1 + a_2$,	ϕ =	41° 09' 01" N.

AI.04.02.3 By computation, as in **11.15**

The latitude may also be found by computation using the formula given in Chapter 11, section **11.15**:

From above, (*SALS* working), L.S.T. of obsn	=	$13^h\ 57^m$	02.2^s
R.A. of Polaris, *SALS*, p. 54	=	2 11	03
H.A. of Polaris = L.S.T. − H.A.,	t =	11 45	59.2
	i.e. t =	176° 29'	48"
Polar distance of Polaris = 90° − δ, from above,	p =	0° 49'	46"
	i.e. p =		2986"

Formula:

$$\phi = h - p.\cos t + \tfrac{1}{2}p^2.\tan h.\sin^2 t.\sin 1"$$
$$= 40° 19' 19" + 2980.4" + 0.1" = 40° 19' 19" + 0° 49' 41"$$
$$= 41° 09' 00" \text{ N}.$$

AI.04.02.4 By observing Polaris at transit

It should be noted that pages 1 through 8 of the *Ephemeris* give the G.M.T.'s of upper culmination of Polaris for every day of the year (refer to **AI.08**). If its altitude is observed once on each face within a period of 4 minutes, from 2^m before to 2^m after culmination, the mean of these corrected for refraction will give the (latitude + polar distance) (see Fig. 11.1). The polar distance, found from the declination given in the same tables, may then be subtracted from this corrected mean altitude to yield the latitude. The result will probably be a few seconds in error because the star culminates instantaneously and is very slightly lower within a couple of minutes on either side of the meridian; also there will inevitably be some observational error. The L.M.T. of culmination, and hence the L.Std.T., for places within the U.S.A. may be found from the G.M.T. of culmination by subtracting the longitude correction (p. 27, *Ephemeris*) in the usual way, and then applying the difference of longitude between local and standard meridians.

Appendix I

Note: the G.M.T. of lower culmination will take place 12 sidereal hours after upper culmination, i.e. 11^h 59^m 02^s mean time-interval afterwards. If Polaris is observed for latitude at lower transit, the polar distance must be *added* to the corrected mean observed altitude.

AI.05 The determination of local time (see Chapter 12 also)

AI.05.01 Using the sun at transit

This method requires the theodolite to be in very good adjustment and set up so that the telescope transits in the plane of the meridian. In other words, when levelled, the instrument's trunnion axis must also be level, and the true astronomical azimuth of some R.O. is needed so that the telescope may be turned from it into the meridian. The higher the altitude of the sun when it transits the meridian, the greater the error will be in the time-determination due to any dislevelment of the trunnion axis (see **7.11**). (This error may be computed and allowed for if the stride-level or plate-level bubble position is recorded at the time of the observation. The effect of an error in setting the telescope in the meridian is discussed in **12.05**.)

The method is described in **12.03** and **12.04**.

EXAMPLE (refer to **AI.08** for appropriate *Ephemeris* tables)

Assume that the mean of the watch times of contact of the E. and W. limbs of the sun with the vertical cross-hair was 12^h 03^m 25.6s p.m. on 1980 May 7 at a place in California whose longitude is 121° 40′ 24″ (8^h 06^m 41.6^s) W. The watch was supposed to be keeping Pacific Standard Time. What was its error? The standard meridian is 8^h W.

L.A.N. (the instant when the sun's centre crosses the observer's meridian) will be 8^h 06^m 41.6^s (the longitude difference) later than G.A.N. The *Ephemeris* gives the Equation of Time at G.A.N., so to get its value at L.A.N. we must interpolate between the values for G.A.N. on 7 and 8 May. The difference between these values, from p. 3 of the *Ephemeris*, is $+3.49^s$. The interpolating factor is $\Delta\lambda/24^h$, or 8^h 06^m $41.6^s/24^h$, or 0.3380.

L.A.T. of L.A.N.	=	12^h 00^m 00^s
Equation of Time at L.A.N. = 3^m 30.99^s +		
(3.49^s × 0.3380)	=	−03 32.2
L.M.T. of L.A.N.	=	11 56 27.8
DUT1	=	0.3
Local time in terms of Coordinated Time	=	11 56 27.5
Standard meridian east of (fast on) local meridian	=	+06 41.6
Pacific Standard Time of observation	=	12 03 09.1
Mean watch time of observation	=	12 03 25.6
∴ Watch fast on P.S.T.	=	16.5^s

Using *SALS*:

Sun's L.H.A. at L.A.N.	=	24^h 00^m 00^s
Longitude of observing station, W.	=	8 06 41.6
Sun's G.H.A. at L.A.N. (and G.H.A. = U.T. + E)	=	32 06 41.6 P
But U.T. + E at U.T. 18^h, May 7 (*SALS*, p. 10)	=	30 03 31.9
∴ Interval between U.T. + E at G.A.N. and U.T. + E at U.T. 18^h	=	2 03 09.7

E at U.T. 18^h, May 7   = 12^h 03^m 31.9^s
ΔE in 2^h 03^m 09.7^s =   +0.3

∴ E at L.A.N.	= 12 03 32.2	=	12	03	32.2 Q
U.T.1 at L.A.N. = P − Q		=	20	03	09.4
DUT1		=			0.3
U.T.C. of L.A.N. (= U.T.1 − DUT1)		=	20	03	09.1
Standard meridian, W.		=	8	00	00
∴ P.S.T. of L.A.N.		=	12	03	09.1
Mean of watch times		=	12	03	25.6
∴ Watch fast on P.S.T.		=			16.5^s

AI.05.02   Using the sun in the east or west

EXAMPLE (refer to **AI.08** for appropriate *Ephemeris* tables)

Consider the example in **AI.03.01**, but imagine that only the altitudes and watch times were recorded. We shall find the watch error. The date was 1980 September 8.

From that example, the three sides of the astronomical triangle were:

$\omega = 54°$ 07′ 54″, from the given latitude;

$p = 84$ 31 59 , from the tabular value of the declination;

$\zeta = 59$ 46 18 , from the corrected observed altitude.

We may find the H.A., t, from:

$$\cos t = (\cos \zeta - \cos \omega . \cos p)/\sin \omega . \sin p$$

From this, L.H.A. (east, because morning obsn),	$t =$	3^h 45^m 11.2^s
Hence L.A.T. of observation = 12^h − east H.A.	=	8 14 48.8 a.m.
Longitude of observing station, W.	=	7 55 13.2
∴ G.A.T. of observation	=	16 10 02.0

Equation of Time at G.A.N.
 (*Ephemeris*, p. 6)   = 2^m 25.96^s
Change, in period since G.A.N.
 (4^h 10^m 02^s)   =   +03.60

Equation of Time at time of obsn   = 2 29.56   =   02 29.6

Appendix I

G.M.T. (= U.T.1) of observation	=	16	07	32.4
DUT1 (from radio time-signals, e.g. WWV)	=			0.0
U.T.C. = U.T.1 − DUT1	=	16	07	32.4
Standard meridian, W.	=	8	00	00
P.S.T. of observation	=	8	07	32.4
Mean of watch times	=	8	07	48.5
∴ Watch fast on P.S.T.	=			16.1^s

(From the data in the example in **AI.03.01**, the watch was, in fact, 16.0^s fast.)

We may use *SALS* to find the watch error:

From above, the east H.A. of the sun	=	3^h	45^m	11.2^s
The west H.A. = 24^h − E.H.A.	=	20	14	48.8
Longitude of observing station, W.	=	7	55	13.2
∴ G.H.A. of sun at time of observation	=	28	10	02.0 A
U.T. + E at U.T. 12^h, Sept 8 = G.H.A. at U.T. 12^h	=	24	02	26.0
Thus, the observation was later than U.T. 12^h by amt	=	4	07	36.0
E at U.T. 12^h, Sept. 8 = 12^h 02^m 26.0^s				
ΔE for 4^h 07^m 36^s = +03.5				
∴ E at time of observation = 12 02 29.5	=	12	02	29.5 B
U.T.1 of observation = A − B = (U.T. + E) − E	=	16	07	32.5
DUT1	=			0.0
U.T.C. of observation (= U.T.1 − DUT1)	=	16	07	32.5
Standard meridian, W.	=	8	00	00.0
∴ P.S.T. of observation	=	8	07	32.5
Mean of watch times	=	8	07	48.5
Thus, watch fast on P.S.T.	=			16.0^s

The observation should be balanced by another in the afternoon.

AI.05.03 Using a star in the east or west

EXAMPLE (refer to **AI.08** for appropriate *Ephemeris* tables)

Consider the example in **AI.03.03**, but imagine that only the altitudes and watch times were recorded. We shall find the watch error. The date was 1980 May 1.

In that example,

$$\omega = 54° \ 07' \ 54''$$
$$p = 84 \ \ 43 \ \ 36$$
$$\zeta = 57 \ \ 00 \ \ 15$$

Using the formula in **AI.05.02** above, $t = 52° \ 32' \ 42''$; we know this to be the west H.A., since the star was observed in the west.

Thus, H.A. of Procyon,	$= t =$		3^h 30^m 10.8^s
Convert to mean time-units	$=$		-00 34.0
H.A. in mean time-units (interval after upper transit)	$=$		3 29 36.8
G.M.T. of transit at Greenwich, May 1	$= 4^h 59^m 06^s$ (*Ephem.*, p.12)		
Sidereal correction for longitude W. (p. 27)	$=$	-01 18	
L.M.T. of local transit, May 1	$= 4$ 57 48 p.m. $=$		16 57 48
Thus, L.M.T. of obsn (in terms of astronomical time)	$=$		20 27 24.8
DUT1 (to bring to U.T.C.)	$=$		-0.3
Difference between local and standard meridians	$=$		-04 46.8
Pacific Standard Time of observation	$=$		20 22 37.7
Mean of watch times	$=$		20 23 09.5
Thus, watch fast on P.S.T.	$=$		31.8^s

The watch was, in fact, 28.0^s fast, and the difference is due largely to the G.M.T.'s of the transits of the stars being given only to 0.1^m (i.e. to 06^s) in the *Ephemeris*. There is bound to be some observational error as well, of course, depending upon the observer's personal equation (see **6.04**).

SALS gives the Right Ascensions of the stars in its catalogue to 0.1^s; therefore we should get a more accurate result by using it:

From above, star's L.H.A.,	$= t =$		3^h 30^m 10.8^s
R.A. of Procyon (No. 212 in *SALS*)	$=$		7 38 15.2
L.S.T. of observation $= t +$ R.A.	$=$		11 08 26.0
Longitude of observing station, W.	$=$		7 55 13.2
G.S.T. of observation $(=$ U.T. $+ R)$	$=$		19 03 39.2
U.T. $+ R$ at U.T. 0^h, May 2	$=$		14 40 14.3
Difference $=$ sidereal interval of obsn after U.T. 0^h, May 2	$=$		4 23 24.9
Convert to a mean time-interval after U.T. 0^h, May 2	$=$		-43.2
Mean time-interval of obsn after U.T. 0^h, May 2	$=$		4 22 41.7
i.e. U.T.1 of observation, May 2	$=$		4 22 41.7
DUT1	$=$		-0.3
U.T.C. of observation, May 2	$=$		4 22 41.4
Standard meridian, W.	$=$		8 00 00
$\therefore$ P.S.T. of observation, May 1	$=$		20 22 41.4
Mean of watch times $= 8^h 23^m 09.5^s$ p.m.	$=$		20 23 09.5
Thus, watch fast on P.S.T.	$=$		28.1^s

Appendix I

For best results, of course, a star near the prime vertical on the other side of the meridian would be observed as soon as possible afterwards, reduced in the same way, and the mean of the two derived watch errors accepted, allowing for the rate of the watch between observations.

AI.05.05   Using a star at transit
The method is described in **12.03** and **12.05** to **12.07**. It is similar to that in **AI.05.01** except that only one time is noted, i.e. when the star transits the vertical cross-hair near the centre of the telescope field. The L.M.T. of transit may be found in the usual way by applying the sidereal correction for longitude (*Ephemeris*, p. 27) to the G.M.T. of transit given in the tables. The L.Std.T. is found from the L.M.T. by application of the difference of longitude between the local and standard meridians, and DUT1.

Note that circumpolar stars will reach lower transit, below the pole but above the horizon, $11^h\ 58^m\ 02^s$ (mean time-interval) after upper transit. An observation at lower transit has the advantage of lower altitude and smaller error due to trunnion axis dislevelment. Such an observation should not be made too near the horizon, however, for fear of lateral refraction. For accuracy, *SALS* should be used rather than the *Ephemeris*, since the latter gives transit times to 06^s only. Unless the theodolite telescope can be set accurately in the meridian, the method is not likely to be very satisfactory (see **12.05**).

AI.06   Determination of longitude (see Chapter 13 also)
To find longitude we must derive two things: the L.M.T. and G.M.T. at the same instant; or the L.A.T. and corresponding G.A.T.; or the L.S.T. and corresponding G.S.T.; or the L.H.A. and corresponding G.H.A. of the object observed. The difference in each case is the longitude. The L.M.T., L.A.T., L.S.T. and L.H.A. are all usually found by local observation of altitude to give the third side of the astronomical triangle which can then be solved for t, the L.H.A. If necessary, the L.H.A. may be transformed into L.M.T., L.A.T. or G.S.T. To get the G.M.T., G.A.T., G.S.T. or G.H.A. we need to know the U.T.C. of observation and DUT1 accurately, and we can get them from radio time-signals, e.g. by listening to **WWV**. A knowledge of U.T.C. and DUT1 allows us to enter published tables to find the Greenwich values we need for the instant of our observation.

AI.06.01   Using the sun

EXAMPLE (refer to **AI.08** for the appropriate *Ephemeris* tables)
Consider again the example in **AI.03.01**, but imagine that we have observed only the altitudes and the watch times corresponding to them. The other things we know are the latitude, the declination and the watch error.

The computation proceeds as in **AI.05.02** to the stage where we find the L.A.T. September $8 = 8^h\ 14^m\ 48.8^s$. After this we use the watch time and its error on P.S.T.:

Astronomy for surveyors

Mean of watch times, Sept. 8	=	8^h	07^m	48.5^s
Watch fast on P.S.T.	=			16.0
PS.T. of observation, Sept. 8	=	8	07	32.5
Longitude of standard meridian, W.	=	8	00	00
U.T.C. of observation, Sept. 8	=	16	07	32.5
DUT1	=			0.0
U.T.1 (= U.T.C. + DUT1), Sept. 8	=	16	07	32.5
Interpolating in table on p. 6 of *Ephemeris*,				
Equation of Time at U.T.1 (G.M.T.) $16^h\ 07^m\ 32.5^s$	=		+02	29.5
∴ G.A.T. of observation, Sept. 8	=	16	10	02.0
But corresponding L.A.T., from above, Sept. 8	=	8	14	48.8
Thus, longitude = L.A.T. − G.A.T.	=	−7	55	13.2
	i.e. λ =	7^h	55^m	13.2^s W.

Or, using *SALS*:				
From **AI.05.02**, L.H.A. sun = 24^h − E.H.A.	=	20^h	14^m	48.8^s P
From **AI.06.01** above, U.T.1 of obsn, Sept. 8	=	16	07	32.5
E at U.T. 12^h, Sept. 8	=	12	02	26.0
ΔE in $4^h\ 07^m\ 32.5^s$	=			+03.5
G.H.A. sun = U.T.1 + E at U.T. 12^h + ΔE	=	28	10	02.0 Q
Thus, longitude = L.H.A. sun − G.H.A. sun = P − Q	=	−7	55	13.2
	i.e. λ =	7	55	13.2 W.

AI.06.02 Using the stars

As we have seen in **AI.05.03**, the watch error determined from observations of a star near the prime vertical differed from the correct value by several seconds, due to transit times of stars being shown in the *Ephemeris* to the nearest 0.1 of a minute. Longitudes determined by using the same tables are also likely to be incorrect by similar amounts because of the direct time–longitude relationship. However, the method of reduction to be used is:

EXAMPLE (refer to **AI.08** for the appropriate *Ephemeris* tables)

In the example in **AI.03.03**, ignore the azimuth aspect, and consider only the altitudes and corrected watch time.

P.S.T. of observation, May 1 (**AI.03.04**)	=	20^h	22^m	41.5^s
Standard meridian, W.	=	8	00	00
Corresponding G.M.T. (U.T.C.)	=	28	22	41.5
DUT1	=			+0.3
U.T.1 of observation	=	28	22	41.8
But G.M.T. of transit of Procyon, May 1 (*Ephem.*, p.12) =		16	59	06
∴ G.H.A. in mean time-units	=	11	23	35.8
Convert to sidereal time-units (*Ephemeris*, p.27)	=		+01	52.3

350

Appendix I

∴ G.H.A. Procyon = 11 25 28.1
But L.H.A. Procyon, from **AI.05.03**, sidereal units = 3 30 10.8

Thus, longitude = L.H.A. − G.H.A. = −7 55 17.3
i.e. λ = $7^h\ 55^m\ 17.3^s$W.

This differs from the originally given value by 04.1^s, or over 01′ arc.

Using *SALS*:
From **AI.03.04**, *SALS* working, G.H.A. Procyon = $11^h\ 25^m\ 24.1^s$
From **AI.05.03**, computed L.H.A. Procyon = 3 30 10.8

Thus, longitude = L.H.A. − G.H.A. = λ = −7 55 13.3
or $7^h\ 55^m\ 13.3^s$W.

AI.06.03 Using transits of the sun or stars
As we have seen in **AI.05**, the instant of transit of the sun's centre across the meridian of a place gives us 12^h L.A.T.; in the case of a star, its instant of transit gives us L.S.T. In both instances the L.H.A. is zero. If the L.Std.T.'s of the instants of transit are recorded and the longitude of the standard meridian of the Zone applied, we have the corresponding U.T.C.'s, and U.T.1's by application of DUT1. Entering tables with these U.T.1's (G.M.T.'s) we may find the G.A.T. (sun) and G.S.T. (star). Then,

$$\text{L.A.T.} - \text{G.A.T.} = \lambda$$

or

$$\text{L.S.T.} - \text{G.S.T.} = \lambda$$

Given an instrument in perfect adjustment, with the telescope rotating in the plane of the meridian, the method of transits would be quick and easy. Unfortunately it is not really possible to achieve this kind of perfection with portable equipment, and the best we can expect from the method is errors of a few seconds of time.

AI.07 Simultaneous determination of latitude and longitude from position lines
The essence of this method is to compare observed altitudes of at least four stars, one in each quadrant azimuth-wise, with the altitudes computed using hour angles obtained from careful timing of these observed altitudes. In deducing the hour angle of a star from corrected watch times, we have seen, as in **AI.03.04**, that we must use the G.M.T. of transit of the star at Greenwich (from tables) in the calculation if we are using the *Ephemeris*. Since this is given to 06^s accuracy only, we cannot compute the altitude (from ω, p and this approximate source of t) to sufficient accuracy to compare with the carefully measured altitudes to get reliable intercepts for plotting or reducing by least squares. The declinations of the stars are also given only to 0.1′ arc, or 06″, which introduces a further approximation.

No example of this method is therefore given in this Appendix, since tables

Astronomy for surveyors

of star positions (or transits) to 0.1^s in R.A. and $01''$ in declination are essential for good results. *SALS*, among other catalogues, has the required specifications.

AI.08 Sample extracts from the *Ephemeris* and *SALS*
Following are full-page facsimiles from *The Ephemeris of the Sun, Polaris and Other Selected Stars with Companion Data and Tables* (1980) published by the United States Department of the Interior, Bureau of Land Management, and from *The Star Almanac for Land Surveyors* published by Her Majesty's Stationery Office, England. (All the tables are slightly reduced in size.)

The author is indebted to the Director of the Bureau of Land Management and the Superintendent of Documents, U.S. Government Printing Office for kind permission to reproduce the material from the *Ephemeris*, and to the Controller of Her Majesty's Stationery Office for kind permission to reproduce the material from the *Star Almanac*.

The order in which the facsimiles appear is pages 353 to 363 show pages 3, 4, 6, 9, 12, 15, 19, 22, 23, 25 and 27 from the *Ephemeris*; and pages 364 and 365 show pages 57 and 58 from the *Star Almanac*.

Appendix I

3   1980

Date		AT GREENWICH APPARENT NOON					POLARIS FOR THE MERIDIAN OF GREENWICH, CIVIL DATE AND MEAN TIME		
		THE SUN'S			Time Semi-diameter Passing Mer.	Equation of Time, Add to Subtract from Apparent Time	Upper Culmination	Elongation Lat. 40°	Declination
		Apparent Declination	Diff. for 1 Hour	Semi-diameter					
		° ′ ″	″	′ ″	s	m s	h m	h m	+89° ′ ″
APRIL, 1980									
Tues.	1	N 4 42 37.7	+57.75	16 01.66	65	3 49.20	1 30.7 P.M.	W.E. 7 27.0 P.M.	10 27.62
Wed.	2	5 05 41.3	57.53	16 01.38	65	3 31.32	1 26.8	7 23.0	10 27.29
Thur.	3	5 28 39.4	57.30	16 01.11	65	3 13.58	1 22.8	7 19.1	10 26.97
Fri.	4	5 51 31.9	57.06	16 00.82	65	2 56.00	1 18.9	7 15.1	10 26.65
Sat.	5	6 14 18.3	56.80	16 00.54	65	2 38.60	1 14.9	7 11.2	10 26.34
Sun.	6	6 36 58.3	56.53	16 00.26	65	2 21.41	1 11.0	7 07.3	10 26.05
Mon.	7	6 59 31.6	56.24	15 59.98	65	2 04.45	1 07.1	7 03.3	10 25.77
Tues.	8	7 21 57.8	55.94	15 59.70	65	1 47.73	1 03.1	6 59.4	10 25.49
Wed.	9	7 44 16.7	55.63	15 59.42	65	1 31.27	12 59.2	6 55.5	10 25.21
Thur.	10	8 06 27.8	55.30	15 59.14	65	1 15.10	12 55.3	6 51.5	10 24.93
Fri.	11	8 28 30.9	54.95	15 58.86	65	0 59.22	12 51.3	6 47.6	10 24.64
Sat.	12	8 50 25.6	54.60	15 58.58	65	0 43.65	12 47.4	6 43.6	10 24.33
Sun.	13	9 12 11.6	54.22	15 58.31	65	0 28.41	12 43.4	6 39.7	10 24.00
Mon.	14	9 33 48.4	53.84	15 58.04	65	0 13.50	12 39.5	6 35.8	10 23.66
Tues.	15	9 55 15.7	53.43	15 57.77	65	0 01.06	12 35.6	6 31.8	10 23.31
Wed.	16	10 16 33.2	53.02	15 57.50	65	0 15.26	12 31.6	6 27.9	10 22.97
Thur.	17	10 37 40.6	52.59	15 57.24	65	0 29.10	12 27.7	6 24.0	10 22.65
Fri.	18	10 58 37.5	52.14	15 56.98	65	0 42.56	12 23.8	6 20.0	10 22.34
Sat.	19	11 19 23.4	51.68	15 56.73	65	0 55.64	12 19.9	6 16.1	10 22.05
Sun.	20	11 39 58.1	51.21	15 56.47	65	1 08.32	12 15.9	6 12.2	10 21.76
Mon.	21	12 00 21.2	50.71	15 56.21	65	1 20.58	12 12.0	6 08.2	10 21.48
Tues.	22	12 20 32.4	50.21	15 55.97	65	1 32.42	12 08.1	6 04.3	10 21.20
Wed.	23	12 40 31.3	49.69	15 55.72	65	1 43.83	12 04.1	6 00.4	10 20.91
Thur.	24	13 00 17.7	49.16	15 55.47	66	1 54.79	12 00.2 P.M.	W.E. 5 56.4 P.M.	10 20.61
Fri.	25	13 19 51.2	48.62	15 55.22	66	2 05.29	11 56.3 A.M.	E.E. 6 00.0 A.M.	10 20.29
Sat.	26	13 39 11.4	48.06	15 54.96	66	2 15.31	11 52.3	5 56.1	10 19.97
Sun.	27	13 58 18.0	47.49	15 54.72	66	2 24.84	11 48.4	5 52.2	10 19.64
Mon.	28	14 17 10.9	46.91	15 54.48	66	2 33.88	11 44.5	5 48.2	10 19.31
Tues.	29	14 35 49.5	46.31	15 54.24	66	2 42.41	11 40.6	5 44.3	10 18.98
Wed.	30	N14 54 13.7	+45.70	15 53.99	66	2 50.41	11 36.6 A.M.	E.E. 5 40.4 A.M.	10 18.65
MAY, 1980									
Thur.	1	N15 12 23.2	+45.08	15 53.75	66	2 57.89	11 32.7 A.M.	E.E. 5 36.5 A.M.	10 18.33
Fri.	2	15 30 17.6	44.45	15 53.51	66	3 04.82	11 28.8	5 32.5	10 18.03
Sat.	3	15 47 56.7	43.80	15 53.27	66	3 11.19	11 24.9	5 28.6	10 17.75
Sun.	4	16 05 20.1	43.14	15 53.04	66	3 17.01	11 20.9	5 24.7	10 17.47
Mon.	5	16 22 27.5	42.47	15 52.80	66	3 22.25	11 17.0	5 20.8	10 17.20
Tues.	6	16 39 18.8	41.79	15 52.56	66	3 26.91	11 13.1	5 16.8	10 16.94
Wed.	7	16 55 53.4	41.09	15 52.34	67	3 30.99	11 09.2	5 12.9	10 16.69
Thur.	8	17 12 11.3	40.39	15 52.10	67	3 34.48	11 05.2	5 09.0	10 16.43
Fri.	9	17 28 12.0	39.66	15 51.88	67	3 37.38	11 01.3	5 05.1	10 16.15
Sat.	10	17 43 55.1	38.93	15 51.65	67	3 39.69	10 57.4	5 01.1	10 15.85
Sun.	11	17 59 20.6	38.18	15 51.43	67	3 41.41	10 53.5	4 57.2	10 15.55
Mon.	12	18 14 28.0	37.43	15 51.22	67	3 42.54	10 49.5	4 53.3	10 15.24
Tues.	13	18 29 17.1	36.66	15 51.01	67	3 43.09	10 45.6	4 49.4	10 14.94
Wed.	14	18 43 47.6	35.87	15 50.81	67	3 43.06	10 41.7	4 45.5	10 14.65
Thur.	15	18 57 59.0	35.08	15 50.60	67	3 42.47	10 37.8	4 41.5	10 14.37
Fri.	16	N19 11 51.3	+34.27	15 50.40	67	3 41.31	10 33.9 A.M.	E.E. 4 37.6 A.M.	10 14.12

The sign + prefixed to the hourly change of declination indicates that $\begin{Bmatrix}\text{north}\\\text{south}\end{Bmatrix}$ declinations are $\begin{Bmatrix}\text{increasing.}\\\text{decreasing.}\end{Bmatrix}$

For time of Western Elongation of Polaris for any date after April 24th, add $5^h\ 56.3^m$ to the time of Upper Culmination of the same date.

353

Astronomy for surveyors

1980

Date		AT GREENWICH APPARENT NOON					POLARIS FOR THE MERIDIAN OF GREENWICH, CIVIL DATE AND MEAN TIME		
		THE SUN'S			Time Semi-diameter Passing Mer.	Equation of Time, Subtract from Add to Apparent Time	Upper Culmination	Elongation Lat. 40°	Declination
		Apparent Declination	Diff. for 1 Hour	Semi-diameter					
MAY, 1980									
		° ′ ″	″	′ ″	s	m s	h m	h m	+89° ′ ″
Fri.	16	N19 11 51.3	+34.27	15 50.40	67	3 41.31	10 33.9 A.M.	E.E. 4 37.6 A.M.	10 14.12
Sat.	17	19 25 24.1	33.45	15 50.22	67	3 39.60	10 29.9	4 33.7	10 13.89
Sun.	18	19 38 37.0	32.62	15 50.02	67	3 37.35	10 26.0	4 29.8	10 13.66
Mon.	19	19 51 30.0	31.78	15 49.84	68	3 34.56	10 22.1	4 25.9	10 13.44
Tues.	20	20 04 02.4	30.92	15 49.66	68	3 31.24	10 18.2	4 22.0	10 13.21
Wed.	21	20 16 14.3	30.06	15 49.48	68	3 27.41	10 14.3	4 18.0	10 12.97
Thur.	22	20 28 05.4	29.19	15 49.32	68	3 23.06	10 10.3	4 14.1	10 12.73
Fri.	23	20 39 35.3	28.30	15 49.15	68	3 18.22	10 06.4	4 10.2	10 12.48
Sat.	24	20 50 43.9	27.41	15 48.97	68	3 12.87	10 02.5	4 06.3	10 12.22
Sun.	25	21 01 31.0	26.51	15 48.81	68	3 07.04	9 58.6	4 02.4	10 11.96
Mon.	26	21 11 56.2	25.59	15 48.65	68	3 00.73	9 54.7	3 58.4	10 11.70
Tues.	27	21 21 59.4	24.67	15 48.49	68	2 53.96	9 50.8	3 54.5	10 11.45
Wed.	28	21 31 40.4	23.74	15 48.33	68	2 46.73	9 46.8	3 50.6	10 11.20
Thur.	29	21 40 59.0	22.81	15 48.19	68	2 39.05	9 42.9	3 46.7	10 10.98
Fri.	30	21 49 55.0	21.86	15 48.04	68	2 30.94	9 39.0	3 42.8	10 10.76
Sat.	31	N21 58 28.3	+20.91	15 47.88	68	2 22.40	9 35.1 A.M.	E.E. 3 38.9 A.M.	10 10.57
JUNE, 1980									
Sun.	1	N22 06 38.7	+19.95	15 47.74	68	2 13.44	9 31.2 A.M.	E.E. 3 35.0 A.M.	10 10.39
Mon.	2	22 14 26.0	18.98	15 47.60	68	2 04.09	9 27.3	3 31.0	10 10.22
Tues.	3	22 21 50.0	18.01	15 47.46	68	1 54.35	9 23.4	3 27.1	10 10.06
Wed.	4	22 28 50.6	17.03	15 47.32	69	1 44.23	9 19.4	3 23.2	10 09.89
Thur.	5	22 35 27.6	16.05	15 47.20	69	1 33.76	9 15.5	3 19.3	10 09.71
Fri.	6	22 41 40.9	15.06	15 47.06	69	1 22.94	9 11.6	3 15.4	10 09.52
Sat.	7	22 47 30.4	14.06	15 46.94	69	1 11.80	9 07.7	3 11.5	10 09.33
Sun.	8	22 52 55.9	13.06	15 46.82	69	1 00.36	9 03.8	3 07.6	10 09.12
Mon.	9	22 57 57.2	12.05	15 46.70	69	0 48.65	8 59.9	3 03.6	10 08.91
Tues.	10	23 02 34.4	11.04	15 46.59	69	0 36.67	8 56.0	2 59.7	10 08.72
Wed.	11	23 06 47.2	10.03	15 46.49	69	0 24.47	8 52.1	2 55.8	10 08.55
Thur.	12	23 10 35.7	9.01	15 46.39	69	0 12.06	8 48.1	2 51.9	10 08.40
Fri.	13	23 13 59.6	7.99	15 46.29	69	0 00.52	8 44.2	2 48.0	10 08.26
Sat.	14	23 16 59.0	6.96	15 46.21	69	0 13.25	8 40.3	2 44.1	10 08.14
Sun.	15	23 19 33.9	5.94	15 46.13	69	0 26.09	8 36.4	2 40.2	10 08.04
Mon.	16	23 21 44.0	4.90	15 46.05	69	0 39.04	8 32.5	2 36.3	10 07.93
Tues.	17	23 23 29.3	3.87	15 45.97	69	0 52.04	8 28.6	2 32.4	10 07.81
Wed.	18	23 24 49.8	2.84	15 45.91	69	1 05.09	8 24.7	2 28.5	10 07.69
Thur.	19	23 25 45.5	1.80	15 45.85	69	1 18.16	8 20.8	2 24.5	10 07.57
Fri.	20	23 26 16.4	+ 0.77	15 45.79	69	1 31.21	8 16.9	2 20.6	10 07.44
Sat.	21	23 26 22.4	− 0.27	15 45.73	69	1 44.24	8 13.0	2 16.7	10 07.30
Sun.	22	23 26 03.6	1.30	15 45.68	69	1 57.21	8 09.0	2 12.8	10 07.16
Mon.	23	23 25 20.0	2.33	15 45.64	69	2 10.11	8 05.1	2 08.9	10 07.03
Tues.	24	23 24 11.7	3.36	15 45.60	69	2 22.90	8 01.2	2 05.0	10 06.91
Wed.	25	23 22 38.5	4.39	15 45.57	69	2 35.57	7 57.3	2 01.1	10 06.80
Thur.	26	23 20 40.7	5.42	15 45.53	69	2 48.10	7 53.4	1 57.2	10 06.71
Fri.	27	23 18 18.4	6.44	15 45.49	69	3 00.47	7 49.5	1 53.3	10 06.64
Sat.	28	23 15 31.4	7.47	15 45.47	69	3 12.66	7 45.6	1 49.4	10 06.59
Sun.	29	23 12 19.9	8.49	15 45.45	69	3 24.65	7 41.7	1 45.5	10 06.55
Mon.	30	23 08 44.1	9.50	15 45.42	69	3 36.42	7 37.8	1 41.6	10 06.52
Tues.	31	N23 04 44.1	−10.51	15 45.41	69	3 47.96	7 33.9 A.M.	E.E. 1 37.6 A.M.	10 06.49

The sign + prefixed to the hourly change of declination indicates that { north / south } declinations are { increasing / decreasing }.

The sign − prefixed to the hourly change of declination indicates that { south / north } declinations are { increasing / decreasing }.

Appendix I

1980

Date		AT GREENWICH APPARENT NOON					POLARIS FOR THE MERIDIAN OF GREENWICH, CIVIL DATE AND MEAN TIME		
		THE SUN'S			Time Semi-diameter Passing Mer.	Equation of Time, Add to Subtract from Apparent Time	Upper Culmination	Elongation Lat. 40°	Declination
		Apparent Declination	Diff. for 1 Hour	Semi-diameter					
		° ′ ″	″	′ ″	s	m s	h m	h m	+89° ′ ″

AUGUST, 1980

Sat.	16	N13 36 05.6	−47.60	15 49.36	65	4 10.76	4 34.2 A.M.	E.E. 10 34.0 P.M.	10 09.32
Sun.	17	13 16 56.8	48.13	15 49.54	65	3 58.08	4 30.2	10 30.1	10 09.47
Mon.	18	12 57 35.4	48.65	15 49.72	65	3 44.87	4 26.3	10 26.2	10 09.63
Tues.	19	12 38 01.7	49.15	15 49.92	65	3 31.16	4 22.4	10 22.3	10 09.80
Wed.	20	12 18 16.1	49.64	15 50.12	65	3 16.94	4 18.5	10 18.4	10 10.00
Thur.	21	11 58 18.9	50.12	15 50.32	65	3 02.24	4 14.6	10 14.5	10 10.21
Fri.	22	11 38 10.4	50.58	15 50.52	65	2 47.06	4 10.7	10 10.6	10 10.43
Sat.	23	11 17 50.8	51.03	15 50.72	65	2 31.42	4 06.8	10 06.6	10 10.67
Sun.	24	10 57 20.8	51.47	15 50.92	65	2 15.35	4 02.9	10 02.7	10 10.92
Mon.	25	10 36 40.3	51.90	15 51.14	65	1 58.85	3 59.0	9 58.8	10 11.17
Tues.	26	10 15 49.7	52.31	15 51.35	65	1 41.94	3 55.1	9 54.9	10 11.41
Wed.	27	9 54 49.5	52.71	15 51.56	65	1 24.65	3 51.2	9 51.0	10 11.63
Thur.	28	9 33 39.8	53.10	15 51.77	65	1 06.99	3 47.3	9 47.1	10 11.84
Fri.	29	9 12 20.9	53.47	15 51.97	64	0 49.00	3 43.3	9 43.2	10 12.04
Sat.	30	8 50 53.2	53.84	15 52.19	64	0 30.67	3 39.4	9 39.3	10 12.25
Sun.	31	N 8 29 16.8	−54.19	15 52.41	64	0 12.04	3 35.5 A.M.	E.E. 9 35.4 P.M.	10 12.46

SEPTEMBER, 1980

Mon.	1	N 8 07 32.2	−54.52	15 52.63	64	0 06.87	3 31.6 A.M.	E.E. 9 31.4 P.M.	10 12.69
Tues.	2	7 45 39.7	54.85	15 52.85	64	0 26.06	3 27.7	9 27.5	10 12.94
Wed.	3	7 23 39.6	55.16	15 53.07	64	0 45.50	3 23.8	9 23.6	10 13.21
Thur.	4	7 01 32.2	55.45	15 53.31	64	1 05.18	3 19.9	9 19.7	10 13.50
Fri.	5	6 39 18.0	55.73	15 53.53	64	1 25.08	3 16.0	9 15.8	10 13.79
Sat.	6	6 16 57.0	56.00	15 53.77	64	1 45.18	3 12.1	9 11.9	10 14.08
Sun.	7	5 54 29.9	56.26	15 54.01	64	2 05.49	3 08.2	9 08.0	10 14.37
Mon.	8	5 31 56.7	56.50	15 54.25	64	2 25.96	3 04.2	9 04.1	10 14.65
Tues.	9	5 09 18.0	56.72	15 54.49	64	2 46.60	3 00.3	9 00.2	10 14.92
Wed.	10	4 46 34.1	56.93	15 54.74	64	3 07.39	2 56.4	8 56.2	10 15.18
Thur.	11	4 23 45.2	57.13	15 54.99	64	3 28.30	2 52.5	8 52.3	10 15.44
Fri.	12	4 00 51.8	57.31	15 55.24	64	3 49.34	2 48.6	8 48.4	10 15.70
Sat.	13	3 37 54.2	57.48	15 55.50	64	4 10.47	2 44.7	8 44.5	10 15.96
Sun.	14	3 14 52.6	57.64	15 55.76	64	4 31.68	2 40.8	8 40.6	10 16.23
Mon.	15	2 51 47.4	57.78	15 56.02	64	4 52.95	2 36.8	8 36.7	10 16.50
Tues.	16	2 28 39.1	57.90	15 56.28	64	5 14.26	2 32.9	8 32.8	10 16.79
Wed.	17	2 05 28.0	58.02	15 56.56	64	5 35.60	2 29.0	8 28.9	10 17.10
Thur.	18	1 42 14.3	58.12	15 56.82	64	5 56.95	2 25.1	8 24.9	10 17.43
Fri.	19	1 18 58.4	58.20	15 57.08	64	6 18.29	2 21.2	8 21.0	10 17.76
Sat.	20	0 55 40.7	58.27	15 57.35	64	6 39.59	2 17.3	8 17.1	10 18.11
Sun.	21	0 32 21.5	58.32	15 57.62	64	7 00.83	2 13.4	8 13.2	10 18.46
Mon.	22	N 0 09 01.2	58.36	15 57.90	64	7 21.99	2 09.4	8 09.3	10 18.80
Tues.	23	S 0 14 19.9	58.39	15 58.17	64	7 43.06	2 05.5	8 05.4	10 19.13
Wed.	24	0 37 41.6	58.41	15 58.44	64	8 03.99	2 01.6	8 01.4	10 19.44
Thur.	25	1 01 03.6	58.41	15 58.71	64	8 24.77	1 57.7	7 57.5	10 19.74
Fri.	26	1 24 25.5	58.40	15 58.97	64	8 45.37	1 53.8	7 53.6	10 20.04
Sat.	27	1 47 46.8	58.38	15 59.24	64	9 05.76	1 49.9	7 49.7	10 20.34
Sun.	28	2 11 07.5	58.34	15 59.51	64	9 25.93	1 45.9	7 45.8	10 20.65
Mon.	29	2 34 27.2	58.29	15 59.78	64	9 45.84	1 42.0	7 41.9	10 20.98
Tues.	30	2 57 45.4	58.22	16 00.05	64	10 05.48	1 38.1	7 37.9	10 21.34
Wed.	31	S 3 21 01.8	−58.14	16 00.32	64	10 24.82	1 34.2 A.M.	E.E. 7 34.0 P.M.	10 21.70

The sign − prefixed to the hourly change of declination indicates that { south / north } declinations are { increasing / decreasing }.

355

SELECTED STARS

Serial Number this list	No. from Nautical Almanac	Constellation Name	North Declination			South Declination		
			R.A.	Decl.	Mag.	R.A.	Decl.	Mag.
			h m	°		h m	°	
1	1	α Andromedae (Alpheratz)	0 07	+29	2.2			
2 x X		β Hydri				0 25	−77	2.9
3	4	β Ceti (Diphda)				0 43	−18	2.2
4 E		α Ursae Minoris (Polaris)	2 11	+89	2.1			
5	6	α Arietis (Hamal)	2 06	+23	2.2			
6	10	α Tauri (Aldebaran)	4 35	+16	1.1			
7	11	β Orionis (Rigel)				5 14	− 8	0.3
8	12	α Aurigae (Capella)	5 15	+46	0.2			
9	16	α Orionis (Betelgeuse)	5 54	+ 7 *	0.5 v			
10 x	17	α Carinae (Canopus)				6 24	−53	−0.9
11	18	α Canis Majoris (Sirius)				6 44	−17	−1.6
12	20	α Canis Minoris (Procyon)	7 38	+ 5	0.5			
13	21	β Geminorum (Pollux)	7 44	+28	1.2			
14	25	α Hydrae (Alphard)				9 27	− 9	2.2
15	26	α Leonis (Regulus)	10 07	+12	1.3			
16	28	β Leonis (Denebola)	11 48	+15	2.2			
17	33	α Virginis (Spica)				13 24	−11	1.2
18 x	36	θ Centauri (Menkent)				14 05	−36	2.3
19	37	α Bootis (Arcturus)	14 15	+19	0.2			
20	41	α Coronae Borealis (Alphecca)	15 34	+27	2.3			
21 x	42	α Scorpii (Antares)				16 28	−26	1.2
22	46	α Ophiuchi (Rasalhague)	17 34	+13	2.1			
23	49	α Lyrae (Vega)	18 36	+39	0.1			
24 x	50	σ Sagittarii (Nunki)				18 54	−26	2.1
25	51	α Aquilae (Altair)	19 50	+ 9	0.9			
26	54	ε Pegasi (Enif)	21 43	+10	2.5			
27 x	56	α Piscis Australis (Fomalhaut)				22 57	−30	1.3
28	57	α Pegasi (Markab)	23 04	+15	2.6			

x: Required for stations south of 30° north latitude.
X: South circumpolar; not included in principal list of stars, Nautical Almanac.
E: Tabulated in this Ephemeris for daily position, including elongation.
v: Varies from 0.1 to 1.2 in magnitude or brilliancy.

α: Alpha. β: Beta. ε: Epsilon. θ: Theta. σ: Sigma. These designations are in the order of brightness, each within its own constellation.

The planets and the bright stars Capella, Canopus, and Vega, are tabulated as an aid in finding stellar positions.

For the south circumpolar star, No. 2 β Hydri, the horizontal angles, and the hour angle at elongation, may be found by the following equations (the lat. and decl. being treated as positive):

At elongation: $\cos t = \cot \delta \tan \phi$; and, $\sin A = \dfrac{\cos \delta}{\cos \phi}$

At any hour angle: $\tan A = \dfrac{\sin t}{\cos \phi \tan \delta - \sin \phi \cos t}$

The product, $\sin \phi \cos t$, is subtracted for hour angles 0° to 90°; added for hour angles 90° to 180°. The hour angle, t, in these equations is in sidereal rate; conversions page 27.

Appendix I

1980

STAR TABLES, MERIDIAN OF GREENWICH, CIVIL DATE AND MEAN TIME

No.	Mag.	Transit May 1	Decl.	Transit May 16	Decl.	Transit June 1	Decl.	Transit June 16	Decl.	Transit Subtract for day of month	
		h m	° ′	h m	′	h m	′	h m	′		m
1- 1	2.2	9 29.5 AM	+28 58.6	8 30.5 AM	58.6	7 27.6 AM	58.7	6 28.6 AM	58.7	1	0.0
2	2.9	9 46.7	-77 21.8	8 47.7	21.7	7 44.8	21.7	6 45.9	21.6	2	3.9
3- 4	2.2	10 04.6	-18 05.8	9 05.6	05.7	8 02.7	05.7	7 03.8	05.6	3	7.9
4	2.1	11 32.7	+89 10.3	10 33.9	10.2	9 31.2	10.2	8 32.5	10.1	4	11.8
5- 6	2.2	11 27.8 AM	+23 22.0	10 28.9 AM	22.0	9 26.0	22.0	8 27.0	22.0	5	15.7
6-10	1.1	1 56.2 PM¹	+16 28.0	12 57.2 PM	28.0	11 54.3 AM	28.1	10 55.3¹	28.1	6	19.7
7-11	0.3	2 34.9	- 8 13.7	1 35.9	13.6	12 33.0 PM	13.6	11 34.0	13.6	7	23.6
8-12	0.2	2 36.5	+45 58.7	1 37.5	58.7	12 34.6	58.6	11 35.6 AM	58.6	8	27.5
9-16	0.5v	3 15.3	+ 7 24.1	2 16.3	24.1	1 13.4	24.1	12 14.4 PM	24.1	9	31.5
10-17	-0.9	3 44.6	-52 41.5	2 45.6	41.4	1 42.7	41.3	12 43.7	41.3	10	35.4
11-18	-1.6	4 05.3	-16 41.6	3 06.3	41.6	2 03.4	41.5	1 04.4	41.5	11	39.3
12-20	0.5	4 59.1	+ 5 16.4	4 00.2	16.4	2 57.3	16.5	1 58.3	16.5	12	43.2
13-21	1.2	5 05.0	+28 04.5	4 06.0	04.5	3 03.1	04.5	2 04.1	04.4	13	47.2
14-25	2.2	6 47.2	- 8 34.5	5 48.2	34.5	4 45.3	34.5	3 46.3	34.5	14	51.1
15-26	1.3	7 27.8²	+12 03.8	6 28.8²	03.8	5 25.9²	03.8	4 26.9²	03.8	15	55.0
16-28	2.2	9 08.3⁴	+14 40.9	8 09.3⁴	41.0	7 06.4²	41.0	6 07.4²	41.0	16	0.0
17-33	1.2	10 44.1	-11 03.6	9 45.1	03.6	8 42.2	03.5	7 43.2	03.5	17	3.9
18-36	2.3	11 25.4	-36 16.4	10 26.4	16.5	9 23.5	16.5	8 24.5	16.5	18	7.9
19-37	0.2	11 34.6 PM	+19 17.1	10 35.6	17.2	9 32.7	17.2	8 33.7	17.2	19	11.8
20-41	2.3	0 57.4 AM {0 02.4 AM / 11 58.4 PM}	+26 46.8	11 54.5 PM	46.9	10 51.6	47.0	9 52.6	47.0	20	15.7
May 15										21	19.7
21-42	1.2	1 51.6 AM	-26 23.3	0 52.6 AM {0 01.5 AM / 11 57.6 PM}	23.3	11 45.8 PM	23.3	10 46.8	23.3	22	23.6
May 29										23	27.5
22-46	2.1	2 57.2	+12 34.4	1 58.3 AM	34.4	0 55.4 AM {0 00.3 AM / 11 56.4 PM}	34.5	11 52.4 PM	34.5	24	31.5
June 15										25	35.4
23-49	0.1	3 59.3	+38 45.7	3 00.3	45.8	1 57.4 AM	45.9	0 58.5 AM {0 03.4 AM / 11 59.5 PM}	45.9	26	39.3
June 30										27	43.2
24-50	2.1	4 17.0	-26 19.2	3 18.1	19.2	2 15.2	19.2	1 16.2 AM	19.2	28	47.2
25-51	0.9	5 12.7	+ 8 48.9	4 13.7	48.9	3 10.8	48.9	2 11.8	49.0	29	51.1
26-54	2.5	7 05.7	+ 9 46.9	6 06.8	46.9	5 03.9	47.0	4 04.9	47.0	30	55.0
27-56	1.3	8 18.9	-29 43.6	7 19.9	43.5	6 17.0	43.5	5 18.0	43.4	31	59.0
28-57	2.6	8 26.1 AM	+15 05.7	7 27.1 AM	05.8	6 24.2 AM	05.8	5 25.2 AM	05.9		
¹Venus	-3.4	2 54 PM	+28	2 28 PM	+27	1 25 PM	+25	11 53 AM	+22		
²Mars	0.8	7 30 PM	+14	6 48 PM	+12	6 08 PM	+ 9	5 34 PM	+ 6		
³Jupiter	-1.6	7 32 PM	+13	6 35 PM	+12	5 37 PM	+12	4 45 PM	+11		
⁴Saturn	1.2	8 49 PM	+ 6	7 49 PM	+ 6	6 46 PM	+ 6	5 49 PM	+ 6		

Footnotes indicate near proximity of a planet to a star in time of transit and declination.

For time of transit subsequent to a star's double transit and until the next tabulated date, subtract for the number of days counting from the second transit of that date. All other interpolations may be taken directly by reference to the day of the month.

Astronomy for surveyors

15   1980
STAR TABLES, MERIDIAN OF GREENWICH, CIVIL DATE AND MEAN TIME

No.	Mag.	Transit Nov. 1	Decl.	Transit Nov. 16	Decl.	Transit Dec. 1	Decl.	Transit Dec. 16	Decl.	Transit Subtract for day of month	
		h m	° ′	h m	′	h m	′	h m	′		m
1– 1	2.2	9 22.1 PM	+28 59.2	8 23.2 PM	59.2	7 24.2 PM	59.2	6 25.2 PM	59.2	1	0.0
2	2.9	9 39.5	–77 22.0	8 40.5	22.0	7 41.5	22.1	6 42.5	22.1	2	3.9
3– 4	2.2	9 57.3	–18 05.6	8 58.3	05.6	7 59.3	05.6	7 00.3	05.7	3	7.9
4	2.1	11 28.6	+89 10.5	10 29.6	10.6	9 30.5	10.7	8 31.3	10.8	4	11.8
5– 6	2.2	11 20.5 PM	+23 22.3	10 21.5 PM	22.3	9 22.6	22.4	8 23.6	22.4	5	15.7
6–10 Nov. 29	1.1	1 52.8 AM	+16 28.2	0 53.8 AM { 0 02.7 AM 11 58.8 PM }	28.2	11 50.9 PM	28.2	10 51.9	28.2	6	19.7
										7	23.6
7–11 Dec. 9	0.3	2 31.5	– 8 13.4	1 32.5 AM	13.4	{ 0 02.1 AM 11 58.1 PM } 0 33.5 AM	13.4	11 30.6	13.5	8	27.5
										9	31.5
8–12 Dec. 9	0.2	2 33.1	+45 58.6	1 34.1	58.6	0 35.2 AM { 0 03.7 AM 11 59.8 PM }	58.7	11 32.3 PM	58.7	10	35.4
										11	39.3
9–16 Dec. 19	0.5v	3 11.9	+ 7 24.2	2 12.9	24.2	1 13.9 AM	24.2	0 14.9 AM { 0 03.2 AM 11 59.2 PM }	24.2	12	43.2
										13	47.2
10–17 Dec. 27	–0.9	3 41.2	–52 40.9	2 42.2	41.0	1 43.3	41.0	0 44.3 AM { 0 01.0 AM 11 57.1 PM }	41.1	14	51.1
										15	55.0
11–18	–1.6	4 01.9	–16 41.2	3 02.9	41.3	2 04.0	41.3	1 05.0 AM	41.4	16	0.0
12–20	0.5	4 55.7	+ 5 16.5	3 56.8	16.5	2 57.8	16.5	1 58.8	16.4	17	3.9
13–21	1.2	5 01.6	+28 04.3	4 02.6	04.3	3 03.6	04.3	2 04.7	04.3	18	7.9
14–25	2.2	6 43.8	– 8 34.3	5 44.8	34.4	4 45.8	34.4	3 46.9	34.5	19	11.8
15–26	1.3	7 24.4	+12 03.7	6 25.4	03.7	5 26.4	03.7	4 27.5	03.6	20	15.7
16–28	2.2	9 04.8 AM	+14 40.9	8 05.8	40.8	7 06.9	40.8	6 07.9	40.7	21	19.7
17–33	1.2	10 40.6 AM [1/4]	–11 03.5	9 41.7 [14/3]	03.5	8 42.7 [14/3]	03.5	7 43.7 [3/4]	03.5	22	23.6
18–36	2.3	11 21.9	–36 16.3	10 22.9	16.3	9 24.0	16.3	8 25.0	16.3	23	27.5
19–37	0.2	11 31.1 AM	+19 17.1	10 32.1	17.1	9 33.2	17.0	8 34.2	16.9	24	31.5
20–41	2.3	12 50.0 PM	+26 47.0	11 51.0 AM	46.9	10 52.0	46.8	9 53.1	46.8	25	35.4
21–42	1.2	1 44.2 [2]	–26 23.3	12 45.2 PM [2]	23.3	11 46.2 AM	23.3	10 47.3 [1]	23.3	26	39.3
22–46	2.1	2 49.8	+12 34.7	1 50.8	34.6	12 51.9 PM	34.6	11 52.9 AM	34.5	27	43.2
23–49	0.1	3 51.9	+38 46.3	2 52.9	46.2	1 53.9	46.2	12 55.0 PM	46.1	28	47.2
24–50	2.1	4 09.6	–26 19.2	3 10.7 [2]	19.2	2 11.7 [2]	19.2	1 12.7 [2]	19.2	29	51.1
25–51	0.9	5 05.3	+ 8 49.2	4 06.3	49.2	3 07.3	49.2	2 08.3	49.1	30	55.0
26–54	2.5	6 58.4	+ 9 47.3	5 59.4	47.3	5 00.4	47.3	4 01.4	47.3	31	59.0
27–56	1.3	8 11.5	–29 43.5	7 12.5	43.6	6 13.6	43.6	5 14.6	43.6		
28–57	2.6	8 18.7 PM	+15 06.2	7 19.7 PM	06.2	6 20.8 PM	06.2	5 21.8 PM	06.2		

[1] Venus	–3.4	9 28 AM	+ 0°	9 36 AM	– 6°	9 47 AM	–13°	10 02 AM	–18°
[2] Mars	1.4	2 09 PM	–23	1 59 PM	–24	1 49 PM	–24	1 41 PM	–23
[3] Jupiter	–1.5	9 22 AM	+ 1	8 33 AM	+ 0	7 43 AM	– 1	6 51 AM	– 2
[4] Saturn	1.1	9 38 AM	+ 0	8 45 AM	– 1	7 51 AM	– 1	6 56 AM	– 1

Footnotes indicate near proximity of a planet to a star in time of transit and declination.

For reduction of time of transit from the Greenwich mean time to the local mean time at the longitude of the station, see table of Sidereal Conversions, page 27.

Appendix I

AZIMUTH OF POLARIS AT ALL HOUR ANGLES, 1980

Mean Declination + 89° 10′ 30″

CORR. TO AZIMUTH

Add for Decl. +89°

	10′20″	10′10″	10′00″

Subtract for Decl. +89°

Mean Time Hour Angle	Latitude										10′40″	10′50″	11′00″
	30°	32°	34°	36°	38°	40°	42°	44°	46°	48°	50°			
h m	′	′	′	′	′	′	′	′	′	′	′	′	′	′
5 55.1	...	...	...	...	...	...	...	...	...	...	77.0	0.3	0.5	0.8
55.4	...	...	...	...	...	...	...	...	...	74.0	...	0.2	0.5	0.7
55.6	...	...	...	...	...	...	...	...	71.3	...	...	0.2	0.5	0.7
5 55.8	...	...	...	...	...	...	...	68.8	...	...	...	0.2	0.5	0.7
56.1	...	...	...	...	...	...	66.6	...	...	...	...	0.2	0.4	0.7
56.3	...	...	...	...	...	64.6	...	...	...	...	...	0.2	0.4	0.7
5 56.4	...	...	...	...	62.8	...	...	...	...	...	...	0.2	0.4	0.6
56.6	...	...	...	61.2	...	...	...	...	...	...	...	0.2	0.4	0.6
56.8	...	...	59.7	...	...	...	...	...	...	...	...	0.2	0.4	0.6
5 57.0	...	58.4	...	...	...	...	...	...	...	...	...	0.2	0.4	0.6
57.1	57.2	...	...	...	...	...	...	...	...	...	...	0.2	0.4	0.6
59.0	57.2	58.4	59.7	61.2	62.8	64.6	66.6	68.8	71.3	74.0	77.0	0.2	0.4	0.7
6 09.0	57.1	58.3	59.6	61.1	62.7	64.5	66.5	68.7	71.1	73.8	76.9	0.2	0.4	0.7
19.0	56.9	58.1	59.4	60.9	62.5	64.3	66.3	68.5	70.9	73.6	76.6	0.2	0.4	0.6
28.9	56.6	57.8	59.1	60.6	62.2	64.0	65.9	68.1	70.5	73.2	76.2	0.2	0.4	0.6
6 38.9	56.2	57.4	58.7	60.1	61.7	63.5	65.4	67.6	70.0	72.6	75.6	0.2	0.4	0.6
48.9	55.7	56.9	58.2	59.6	61.2	62.9	64.8	67.0	69.3	72.0	74.9	0.2	0.4	0.6
58.9	55.1	56.2	57.5	58.9	60.5	62.2	64.1	66.2	68.6	71.2	74.0	0.2	0.4	0.6
7 08.8	54.4	55.5	56.8	58.2	59.7	61.4	63.3	65.4	67.7	70.2	73.1	0.2	0.4	0.6
18.8	53.6	54.7	55.9	57.3	58.8	60.5	62.3	64.4	66.6	69.1	71.9	0.2	0.4	0.6
28.8	52.6	53.7	55.0	56.3	57.8	59.4	61.2	63.2	65.5	67.9	70.7	0.2	0.4	0.6
7 38.7	51.6	52.7	53.9	55.2	56.7	58.3	60.0	62.0	64.2	66.6	69.3	0.2	0.4	0.6
48.7	50.5	51.6	52.7	54.0	55.4	57.0	58.7	60.6	62.8	65.1	67.8	0.2	0.4	0.6
58.7	49.3	50.3	51.5	52.7	54.1	55.6	57.3	59.2	61.3	63.6	66.1	0.2	0.4	0.6
8 08.7	48.0	49.0	50.1	51.3	52.7	54.1	55.8	57.6	59.6	61.9	64.4	0.2	0.4	0.5
18.6	46.6	47.6	48.6	49.8	51.1	52.6	54.2	55.9	57.9	60.0	62.5	0.2	0.4	0.5
28.6	45.1	46.1	47.1	48.2	49.5	50.9	52.4	54.1	56.0	58.1	60.5	0.2	0.3	0.5
8 38.6	43.6	44.5	45.5	46.6	47.8	49.1	50.6	52.2	54.1	56.1	58.3	0.2	0.3	0.5
48.6	41.9	42.8	43.7	44.8	46.0	47.3	48.7	50.3	52.0	54.0	56.1	0.2	0.3	0.5
58.5	40.2	41.0	41.9	42.9	44.1	45.3	46.7	48.2	49.9	51.7	53.8	0.2	0.3	0.5
9 08.5	38.4	39.2	40.1	41.0	42.1	43.3	44.6	46.0	47.6	49.4	51.4	0.1	0.3	0.4
18.5	36.5	37.3	38.1	39.0	40.0	41.2	42.4	43.8	45.3	47.0	48.9	0.1	0.3	0.4
28.4	34.6	35.3	36.1	36.9	37.9	39.0	40.1	41.4	42.9	44.5	46.3	0.1	0.3	0.4
9 38.4	32.6	33.2	34.0	34.8	35.7	36.7	37.8	39.0	40.4	41.9	43.6	0.1	0.2	0.4
48.4	30.5	31.1	31.8	32.6	33.4	34.4	35.4	36.5	37.8	39.2	40.8	0.1	0.2	0.3
58.4	28.4	29.0	29.6	30.3	31.1	32.0	32.9	34.0	35.2	36.5	37.9	0.1	0.2	0.3
10 08.3	26.2	26.7	27.3	28.0	28.7	29.5	30.4	31.4	32.5	33.7	35.0	0.1	0.2	0.3
18.3	24.0	24.5	25.0	25.6	26.3	27.0	27.8	28.7	29.7	30.8	32.0	0.1	0.2	0.3
28.3	21.7	22.2	22.6	23.2	23.8	24.5	25.2	26.0	26.9	27.9	29.0	0.1	0.2	0.2
10 38.3	19.4	19.8	20.2	20.7	21.3	21.9	22.5	23.2	24.0	24.9	25.9	0.1	0.1	0.2
48.2	17.1	17.4	17.8	18.2	18.7	19.2	19.8	20.4	21.1	21.9	22.8	0.1	0.1	0.2
58.2	14.7	15.0	15.3	15.7	16.1	16.5	17.0	17.6	18.2	18.9	19.6	0.1	0.1	0.2
11 08.2	12.3	12.5	12.8	13.1	13.4	13.8	14.2	14.7	15.2	15.8	16.4	0.0	0.1	0.1
18.1	9.8	10.0	10.3	10.5	10.8	11.1	11.4	11.8	12.2	12.6	13.2	0.0	0.1	0.1
28.1	7.4	7.6	7.7	7.9	8.1	8.3	8.6	8.9	9.2	9.5	9.9	0.0	0.1	0.1
11 38.1	4.9	5.0	5.2	5.3	5.4	5.6	5.7	5.9	6.1	6.3	6.6	0.0	0.0	0.1
48.1	2.5	2.5	2.6	2.6	2.7	2.8	2.9	3.0	3.1	3.2	3.3	0.0	0.0	0.0
58.0	0.0	0.0	0.0	0.0	0.0	0.0	0.0	0.0	0.0	0.0	0.0	0.0	0.0	0.0

$$\tan A = \frac{\sin t}{\cos \phi \tan \delta - \sin \phi \cos t}$$

The product, $\sin \phi \cos t$, is subtracted for hour angles 0° to 90° and added for hour angles from 90° to 180°.

Astronomy for surveyors

1980

Lat.	Correction to the Time of Elongation	AZIMUTH OF POLARIS AT ELONGATION, 1980 Decl. +89°						
		10' 00"	10' 10"	10' 20"	10' 30"	10' 40"	10' 50"	11' 00"
° '	W.E. ^{m} E.E.	° ' "	° ' "	° ' "	° ' "	° ' "	° ' "	° ' "
10 00	+2.2−	0 50 46	0 50 36	0 50 26	0 50 16	0 50 06	0 49 56	0 49 45
12 00	2.1	0 51 07	0 50 57	0 50 47	0 50 36	0 50 26	0 50 16	0 50 06
14 00	2.0	0 51 32	0 51 22	0 51 11	0 51 01	0 50 51	0 50 40	0 50 30
16 00	1.8	0 52 01	0 51 51	0 51 40	0 51 30	0 51 19	0 51 09	0 50 58
18 00	1.7	0 52 34	0 52 24	0 52 13	0 52 03	0 51 52	0 51 42	0 51 31
20 00	1.6	0 53 13	0 53 02	0 52 51	0 52 41	0 52 30	0 52 19	0 52 09
21 00	1.5	0 53 33	0 53 23	0 53 12	0 53 01	0 52 51	0 52 40	0 52 29
22 00	1.4	0 53 56	0 53 45	0 53 34	0 53 23	0 53 12	0 53 02	0 52 51
23 00	1.4	0 54 19	0 54 08	0 53 57	0 53 47	0 53 36	0 53 25	0 53 14
24 00	1.3	0 54 44	0 54 33	0 54 22	0 54 11	0 54 00	0 53 49	0 53 38
25 00	1.2	0 55 10	0 54 59	0 54 48	0 54 37	0 54 26	0 54 15	0 54 04
26 00	1.2	0 55 38	0 55 27	0 55 16	0 55 04	0 54 53	0 54 42	0 54 31
27 00	1.1	0 56 07	0 55 56	0 55 45	0 55 33	0 55 22	0 55 11	0 55 00
28 00	1.0	0 56 38	0 56 26	0 56 15	0 56 04	0 55 52	0 55 41	0 55 30
29 00	0.9	0 57 10	0 56 59	0 56 47	0 56 36	0 56 24	0 56 13	0 56 01
30 00	0.9	0 57 44	0 57 33	0 57 21	0 57 10	0 56 58	0 56 46	0 56 35
31 00	0.8	0 58 20	0 58 08	0 57 57	0 57 45	0 57 33	0 57 22	0 57 10
32 00	0.7	0 58 58	0 58 46	0 58 34	0 58 22	0 58 10	0 57 59	0 57 47
33 00	0.6	0 59 37	0 59 25	0 59 13	0 59 01	0 58 49	0 58 38	0 58 26
34 00	0.5	1 00 19	1 00 07	0 59 55	0 59 43	0 59 30	0 59 18	0 59 06
35 00	0.5	1 01 02	1 00 50	1 00 38	1 00 26	1 00 14	1 00 01	0 59 49
36 00	0.4	1 01 48	1 01 36	1 01 24	1 01 11	1 00 59	1 00 46	1 00 34
37 00	0.3	1 02 36	1 02 24	1 02 11	1 01 59	1 01 46	1 01 34	1 01 21
38 00	0.2	1 03 27	1 03 14	1 03 02	1 02 49	1 02 36	1 02 24	1 02 11
39 00	+0.1−	1 04 20	1 04 07	1 03 55	1 03 42	1 03 29	1 03 16	1 03 03
40 00	0.0	1 05 16	1 05 03	1 04 50	1 04 37	1 04 24	1 04 11	1 03 58
41 00	−0.1+	1 06 15	1 06 02	1 05 49	1 05 35	1 05 22	1 05 09	1 04 56
42 00	0.2	1 07 17	1 07 04	1 06 50	1 06 37	1 06 23	1 06 10	1 05 56
43 00	0.3	1 08 22	1 08 08	1 07 55	1 07 41	1 07 27	1 07 14	1 07 00
44 00	0.4	1 09 31	1 09 17	1 09 03	1 08 49	1 08 35	1 08 21	1 08 07
45 00	0.5	1 10 43	1 10 29	1 10 15	1 10 00	1 09 46	1 09 32	1 09 18
46 00	0.7	1 11 59	1 11 44	1 11 30	1 11 16	1 11 01	1 10 47	1 10 32
47 00	0.8	1 13 19	1 13 04	1 12 50	1 12 35	1 12 20	1 12 06	1 11 51
48 00	0.9	1 14 44	1 14 29	1 14 14	1 13 59	1 13 44	1 13 29	1 13 14
49 00	1.0	1 16 13	1 15 58	1 15 42	1 15 27	1 15 12	1 14 57	1 14 42
50 00	1.2	1 17 47	1 17 32	1 17 16	1 17 01	1 16 45	1 16 30	1 16 14
51 00	1.3	1 19 27	1 19 11	1 18 56	1 18 40	1 18 24	1 18 08	1 17 52
52 00	1.5	1 21 13	1 20 57	1 20 41	1 20 24	1 20 08	1 19 52	1 19 36
53 00	1.6	1 23 05	1 22 49	1 22 32	1 22 15	1 21 59	1 21 42	1 21 26
54 00	1.8	1 25 04	1 24 47	1 24 30	1 24 13	1 23 56	1 23 39	1 23 22
55 00	1.9	1 27 11	1 26 53	1 26 36	1 26 18	1 26 01	1 25 44	1 25 26
55 30	2.0	1 28 17	1 27 59	1 27 42	1 27 24	1 27 06	1 26 49	1 26 31
56 00	2.1	1 29 25	1 29 07	1 28 50	1 28 32	1 28 14	1 27 56	1 27 38
56 30	2.2	1 30 36	1 30 18	1 30 00	1 29 41	1 29 23	1 29 05	1 28 47
57 00	2.3	1 31 49	1 31 30	1 31 12	1 30 54	1 30 35	1 30 17	1 29 59
57 30	2.4	1 33 04	1 32 45	1 32 27	1 32 08	1 31 49	1 31 31	1 31 12
58 00	2.5	1 34 22	1 34 03	1 33 44	1 33 25	1 33 06	1 32 47	1 32 28
58 30	2.6	1 35 42	1 35 23	1 35 04	1 34 45	1 34 26	1 34 06	1 33 47
59 00	2.7	1 37 05	1 36 46	1 36 27	1 36 07	1 35 48	1 35 28	1 35 09
59 30	2.8	1 38 31	1 38 12	1 37 52	1 37 32	1 37 13	1 36 53	1 36 33
60 00	3.0	1 40 01	1 39 41	1 39 21	1 39 01	1 38 41	1 38 21	1 38 01
60 30	3.1	1 41 33	1 41 13	1 40 52	1 40 32	1 40 12	1 39 51	1 39 31
61 00	3.2	1 43 09	1 42 48	1 42 27	1 42 07	1 41 46	1 41 26	1 41 05
61 30	3.3	1 44 48	1 44 27	1 44 06	1 43 45	1 43 24	1 43 03	1 42 42
62 00	3.4	1 46 31	1 46 10	1 45 48	1 45 27	1 45 06	1 44 44	1 44 23
62 30	3.6	1 48 18	1 47 56	1 47 35	1 47 13	1 46 51	1 46 30	1 46 08
63 00	3.7	1 50 09	1 49 47	1 49 25	1 49 03	1 48 41	1 48 19	1 47 57
63 30	3.9	1 52 04	1 51 42	1 51 20	1 50 57	1 50 35	1 50 12	1 49 50
64 00	4.0	1 54 05	1 53 42	1 53 19	1 52 56	1 52 33	1 52 10	1 51 48
64 30	4.2	1 56 10	1 55 46	1 55 23	1 55 00	1 54 37	1 54 13	1 53 50
65 00	4.3	1 58 20	1 57 56	1 57 32	1 57 09	1 56 45	1 56 21	1 55 58
65 30	4.5	2 00 35	2 00 11	1 59 47	1 59 23	1 58 59	1 58 35	1 58 11
66 00	4.7	2 02 57	2 02 32	2 02 08	2 01 43	2 01 19	2 00 54	2 00 29
66 30	4.8	2 05 25	2 05 00	2 04 35	2 04 10	2 03 45	2 03 19	2 02 54
67 00	5.0	2 07 59	2 07 34	2 07 08	2 06 43	2 06 17	2 05 51	2 05 26
67 30	5.2	2 10 41	2 10 15	2 09 49	2 09 23	2 08 56	2 08 30	2 08 04
68 00	5.4	2 13 30	2 13 03	2 12 37	2 12 10	2 11 43	2 11 17	2 10 50
68 30	5.6	2 16 27	2 16 00	2 15 33	2 15 05	2 14 38	2 14 11	2 13 44
69 00	5.8	2 19 33	2 19 05	2 18 37	2 18 10	2 17 42	2 17 14	2 16 46
69 30	6.1	2 22 49	2 22 20	2 21 51	2 21 23	2 20 54	2 20 26	2 19 57
70 00	−6.3+	2 26 14	2 25 44	2 25 15	2 24 46	2 24 17	2 23 47	2 23 18

Appendix I

TO FIND THE LATITUDE BY AN ALTITUDE OBSERVATION OF POLARIS AT ANY HOUR ANGLE, 1980

Mean Time Hour Angle	Primary Adjustment to Elevation of Pole								Mean Time Hour Angle
	Subtract — Polaris Above the Pole — Decl. +89°				Add — Polaris Below the Pole — Decl. +89°				
	10' 00"	10' 20"	10' 40"	11' 00"	10' 00"	10' 20"	10' 40"	11' 00"	
h m	° ' "	° ' "	° ' "	° ' "	° ' "	° ' "	° ' "	° ' "	h m
0 00.0	0 50 00	0 49 40	0 49 20	0 49 00	0 00 00	0 00 00	0 00 00	0 00 00	5 57.4
					0 00 22	0 00 22	0 00 21	0 00 21	5 59.0
0 12.0	0 49 56	0 49 36	0 49 16	0 48 56	0 01 40	0 01 40	0 01 39	0 01 38	6 05.0
23.9	49 43	49 23	49 04	48 44	02 59	02 57	02 56	02 55	11.0
35.9	49 23	49 03	48 43	48 23	04 17	04 15	04 13	04 12	17.0
47.9	48 54	48 34	48 14	47 55	05 35	05 33	05 30	05 28	23.0
59.8	48 16	47 57	47 38	47 18	06 53	06 50	06 47	06 44	28.9
1 11.8	0 47 31	0 47 12	0 46 53	0 46 34	0 08 07	0 08 07	0 08 04	0 08 00	6 34.9
23.8	46 38	46 19	46 01	45 42	09 28	09 24	09 20	09 16	40.9
35.7	45 37	45 19	45 01	44 42	10 45	10 40	10 36	10 31	46.9
47.7	44 29	44 11	43 53	43 35	12 01	11 56	11 51	11 46	52.9
59.7	43 13	42 55	42 38	42 21	13 17	13 11	13 06	13 01	58.9
2 11.6	0 41 50	0 41 33	0 41 16	0 41 00	0 14 32	0 14 26	0 14 20	0 14 14	7 04.8
23.6	40 20	40 03	39 47	39 31	15 47	15 40	15 34	15 28	10.8
35.6	38 43	38 27	38 12	37 57	17 01	16 54	16 47	16 40	16.8
47.5	37 00	36 45	36 30	36 16	18 14	18 07	17 59	17 52	22.8
59.5	35 10	34 56	34 42	34 28	19 27	19 19	19 11	19 03	28.8
3 05.5	0 34 14	0 34 00	0 33 46	0 33 33	0 20 38	0 20 30	0 20 22	0 20 13	7 34.8
11.5	33 15	33 02	32 49	32 36	21 49	21 41	21 32	21 23	40.7
17.5	32 16	32 03	31 50	31 37	22 59	22 50	22 41	22 31	46.7
23.4	31 15	31 02	30 50	30 38	24 08	23 59	23 49	23 39	52.7
29.4	30 13	30 01	29 49	29 37	25 16	25 06	24 56	24 46	58.7
3 35.4	0 29 09	0 28 58	0 28 46	0 28 34	0 26 23	0 26 13	0 26 02	0 25 51	8 04.7
41.4	28 04	27 53	27 42	27 31	27 29	27 18	27 07	26 56	10.7
47.4	26 59	26 48	26 37	26 27	28 34	28 23	28 11	28 00	16.6
53.4	25 52	25 41	25 31	25 21	29 38	29 26	29 14	29 02	22.6
59.3	24 44	24 34	24 24	24 14	30 40	30 28	30 15	30 03	28.6
4 05.3	0 23 35	0 23 25	0 23 16	0 23 07	0 31 41	0 31 28	0 31 16	0 31 03	8 34.6
11.3	22 25	22 16	22 07	21 58	32 41	32 28	32 15	32 02	40.6
17.3	21 14	21 05	20 57	20 49	33 40	33 26	33 12	32 59	46.6
23.3	20 02	19 54	19 46	19 38	34 37	34 23	34 09	33 55	52.5
29.3	18 49	18 42	18 35	18 27	35 32	35 18	35 04	34 49	58.5
4 35.2	0 17 36	0 17 29	0 17 22	0 17 15	0 37 19	0 37 04	0 36 49	0 36 34	9 10.5
41.2	16 22	16 16	16 09	16 03	39 00	38 44	38 29	38 13	22.5
47.2	15 07	15 01	14 56	14 50	40 35	40 18	40 02	39 46	34.4
53.2	13 52	13 47	13 41	13 36	42 03	41 46	41 29	41 12	46.4
59.2	12 36	12 31	12 26	12 21	43 24	43 06	42 49	42 31	58.4
5 05.2	0 11 20	0 11 15	0 11 11	0 11 07	0 44 38	0 44 20	0 44 02	0 43 44	10 10.3
11.1	10 03	09 59	09 55	09 51	45 44	45 26	45 08	44 49	22.3
17.1	08 46	08 42	08 39	08 36	46 44	46 25	46 06	45 47	34.3
23.1	07 28	07 25	07 22	07 20	47 35	47 16	46 57	46 38	46.2
29.1	06 10	06 08	06 06	06 03	48 19	48 00	47 41	47 21	58.2
5 35.1	0 04 52	0 04 50	0 04 48	0 04 47	0 48 55	0 48 36	0 48 16	0 47 57	11 10.2
41.1	03 34	03 32	03 31	03 30	49 24	49 04	48 44	48 24	22.1
47.0	02 15	02 15	02 14	02 13	49 44	49 24	49 04	48 44	34.1
53.0	00 57	00 57	00 56	00 56	49 56	49 36	49 16	48 56	46.1
57.4	00 00	00 00	00 00	00 00	50 00	49 40	49 20	49 00	58.0

Mean Time Hour Angle	Supplemental Correction to be Applied to the Vertical Angle Reading — Altitude												Mean Time Hour Angle
	10°	15°	20°	25°	30°	35°	40°	45°	50°	55°	60°	65°	70°	
h m	"	"	"	"	"	"	"	"	"	"	"	"	"	h m
0 00	0	0	0	0	0	0	0	0	0	0	0	0	0	11 58
1 00	− 1	− 1	− 1	− 1	− 1	0	− 0	0	+ 0	+ 1	+ 1	+ 2	+ 3	10 58
2 00	4	4	3	3	2	− 2	− 1	0	1	2	4	6	9	9 58
3 00	9	8	7	6	5	3	2	0	2	5	8	12	19	8 58
3 59	13	12	10	9	7	5	3	0	3	7	12	18	28	7 59
4 59	16	15	13	11	8	6	3	0	4	9	15	23	35	6 59
5 59	−18	−16	−14	−11	− 9	− 6	− 3	0	+ 4	+ 9	+16	+24	+37	5 59

361

Astronomy for surveyors

25

MEAN REFRACTIONS IN ZENITH DISTANCE AND SUN'S PARALLAX IN ALTITUDE
Bar. : 29.6 ins.   Temp. : 50° F.

Apparent Altitude	Refraction	Sun's Par.	Apparent Altitude	Refraction	Sun's Par.	Apparent Altitude	Refraction	Sun's Par.
° ′	′ ″	″	° ′	′ ″	″	°	′ ″	″
7 30	6 53	8.8	12 0	4 25	8.7	25	2 3	8.1
7 40	6 45	8.8	12 30	4 15	8.7	26	1 58	8.0
7 50	6 37	8.8	13 0	4 5	8.7	27	1 53	7.9
8 0	6 30	8.8	13 30	3 56	8.7	28	1 48	7.9
8 10	6 22	8.8	14 0	3 47	8.6	29	1 44	7.8
8 20	6 15	8.8	14 30	3 39	8.6	30	1 40	7.7
8 30	6 8	8.8	15 0	3 32	8.6	32	1 32	7.6
8 40	6 2	8.8	15 30	3 25	8.6	34	1 25	7.4
8 50	5 55	8.8	16 0	3 19	8.6	36	1 19	7.2
9 0	5 49	8.8	16 30	3 13	8.5	38	1 14	7.0
9 10	5 43	8.8	17 0	3 7	8.5	40	1 9	6.8
9 20	5 38	8.8	17 30	3 1	8.5	42	1 4	6.6
9 30	5 32	8.8	18 0	2 56	8.5	44	1 0	6.4
9 40	5 26	8.8	18 30	2 51	8.4	46	0 56	6.2
9 50	5 21	8.8	19 0	2 46	8.4	48	0 52	6.0
10 0	5 16	8.8	19 30	2 42	8.4	50	0 48	5.7
10 20	5 6	8.8	20 0	2 37	8.4	55	0 40	5.1
10 40	4 57	8.7	21 0	2 29	8.3	60	0 33	4.4
11 0	4 48	8.7	22 0	2 22	8.2	65	0 27	3.8
11 20	4 40	8.7	23 0	2 15	8.2	70	0 21	3.0
11 40	4 32	8.7	24 0	2 9	8.1	80	0 10	1.5
12 0	4 25	8.7	25 0	2 3	8.1	90	0 0	0.0

Apparent altitude (observed vertical angle) : v
Sun's true altitude : h = v − refraction + parallax
Star's true altitude : h = v − refraction

True refraction : mean refraction × coefficient for barometric pressure × coefficient for temperature.

COEFFICIENTS TO APPLY TO MEAN REFRACTIONS FOR VARIATIONS IN BAROMETER AND TEMPERATURE

Barometer (Ins.)	Elevation above sea level (Feet)	Coefficient	Barometer (Ins.)	Elevation above sea level (Feet)	Coefficient	Barometer (Ins.)	Elevation above sea level (Feet)	Coefficient	Temperature (Fahr.)	Coefficient
30.5	−451	1.03	27.2	2,670	0.92	23.6	6,538	0.80	−10°	1.13
30.2	−181	1.02	26.9	2,972	.91	23.3	6,887	.79	0°	1.11
30.0	00	1.01	26.6	3,277	.90	23.0	7,239	.78	+10°	1.08
29.9	+ 91	1.01	26.3	3,586	.89	22.7	7,597	.77	20°	1.06
29.6	366	1.00	26.0	3,899	.88	22.4	7,960	.76	30°	1.04
29.3	643	.99	25.7	4,215	.87	22.1	8,327	.75	40°	1.02
29.0	924	.98	25.4	4,535	.86	21.8	8,700	.74	50°	1.00
28.7	1,207	.97	25.1	4,859	.85	21.5	9,077	.73	60°	.98
28.4	1,493	.96	24.8	5,186	.84	21.2	9,460	.72	70°	.96
28.1	1,783	.95	24.5	5,518	.83	20.9	9,848	.71	80°	.94
27.8	2,075	.94	24.2	5,854	.82	20.6	10,242	.70	90°	.93
27.5	2,371	.93	23.9	6,194	.81	20.3	10,642	.69	100°	.91
27.2	2,670	.92	23.6	6,538	.80	20.0	11,047	.68	+110°	.90

Appendix I

CONVERSION OF ARC TO TIME

° '	h m / m s	° '	h m / m s	°	h m	°	h m	°	h m	°	h m	"	s
0	0 00	30	2 00	60	4 00	90	6 00	120	8 00	150	10 00	0	0,00
1	0 04	31	2 04	61	4 04	91	6 04	121	8 04	151	10 04	2	0,13
2	0 08	32	2 08	62	4 08	92	6 08	122	8 08	152	10 08	4	0,27
3	0 12	33	2 12	63	4 12	93	6 12	123	8 12	153	10 12	6	0,40
4	0 16	34	2 16	64	4 16	94	6 16	124	8 16	154	10 16	8	0,53
5	0 20	35	2 20	65	4 20	95	6 20	125	8 20	155	10 20	10	0,67
6	0 24	36	2 24	66	4 24	96	6 24	126	8 24	156	10 24	12	0,80
7	0 28	37	2 28	67	4 28	97	6 28	127	8 28	157	10 28	14	0,93
8	0 32	38	2 32	68	4 32	98	6 32	128	8 32	158	10 32	16	1,07
9	0 36	39	2 36	69	4 36	99	6 36	129	8 36	159	10 36	18	1,20
10	0 40	40	2 40	70	4 40	100	6 40	130	8 40	160	10 40	20	1,33
11	0 44	41	2 44	71	4 44	101	6 44	131	8 44	161	10 44	22	1,47
12	0 48	42	2 48	72	4 48	102	6 48	132	8 48	162	10 48	24	1,60
13	0 52	43	2 52	73	4 52	103	6 52	133	8 52	163	10 52	26	1,73
14	0 56	44	2 56	74	4 56	104	6 56	134	8 56	164	10 56	28	1,87
15	1 00	45	3 00	75	5 00	105	7 00	135	9 00	165	11 00	30	2,00
16	1 04	46	3 04	76	5 04	106	7 04	136	9 04	166	11 04	32	2,13
17	1 08	47	3 08	77	5 08	107	7 08	137	9 08	167	11 08	34	2,27
18	1 12	48	3 12	78	5 12	108	7 12	138	9 12	168	11 12	36	2,40
19	1 16	49	3 16	79	5 16	109	7 16	139	9 16	169	11 16	38	2,53
20	1 20	50	3 20	80	5 20	110	7 20	140	9 20	170	11 20	40	2,67
21	1 24	51	3 24	81	5 24	111	7 24	141	9 24	171	11 24	42	2,80
22	1 28	52	3 28	82	5 28	112	7 28	142	9 28	172	11 28	44	2,93
23	1 32	53	3 32	83	5 32	113	7 32	143	9 32	173	11 32	46	3,07
24	1 36	54	3 36	84	5 36	114	7 36	144	9 36	174	11 36	48	3,20
25	1 40	55	3 40	85	5 40	115	7 40	145	9 40	175	11 40	50	3,33
26	1 44	56	3 44	86	5 44	116	7 44	146	9 44	176	11 44	52	3,47
27	1 48	57	3 48	87	5 48	117	7 48	147	9 48	177	11 48	54	3,60
28	1 52	58	3 52	88	5 52	118	7 52	148	9 52	178	11 52	56	3,73
29	1 56	59	3 56	89	5 56	119	7 56	149	9 56	179	11 56	58	3,87
30	2 00	60	4 00	90	6 00	120	8 00	150	10 00	180	12 00	60	4,00

Degrees of arc convert to hours and minutes of time ; minutes of arc convert to minutes and seconds of time. See right hand column for conversion of seconds of arc to seconds of time.

SIDEREAL CONVERSIONS

		Longitude						
		0° 00'	2° 30'	5° 00'	7° 30'	10° 00'	12° 30'	15° 00'
		Minutes						
Long.	Hours	0	10	20	30	40	50	60
°		m s	m s	m s	m s	m s	m s	m s
0	0	0 00	0 02	0 03	0 05	0 07	0 08	0 10
15	1	0 10	0 11	0 13	0 15	0 16	0 18	0 20
30	2	0 20	0 21	0 23	0 25	0 26	0 28	0 30
45	3	0 30	0 31	0 33	0 34	0 36	0 38	0 39
60	4	0 39	0 41	0 43	0 44	0 46	0 48	0 49
75	5	0 49	0 51	0 53	0 54	0 56	0 57	0 59
90	6	0 59	1 01	1 02	1 04	1 06	1 07	1 09
105	7	1 09	1 11	1 12	1 14	1 15	1 17	1 19
120	8	1 19	1 20	1 22	1 24	1 25	1 27	1 29
135	9	1 29	1 30	1 32	1 34	1 35	1 37	1 38
150	10	1 38	1 40	1 42	1 43	1 45	1 47	1 48
165	11	1 48	1 50	1 52	1 53	1 55	1 56	1 58
180	12	1 58	2 00	2 01	2 03	2 05	2 06	2 08

Sidereal into mean solar time : To be subtracted from a sidereal time interval :
Argument, hours and minutes of sidereal interval.

Mean solar into sidereal time : To be added to a mean time interval :
Argument, hours and minutes of mean time interval.

For any stellar observation : Amount to be subtracted from the Greenwich mean time of transit or elongation to obtain local mean time :
Argument, longitude west from Greenwich ; add for longitudes east from Greenwich.

Astronomy for surveyors

POLE STAR TABLE, 1980

L.S.T.	6^h		7^h		8^h		9^h		10^h		11^h	
	a_0	b_0	a_0	b_0	a_0	b_0	a_0	b_0	a_0	b_0	a_0	b_0
m	′	′	′	′	′	′	′	′	′	′	′	′
0	−26.7	−41.9	−14.9	−47.3	− 2.2	−49.5	+10.7	−48.2	+22.8	−43.8	+33.4	−36.4
3	26.1	42.2	14.3	47.5	1.5	49.5	11.3	48.1	23.4	43.5	33.8	35.9
6	25.6	42.6	13.7	47.7	0.9	49.5	12.0	47.9	23.9	43.1	34.3	35.5
9	25.0	42.9	13.0	47.9	− 0.2	49.5	12.6	47.8	24.5	42.8	34.8	35.0
12	24.4	43.2	12.4	48.0	+ 0.4	49.5	13.2	47.6	25.1	42.5	35.2	34.6
15	−23.9	−43.6	−11.8	−48.2	+ 1.1	−49.5	+13.8	−47.4	+25.6	−42.2	+35.7	−34.1
18	23.3	43.9	11.2	48.3	1.7	49.5	14.4	47.2	26.2	41.8	36.1	33.6
21	22.7	44.2	10.5	48.5	2.4	49.4	15.1	47.0	26.7	41.5	36.5	33.2
24	22.1	44.4	9.9	48.6	3.0	49.4	15.7	46.8	27.3	41.1	37.0	32.7
27	21.6	44.7	9.2	48.7	3.7	49.3	16.3	46.6	27.8	40.8	37.4	32.2
30	−21.0	−45.0	− 8.6	−48.8	+ 4.3	−49.3	+16.9	−46.4	+28.3	−40.4	+37.8	−31.7
33	20.4	45.3	8.0	48.9	5.0	49.2	17.5	46.2	28.9	40.0	38.2	31.2
36	19.8	45.5	7.3	49.0	5.6	49.1	18.1	45.9	29.4	39.6	38.6	30.7
39	19.2	45.8	6.7	49.1	6.2	49.1	18.7	45.7	29.9	39.3	39.0	30.2
42	18.6	46.0	6.0	49.2	6.9	49.0	19.3	45.4	30.4	38.9	39.4	29.7
45	−18.0	−46.3	− 5.4	−49.3	+ 7.5	−48.9	+19.9	−45.2	+30.9	−38.5	+39.8	−29.2
48	17.4	46.5	4.8	49.3	8.2	48.8	20.5	44.9	31.4	38.1	40.2	28.7
51	16.8	46.7	4.1	49.4	8.8	48.6	21.1	44.6	31.9	37.6	40.6	28.2
54	16.1	46.9	3.5	49.4	9.4	48.5	21.7	44.3	32.4	37.2	40.9	27.6
57	15.5	47.1	2.8	49.4	10.1	48.4	22.2	44.1	32.9	36.8	41.3	27.1
60	−14.9	−47.3	− 2.2	−49.5	+10.7	−48.2	+22.8	−43.8	+33.4	−36.4	+41.6	−26.6

Lat.	a_1	b_1	a_1	b_1	a_1	b_1	a_1	b_1	a_1	b_1	a_1	b_1
°	′	′	′	′	′	′	′	′	′	′	′	′
0	−.3	+.3	−.4	+.2	−.4	−.1	−.4	−.3	−.3	−.4	−.2	−.4
10	−.3	+.3	−.3	+.1	−.4	−.1	−.3	−.2	−.2	−.3	−.2	−.4
20	−.2	+.2	−.3	+.1	−.3	.0	−.3	−.2	−.2	−.3	−.1	−.3
30	−.2	+.2	−.2	+.1	−.2	.0	−.2	−.1	−.1	−.2	−.1	−.2
40	−.1	+.1	−.1	.0	−.1	.0	−.1	−.1	−.1	−.1	−.1	−.1
45	−.1	+.1	−.1	.0	−.1	.0	−.1	.0	.0	−.1	.0	−.1
50	.0	.0	.0	.0	.0	.0	.0	.0	.0	.0	.0	.0
55	+.1	−.1	+.1	.0	+.1	.0	+.1	+.1	+.1	+.1	.0	+.1
60	+.2	−.1	+.2	−.1	+.2	.0	+.2	+.1	+.2	+.2	+.1	+.2
62	+.2	−.2	+.2	−.1	+.2	.0	+.2	+.2	+.2	+.2	+.1	+.2
64	+.2	−.2	+.3	−.1	+.3	.0	+.3	+.2	+.2	+.3	+.1	+.3
66	+.3	−.3	+.4	−.1	+.4	+.1	+.3	+.2	+.3	+.4	+.2	+.4

Month	a_2	b_2	a_2	b_2	a_2	b_2	a_2	b_2	a_2	b_2	a_2	b_2
Jan.	.0	+.1	−.1	+.1	−.1	+.1	−.1	+.1	−.2	.0	−.2	.0
Feb.	+.1	+.1	+.1	+.1	.0	+.1	.0	+.1	.0	+.1	−.1	+.1
Mar.	+.2	−.1	+.2	.0	+.2	.0	+.2	+.1	+.1	+.1	+.1	+.2
Apr.	+.2	−.2	+.2	−.2	+.2	−.1	+.3	.0	+.3	.0	+.2	+.1
May	+.1	−.3	+.1	−.3	+.2	−.2	+.3	−.2	+.3	−.1	+.3	.0
June	−.1	−.4	.0	−.4	+.1	−.4	+.2	−.3	+.3	−.2	+.3	−.2
July	−.2	−.3	−.1	−.4	.0	−.4	+.1	−.4	+.2	−.4	+.3	−.3
Aug.	−.4	−.2	−.3	−.3	−.2	−.4	−.1	−.4	.0	−.4	+.1	−.4
Sept.	−.4	.0	−.4	−.1	−.4	−.2	−.3	−.3	−.2	−.4	−.1	−.4
Oct.	−.4	+.2	−.5	.0	−.4	−.1	−.4	−.2	−.3	−.3	−.3	−.4
Nov.	−.4	+.3	−.4	+.2	−.5	+.1	−.5	.0	−.5	−.1	−.4	−.3
Dec.	−.2	+.4	−.3	+.4	−.4	+.3	−.5	+.2	−.5	.0	−.5	−.1

Latitude = Corrected observed altitude of *Polaris* + $a_0 + a_1 + a_2$
Azimuth of *Polaris* = $(b_0 + b_1 + b_2)$ sec (latitude)

Appendix I

POLE STAR TABLE, 1980

L.S.T.	12^h		13^h		14^h		15^h		16^h		17^h	
	a_0	b_0	a_0	b_0	a_0	b_0	a_0	b_0	a_0	b_0	a_0	b_0
m	′	′	′	′	′	′	′	′	′	′	′	′
0	+41·6	−26·6	+47·1	−15·0	+49·4	− 2·5	+48·4	+10·1	+44·2	+22·1	+37·0	+32·7
3	42·0	26·0	47·3	14·4	49·5	1·9	48·3	10·7	43·9	22·7	36·5	33·2
6	42·3	25·5	47·5	13·8	49·5	1·3	48·2	11·4	43·6	23·3	36·1	33·6
9	42·7	24·9	47·7	13·2	49·5	− 0·6	48·0	12·0	43·3	23·8	35·7	34·1
12	43·0	24·4	47·8	12·6	49·5	0·0	47·8	12·6	43·0	24·4	35·2	34·6
15	+43·3	−23·8	+48·0	−12·0	+49·5	+ 0·6	+47·7	+13·2	+42·7	+24·9	+34·8	+35·0
18	43·6	23·3	48·2	11·4	49·5	1·3	47·5	13·8	42·3	25·5	34·3	35·5
21	43·9	22·7	48·3	10·7	49·5	1·9	47·3	14·4	42·0	26·0	33·8	35·9
24	44·2	22·1	48·4	10·1	49·4	2·5	47·1	15·0	41·6	26·6	33·4	36·4
27	44·5	21·6	48·6	9·5	49·4	3·2	46·9	15·7	41·3	27·1	32·9	36·8
30	+44·8	−21·0	+48·7	− 8·9	+49·4	+ 3·8	+46·7	+16·3	+40·9	+27·6	+32·4	+37·2
33	45·0	20·4	48·8	8·2	49·3	4·5	46·5	16·9	40·6	28·2	31·9	37·6
36	45·3	19·8	48·9	7·6	49·2	5·1	46·3	17·5	40·2	28·7	31·4	38·1
39	45·5	19·2	49·0	7·0	49·2	5·7	46·0	18·1	39·8	29·2	30·9	38·5
42	45·8	18·6	49·1	6·4	49·1	6·4	45·8	18·6	39·4	29·7	30·4	38·9
45	+46·0	−18·1	+49·2	− 5·7	+49·0	+ 7·0	+45·5	+19·2	+39·0	+30·2	+29·9	+39·3
48	46·3	17·5	49·2	5·1	48·9	7·6	45·3	19·8	38·6	30·7	29·4	39·6
51	46·5	16·9	49·3	4·5	48·8	8·2	45·0	20·4	38·2	31·2	28·9	40·0
54	46·7	16·3	49·4	3·8	48·7	8·9	44·8	21·0	37·8	31·7	28·3	40·4
57	46·9	15·7	49·4	3·2	48·6	9·5	44·5	21·6	37·4	32·2	27·8	40·8
60	+47·1	−15·0	+49·4	− 2·5	+48·4	+10·1	+44·2	+22·1	+37·0	+32·7	+27·3	+41·1

Lat.	a_1	b_1	a_1	b_1	a_1	b_1	a_1	b_1	a_1	b_1	a_1	b_1
°												
0	−·1	−·3	·0	−·2	·0	+·1	·0	+·3	−·1	+·4	−·2	+·4
10	−·1	−·3	·0	−·1	·0	+·1	·0	+·2	−·1	+·3	−·2	+·4
20	−·1	−·2	·0	−·1	·0	·0	·0	+·2	−·1	+·3	−·2	+·3
30	·0	−·2	·0	−·1	·0	·0	·0	+·1	−·1	+·2	−·1	+·2
40	·0	−·1	·0	·0	·0	·0	·0	+·1	·0	+·1	−·1	+·1
45	·0	−·1	·0	·0	·0	·0	·0	·0	·0	+·1	·0	+·1
50	·0	·0	·0	·0	·0	·0	·0	·0	·0	·0	·0	·0
55	·0	+·1	·0	·0	·0	·0	·0	−·1	·0	−·1	·0	−·1
60	·0	+·1	·0	+·1	·0	·0	·0	−·1	+·1	−·2	+·1	−·2
62	·0	+·2	·0	+·1	·0	·0	·0	−·2	+·1	−·2	+·1	−·2
64	+·1	+·2	·0	+·1	·0	·0	·0	−·2	+·1	−·3	+·2	−·3
66	+·1	+·3	·0	+·1	·0	−·1	·0	−·2	+·1	−·4	+·2	−·4

Month	a_2	b_2	a_2	b_2	a_2	b_2	a_2	b_2	a_2	b_2	a_2	b_2
Jan.	−·1	·0	−·1	−·1	−·1	−·1	−·1	−·1	·0	−·2	·0	−·2
Feb.	−·1	+·1	−·1	+·1	−·1	·0	−·1	·0	−·1	·0	−·1	−·1
Mar.	+·1	+·2	·0	+·2	·0	+·2	−·1	+·2	−·1	+·1	−·2	+·1
Apr.	+·2	+·2	+·2	+·2	+·1	+·2	·0	+·3	·0	+·3	−·1	+·2
May	+·3	+·1	+·3	+·1	+·2	+·2	+·2	+·3	+·1	+·3	·0	+·3
June	+·4	−·1	+·4	·0	+·4	+·1	+·3	+·2	+·2	+·3	+·2	+·3
July	+·3	−·2	+·4	−·1	+·4	·0	+·4	+·1	+·4	+·2	+·3	+·3
Aug.	+·2	−·4	+·3	−·3	+·4	−·2	+·4	−·1	+·4	·0	+·4	+·1
Sept.	·0	−·4	+·1	−·4	+·2	−·4	+·3	−·3	+·4	−·2	+·4	−·1
Oct.	−·2	−·4	·0	−·5	+·1	−·4	+·2	−·4	+·3	−·3	+·4	−·3
Nov.	−·3	−·4	−·2	−·4	−·1	−·5	·0	−·5	+·1	−·5	+·3	−·4
Dec.	−·4	−·2	−·4	−·3	−·3	−·4	−·2	−·5	·0	−·5	+·1	−·5

Latitude = Corrected observed altitude of *Polaris* + $a_0 + a_1 + a_2$
Azimuth of *Polaris* = $(b_0 + b_1 + b_2)$ sec (latitude)

APPENDIX II

SOME DESK-TOP AND MICROCOMPUTER PROGRAMS FOR USE IN FIELD ASTRONOMY

AII.01 Introduction

The advance in computer technology since 1970 has been phenomenal. In the early part of this period small hand-held calculators with "hard-wired" mathematical functions put paid for ever to tables of logarithms and to those of natural functions used for so long with hand-cranked mechanical calculators, now unknown to many new-generation surveyors. Lengthier computations such as those required for survey adjustment could be done by using large "main-frame" computers, but generally these were available only to surveyors and students working in the larger institutions. The flexibility and memory capacity of pocket electronic calculators has increased, however, to the point where they can cope with a great many survey computations. The advent of small attachable modules, such as printers, extra memory and program packs, and the ability of many modern calculators to retain programs in memory after turning off the display, are but some of the features of this new technology.

Large-scale integration techniques in electronics made it possible to manufacture powerful minicomputers to take over some of the tasks of the main-frame installations, and the appearance of the microprocessor has led to the development of the microcomputer—a new breed with all the earmarks of main-frame and minicomputers, such as disk storage and high-level programming language capabilities. It is now interesting to see the microcomputer living up to its name and approaching in size some of the pocket electronic calculators without sacrificing any of its capability. The built-in scanning "window" does away with the full-sized screen of the familiar, TV-like visual display unit.

This Appendix does not aim to present a large range of programs for all kinds of computers. There are two main sections: the first has four programs using **RPN** (Reverse Polish Notation) logic and designed specifically for the Hewlett-Packard 67/97 series of calculators, while the second has two relatively long and comprehensive programs written in BASIC language for a 16K microcomputer incorporating a Z-80 microprocessor. A third, shorter program is presented for computing the ephemeris of the sun given the year, month, day and U.T.

Considerable care is needed in following the program instructions faithfully and accurately; do not blame the program unless you are completely sure you have entered the data in the correct form and that you have made the decisions exactly as required by the instructions! Anyone familiar with electronic computers will know that only one very small error can produce surprising results. If some of the programming techniques do not appeal to you, you can use the listings as bases for modification.

Appendix II

Although the RPN programs were designed for the HP-67/69 calculators, they should be easily enough adapted for others using this type of logic. The microcomputer programs were written for a Model 1 TRS-80 with 16K of RAM and Level II BASIC in a 12K ROM (TRS-80 is a trade mark of the Tandy Corporation, U.S.A.).

All programs are structured around the *Star Almanac for Land Surveyors* (*SALS*) published by H.M.S.O. This set of tables with its comprehensive coverage and excellent presentation is used in many countries. It is still reasonably cheap too.

Notation used in the instructions is that given at the beginning of this book (pp. xi–xii).

AII.02   The Reverse Polish Notation programs

There are four programs in this series:

> Azimuth by altitude of the sun,
> Azimuth by altitude of a star,
> Azimuth by hour angle of the sun,
> Azimuth by hour angle of a star.

The first two are designated A and B respectively and they may be stored consecutively on a single magnetic storage card. The third and fourth may similarly be combined as A and B on one magnetic card.

Throughout all RPN programs, the following conventions are used:

> North latitude is $+$, south is $-$.
> East longitude is $+$, west is $-$.
> North declination is $+$, south is $-$.

Altitudes of observed objects are always entered as true elevations; they will have to be deduced from the vertical circle readings before being entered as data.

When FL and FR horizontal circle readings on the R.M. and object are to be entered (in azimuth determinations) the degrees of the face right readings must be put into terms of those of face left before they are entered as data; e.g. actual readings may be:

> FL   79° 07′ 47″
> FR 258   23   46;

these are entered as 79° 07′ 47″ and 78° 23′ 46″.

When quantities such as the declination of the sun and its 6-hourly differences, which are given in *SALS* to decimals of an arc-minute, are required, they are always entered in the RPN programs as degrees, minutes and seconds: e.g. $-19°$ 42.6′ is entered as -19.4236; and a 6-hourly difference of 5.9 is entered as .0554. (This does not apply to the BASIC programs which will accommodate the minutes and fractions.) Small quantities such as DUT1 must be entered as hours, minutes and seconds: e.g. if DUT1 $= -0.2^s$ it is entered as $-.00002$.

Note that after entering data the "R/S" key must be used to make the program resume operation.

AII.02.01 Azimuth by altitude of the sun

This program is designed to deal with two pointings on opposite limbs of the sun as set out in the examples in **10.21**(a), (i) and (ii) of Chapter 10, and in **AI.03.01** of Appendix I, the observational data from which are used in the column headed "Example".

Comments on the program steps:

Step 5: enter hours (and decimal if necessary, e.g. -7.5).

Steps 11 and 13: δ_1 is the sun's declination at U.T. 0^h on the U.T. date which has the same number as the local date, and δ_2 is its declination at U.T. 0^h on the next day. In the example,

$$\delta_1 = -5° \ 43.2', \ 1980 \ \text{September 8, U.T. } 0^h$$
$$\delta_2 = -5 \ \ 20.6 \ , \ 1980 \ \text{September 9, U.T. } 0^h$$

Note that they are entered in the reverse order, δ_2 first.

Step 82: If the sun was observed west of the meridian, the sign of the figure in the display is to be reversed and 360 added. This is done manually by keying CHS, 3, 6, 0, +, then R/S. If no action required, press R/S to continue.

Steps 90 and 91: For ΣL, the two left-hand plate level bubble end readings are added mentally and entered; for ΣR, the two right-hand readings are summed and entered.

Step 114: If the azimuth displayed is negative, convert to decimal, add 360 manually, and convert back to degrees, minutes and seconds. If it is greater than 360, subtract 360 from the degrees.

The formula used for computing the zenith angle is:

$$\cos Z = (\sin \delta - \sin h . \sin \phi)/\cos h . \cos \phi;$$

and for the refraction correction:

$$+16.3''.\tan \zeta.(P/T) \text{ to zenith distances,}$$

with P in mb and T in °K (Bomford, 1980). This translates to

$$-0.004528.\cot h.(P/T)$$

to obtain the correction in degrees to altitudes.

For parallax, the formula for the correction is:

$$8.8''.\cos h, \text{ or}$$
$$0.00244.\cos h \text{ degrees.}$$

Appendix II

Program listing:

Step	Instruction	Example	Step	Instruction	Example
1	LBL A	(Press A)	40	TAN	
2	DSP 5		41	1/X	
3	R/S:Key approx.Zone Time	8.0700	42	×	
4	f H.MS→		43	4	
5	R/S: Key Zone (hours)	-8	44	.	
6	-		45	5	
7	2		46	2	
8	4		47	8	
9	÷		48	EEX	
10	STO 1		49	CHS	
11	R/S: Key δ_2	5.2036	50	3	
12	f H.MS→		51	×	
13	R/S: Key δ_1	5.4312	52	CHS	
14	f H.MS→		53	.	
15	-		54	0	
16	f LAST X		55	0	
17	X⇄Y		56	2	
18	RCL 1		57	4	
19	×		58	4	
20	+		59	RCL 3	
21	STO 2		60	COS	
22	R/S:Key obs. alt. FL	30.2320	61	×	
23	R/S:Key obs. alt. FR	30.0700	62	+	
24	f H.MS→		63	STO+3	
25	f H.MS→		64	R/S: Key latitude	35.5206
26	2		65	f H.MS→	
27	÷		66	STO 4	
28	STO 3		67	SIN	
29	R/S:Key pressure, mb	1016 (30in.)	68	RCL 3	
30	R/S:Key temp.,°C	22.2 (72°F)	69	SIN	
31	2		70	×	
32	7		71	CHS	
33	3		72	RCL 2	
34	.		73	SIN	
35	1		74	+	
36	6		75	RCL 4	
37	+		76	COS	
38	÷		77	RCL 3	
39	RCL 3		78	COS	

Step	Instruction	Example	Step	Instruction	Example
79	×		98	RCL 3	
80	÷		99	TAN	
81	f COS		100	×	
82	R/S:If Sun west, CHS and add 360	(106°.572)	101	+	
83	STO 5		102	STO 6	
84	R/S:Key circle Sun,FL	112.1920	103	R/S:Key circle RM,FL	25.4220
85	R/S:Key circle Sun,FR	112.1800	104	R/S:Key circle RM,FR	25.4200
86	f H.MS→		105	f H.MS→	
87	f H.MS→		106	f H.MS→	
88	2		107	2	
89	÷		108	÷	
90	R/S:Key ΣL level	7	109	RCL 5	
91	R/S:Key ΣR level	9	110	+	
92	−		111	RCL 6	
93	4		112	−	
94	÷		113	f→H.MS	
95	R/S:Key secs./lev.div.	.0030	114	R/S: (See azimuth RM)	19.5759
96	f H.MS→		115	RTN	
97	×				

Note: Example is in AI.03.01 of Appendix I.

AII.02.02 Azimuth by altitude of a star

This program is very similar to the previous one. However, it is designed to compute the azimuth from one pointing on the star and one on the R.M., both on one face of the instrument. The pointings on the other face may then be dealt with in a second run, and the mean azimuth obtained from the two results, substantially free of instrument error. This provides a useful check that the pointings are consistent.

For a star, the declination does not have to be interpolated from fairly rapidly changing values as for the sun; it is taken from the star tables in *SALS* by inspection and entered directly. No parallax correction is applied, of course.

Comments on the program steps:

Steps 56 and 57: ΣL and ΣR level readings in this case relate to one only on each end of the bubble.

Step 75: the display stops to show the azimuth in decimal degrees, in case 360 is to be subtracted or added to get it within the range 0–360; this is done manually. Do not forget to press R/S to conclude the run, so that the azimuth is converted to degrees, minutes and seconds.

The figures in the "Example" column are from the data in **AI.03.03** of Appendix I.

Program listing:

Appendix II

Step	Instruction	Example	Step	Instruction	Example
1	LBL B	(Press B)	40	×	
2	DSP 5		41	CHS	
3	R/S: Key Dec.	5.1624	42	RCL 2	
4	f H.MS→		43	SIN	
5	STO 2		44	+	
6	R/S:Key obs. alt. FL	33.0801	45	RCL 4	
7	f H.MS→		46	COS	
8	STO 3		47	RCL 3	
9	R/S:Key pressure, mb	1009(29.8in)	48	COS	
10	R/S:Key temp.,°C	17.8(64°F)	49	×	
11	2		50	÷	
12	7		51	f COS	
13	3		52	R/S:If star West, CHS and add 360	(250°.35968)
14	.		53	STO 5	
15	1		54	R/S:Key circle star,FL	241.2147
16	6		55	f H.MS→	
17	+		56	R/S:Key ΣL, level	8
18	÷		57	R/S:Key ΣR, level	6
19	RCL 3		58	−	
20	TAN		59	2	
21	1/X		60	÷	
22	×		61	R/S:Key secs./lev.div.	.0030
23	4		62	f H.MS→	
24	.		63	×	
25	5		64	RCL 3	
26	2		65	TAN	
27	8		66	×	
28	EEX		67	+	
29	CHS		68	STO 6	
30	3		69	R/S:Key circle RM, FL	10.5830
31	×		70	f H.MS→	
32	CHS		71	RCL 5	
33	STO+3		72	+	
34	R/S:Key latitude	35.5206	73	RCL 6	
35	f H.MS→		74	−	
36	STO 4		75	R/S	(19°.96619)
37	SIN		76	f→H.MS	19.5758
38	RCL 3		77	RTN	
39	SIN				

Repeat for right face pointings on star and RM.
Example is in AI.03.03, Appendix I.

AII.02.03 Azimuth by hour angle of the sun

This particular program is based on one given by G. G. Bennett in "Sun Observations for Azimuth" in *The Australian Surveyor*, **26**(1), March 1974. He comments:

"A suggestion for the observer who wants to improve the accuracy of the pointing to the sun's side limbs without using a Roelofs solar prism attachment, is to observe only that limb which is about to leave the vertical hair,

Astronomy for surveyors

otherwise it will be found when observing the other limb that the vertical hair is usually invisible and the observer must of necessity either anticipate the tangency of the disc and hair or observe slightly late."

No plate level correction is included in the program, because Bennett recommends careful preliminary levelling of the instrument, preferably by using the alidade level which often has a split image for greater accuracy in setting.

In the program, one timed pointing to the sun's limb and a pointing to the R.M., on one face, are used to obtain the azimuth of the R.M. The program should be run again using the data from a similar pair of pointings on the other face. The mean of the two computed azimuths will yield the azimuth of the R.M. free of most instrument error. In the computation the sun's semi-diameter (S.D.), corrected for altitude, is applied to the computed azimuth of its centre to give the azimuth of the limb at which tangency was made. If Bennett's suggestion of observing the limb which is about to leave the vertical hair is followed, then in the northern hemisphere with the sun to the south, this will be the true left limb; in the southern hemisphere, with the sun to the north, it will be the true right limb. Thus, in those northern hemisphere conditions the S.D. will have to be subtracted and in the southern, added.

Comments on the program steps:

Step 8: if the watch is fast on Zone time, the sign is $-$. The DUT1 correction is to be combined with the watch correction.

Steps 26 and 28: δ_1 and δ_2 as in **AII.02.01**.

Steps 39 and 41: E_1 is the value of E at U.T. 0^h on the U.T. date which has the same number as the local date of the observation, in this case 26 February 1975; E_2 is the value of E at U.T. 0^h on the following day. Note that E_2 is keyed in before E_1.

Step 80: If you are in the southern hemisphere and follow Bennett's advice, the sign will be $+$; if in the northern, it will be $-$. If, however, you do observe the opposite limb for the separate calculation of the azimuth on the other face, you will have to change the sign at step 80 before starting on the second run. If you wish to use the means of the pointings to R.M. and sun to eliminate the need for the S.D. correction, you will have to delete steps 23-25, 37-38, 68 and 76-80.

Step 87: If the value of the azimuth, which is displayed in degrees and decimals, does not lie in the range 0-360, then it is to be brought into this range by either adding or subtracting multiples of 360. Do not forget to press R/S after this step so that the azimuth is displayed in degrees, minutes and seconds.

The data in the column headed "Example" are from (b), the afternoon observation on 26 February 1975, in **10.25** of Chapter 10.

Program listing:

Appendix II

Step	Instruction	Example	Step	Instruction	Example
1	LBL A	(Press A)	39	R/S:Key E_2	11.47038
2	f CL REG		40	f H.MS→	
3	f P⇄S		41	R/S:Key E_1	11.4654
4	f CL REG		42	f H.MS→	
5	DSP 5		43	−	
6	R/S:Key watch time	17.16573	44	f LAST X	
7	f H.MS→		45	X⇄Y	
8	R/S:Key watch corrn. including DUT1	−.01561	46	RCL 1	
			47	×	
9	f H.MS→		48	+	
10	+		49	f P⇄S	
11	R/S:Key Zone	12	50	RCL 4	
12	f H.MS→		51	+	
13	−		52	f P⇄S	
14	Σ+		53	1	
15	f LAST x		54	5	
16	2		55	×	
17	4		56	1	
18	÷		57	→R	
19	STO 1		58	R/S:Key latitude	−45.5215
20	R/S:Key longitude	11.22039	59	f H.MS→	
21	f H.MS→		60	1	
22	Σ+		61	→R	
23	R/S:Key semi-diam.	−.1612	62	R↓	
24	f H.MS→		63	×	
25	STO 2		64	RCL 3	
26	R/S:Key δ_2	−8.3918	65	TAN	
27	f H.MS→		66	R↓	
28	R/S:Key δ_1	−9.0154	67	R↓	
29	f H.MS→		68	STO+2	
30	−		69	CHS	
31	f LAST X		70	R↓	
32	X⇄Y		71	×	
33	RCL 1		72	X⇄Y	
34	×		73	−	
35	+		74	→P	
36	STO 3		75	R↓	
37	COS		76	SIN	
38	STO+2		77	STO×2	

Step	Instruction	Example	
78	f LAST X		
79	RCL 2		
80	+ (True R limb +) (True L limb −)	Southern Hemisphere	Opposite sense in N. Hemisphere
81	R/S:Key circle RM	271.0544	
82	f H.MS→		
83	+		
84	R/S:Key circle Sun	281.3519 *	Corrected for mis-levelment
85	f H.MS→		
86	−		
87	R/S:If <0, add 360 If >0, less 360	271.°08137	
88	f→H.MS (See azimuth RM)	271.0453	
89	RTN		

Repeat for right face pointings and take the mean of the two computed azimuths.

The example is (b) afternoon observation on 26 Feb. 1975, Sec.10.25, Chapter 10.

AII.02.04 Azimuth by hour angle of a star

This program is designed to calculate the azimuth of the R.M. by taking the mean of two-face observations on star and R.M. The star's declination and R.A. are interpolated by inspection from *SALS* and entered at steps 26 and 43. R is interpolated within the program from the two values entered at steps 32 and 34. See the comments in **AII.02.01** for the manner of the plate level entries.

When the observation is being made, remember to note the approximate mean altitude for keying-in at step 95 so that the plate level correction to the star's azimuth may be computed.

Comments on the program steps:

Steps 32 and 34: R_1 is the value of R at U.T. 0^h on the U.T. date which has the same number as the local date of the observation (in this example 1980 May 1). R_2 is the value of R at U.T. 0^h on the following day, May 2 in this case:

$$R_1 = R \text{ at U.T. } 0^h \text{ 1980 May 1} = 14^h \ 36^m \ 17.8^s$$
$$R_2 = R \text{ at U.T. } 0^h \text{ 1980 May 2} = 14 \quad 40 \quad 14.3$$

Remember that R_2 is keyed in before R_1.

Step 105: the computed azimuth must be brought into the range 0–360 by adding or subtracting (manually) $n \times 360$; but remember to press R/S after doing so.

The data in the column headed "Example" are from **AI.03.03** of Appendix I. The formula used for computing the zenith angle in this and the previous program in **AII.02.03**, is:

$$\tan Z = \sin t / (\tan \delta . \cos \phi - \sin \phi . \cos t)$$

Appendix II

Program listing:

Step	Instruction	Example	Step	Instruction	Example
1	LBL B	(Press B)	39	RCL 1	
2	f CL REG		40	×	
3	f P⇄S		41	+	
4	f CL REG		42	Σ+	
5	DSP 5		43	R/S:Key R.A.	7.38152
6	R/S:Key watch time FL	20.22345	44	f H.MS→	
7	R/S:Key watch time FR	20.23445	45	CHS	
8	f H.MS+		46	Σ+	
9	f H.MS→		47	f P⇄S	
10	2		48	RCL 4	
11	÷		49	f P⇄S	
12	R/S:Key watch corrn.	-.0028	50	1	
13	f H.MS→		51	5	
14	R/S:Key DUT1	.00003	52	×	
15	f H.MS→		53	1	
16	+		54	→R	
17	+		55	R/S:Key latitude	35.5206
18	R/S:Key Zone	-8	56	f H.MS→	
19	-		57	1	
20	Σ+		58	→R	
21	f LAST X		59	R↓	
22	2		60	×	
23	4		61	RCL 3	
24	+		62	TAN	
25	STO 1		63	R↓	
26	R/S:Key Dec.	5.1624	64	R↓	
27	f H.MS→		65	CHS	
28	STO 3		66	R↓	
29	R/S:Key longitude	-7.55132	67	×	
30	f H.MS→		68	X⇄Y	
31	Σ+		69	-	
32	R/S:Key R₂	14.40143	70	→P	
33	f H.MS→		71	R↓	
34	R/S:Key R₁	14.36178	72	STO 4	
35	f H.MS→		73	R/S:Key circle RM, FL	10.5830
36	-		74	R/S:Key circle RM, FR	10.5800
37	f LAST X		75	f H.MS+	
38	X⇄Y		76	f H.MS→	

Astronomy for surveyors

Step	Instruction	Example	Step	Instruction	Example
77	2		93	f H.MS→	
78	÷		94	×	
79	STO 5		95	R/S:Key approx. alt.	33.01
80	R/S:Key circle star,FL	241.2147	96	f H.MS→	
81	R/S:Key circle star,FR	241.3453	97	TAN	
82	f H.MS→		98	×	
83	f H.MS→		99	+	
84	2		100	RCL 4	
85	÷		101	RCL 5	
86	STO 6		102	+	
87	R/S:Key ΣL, level	8	103	x⇌y	
88	R/S:Key ΣR, level	6	104	−	
89	−		105	R/S:If <0, add 360	−340°03074
90	4			If >360, less 360	
91	÷		106	f→H.MS: (Azimuth RM)	19.5809
92	R/S:Key secs./lev.div.	.0030	107	RTN	

The example is in AI.03.03 of Appendix I.

AII.03 The BASIC programs

These are in interactive form, so not a great deal of instruction in their use is necessary. However, there are some warnings and precautions. Firstly, these programs are copyright, like *Astronomy for Surveyors*, and therefore they cannot be copied on to tape or disc and freely passed around, or sold, in recorded form. They may be stored magnetically for use in an institution or survey firm. The question of making the programs available on cassette tapes or floppy disks was considered but abandoned because of a number of problems, of which compatibility with other popular microcomputers was not the least.

Secondly, there are some conventions adopted in the programs:

After the program has been loaded into the microcomputer, the usual procedure for starting is to type RUN and then press the ENTER key. From this point onwards, after each question is answered with the appropriate data, ENTER is pressed. When, H,M,S or D,M,S appear in a question they mean that the data is in the form Hours, Minutes and Seconds or Degrees, Minutes and Seconds respectively. If any of the quantities is zero, it *must* be entered by typing a zero. Failure to do this before pressing ENTER will result in a query (?) which cannot be answered by re-typing the data; the program must be re-started by typing BREAK and then RUN. The three quantities must be separated by commas. For example, a time like $14^h\ 00^m\ 03.6^s$ is entered as 14,0,3.6. The common survey practice of writing zeros in the minutes and seconds, as in 28° 09′ 07.5″ is ignored when entering these figures as data—in this case, 28,9,7.5.

In some cases only degrees and minutes are required, such as in the sun's declination taken from *SALS*; e.g. 16° 42.7′, in which D,M only are asked for, is entered as 16,42.7.

Other instances of two quantities being called for in one entry are in the

Appendix II

temperature–pressure data for finding the refraction correction to altitudes, e.g. T = 16 °C, p = 1019 mb, which are entered as 16,1019. The units of degrees Celsius and millibars are mandatory for these programs. The refraction is calculated with the aid of Saastamoinen's formula ((7.2) in Chapter 7), ignoring the partial water-vapour pressure. Use of this formula may give answers slightly different from those obtained using the *SALS* tables.

Another case of two quantities in one entry involves the sum of the left ends and the sum of the right ends of the plate level bubble (divisions) as in 7.8,4.2.

In some cases, seconds of time or arc-minutes are called for, but the units are designated in the question requesting the data; DUT1, and the 6-hourly changes in the quantity E and the sun's declination are in this category.

One question which asks for the difference between the clock time and Zone time requires H,M,S—i.e. if the difference is only 16.7^s it must be entered as 0,0,16.7, the appropriate zeros being inserted.

Where D,M,S or H,M,S are for negative quantities, they are nevertheless always entered as positive; their negative sign is looked after by other questions. This avoids having to test the degrees or hours for sign and then make the minutes and seconds correspond—a slight problem with computers which work in radians.

The method of feeding-in timing information perhaps deserves some explanation. The distinction is made between a simple stop-watch and one with lap-time facility. In using a simple stop-watch, the observer usually starts the watch when the star or sun is in contact with a cross-hair, and then stops it at a recorded clock time. The stop-watch time is subtracted from that time to give the clock time of the event. Both the clock time and stop-watch reading are entered into the program for each pointing to the object. If a lap-time stop-watch is used, it is usually started on a known instant of time signal or clock time and the events are recorded as the watch runs on. It is usually easy to deduce the times of the events by inspection —adding the clock start time to the "lap" times. In this case, only the final time of each event need be entered into the program.

The answers to some questions require single letters or short words instead of figures, e.g. N, S, E, W, U, L, Y, N, SUN, STAR, the first eight standing for North, South, East, West, Upper, Lower, Yes, No, respectively. The context makes them fairly obvious.

In some parts of the programs there are pauses to enable the user to read some of the text; these are set up by simple counting loops, but none of them is very long and a little patience will soon see a resumption of activity. In all three programs, after introductory comments, a series of questions is put to collect basic data, after which the information required for each particular type of solution is sought.

The best advice the author can give to users trying the programs for the first time is to be *very* careful to enter the data in the correct form.

Astronomy for surveyors

AII.03.01 A program for calculating azimuth, latitude, local time (watch error) and longitude

For azimuth, altitude and hour angle, methods are applied to observations of the sun or stars in the east and west.

For latitude, the circum-meridian method of observations of sun or stars is used, and for local time and longitude the traditional method of east–west observations to sun or stars.

All these are described in detail in Chapters 10, 11, 12, and 13.

The program, approximately 14 500 bytes in length, is listed in the print-out at the end of the following example. The latter takes the operator through the preliminary section of the program, listing the questions and the answers applicable to part of the example given in **10.25(A)(ii)(a)** in Chapter 10. After the preliminary questions, the program branches to "Azimuth by hour angle" and the remainder of the data is requested. The method computes the azimuth of the R.M. from the FL pointings to sun and R.M., applying the semi-diameter (S.D.) of the sun so that the azimuth of its observed true right limb is first computed, then applying the included angle to this to give the azimuth of the R.M. The illuminated cross-hair is timed across the sun's limb in this FL observation as recommended by G. G. Bennett (see **AII.02.03** earlier). In completing this method, FR observations should be made to the same (true right in this case) limb and the R.M., and the S.D. again applied. The mean of the two computed azimuths will then eliminate most instrument errors. Of course, in the example from Chapter 10, the observation is completed on FR by timing tangency with the true left limb to eliminate the need for the S.D. correction, as well as to get rid of instrumental errors. The program will deal with this second part of the observation to provide an azimuth computed from the FR pointings.

EXAMPLE: Use of the program, using FL data from **10.25(A)(ii)(a)** of Chapter 10 (the bold-face type shows the answers to the questions generated by the computer program).

LOAD PROGRAM   Type RUN   Press ENTER

Press ENTER after each question has been answered.

1.01   Charles Griffin Software by J. B. Mackie, "Astronomy for Surveyors"

This is a comprehensive program for processing the data from field astronomy observations to yield azimuth, latitude, time and longitude by methods explained in "Astronomy for Surveyors" by J. B. Mackie, published by Charles Griffin & Co. Ltd, Crendon Street, High Wycombe, England.

The program is copyright.

Please read the information in the book before using the program.

Name of station? **ARTS BLOCK F**
Name of R.M. (if none, type X)? **RED LIGHT 1 COBDEN STREET**
When prompted, enter the latitude and longitude of the observing station (both as positive values).

378

Appendix II

If either of these is not required, enter 0,0,0 for D (or H), M,S.
Latitude of stn., D,M,S, (always +)? **45,52,15**
Longitude of stn., H,M,S, (always +)? **11,22,3.9**
Is longitude east or west; type E or W? **E**
Longitude of Zone meridian (type 0 if not reqd.),+/−H, e.g. −7.5? **12**
Answer next question with X if circum-merid. latitude.
Was object observed E or W of meridian; type E or W? **E**
Was sun or star observed; type SUN or STAR? **SUN**
Was declination of sun or star north or south, type N or S? **S**
Is observing station in N or S hemisphere; enter N or S? **S**
Number of pointings on sun/star? **1**

Menu

Select from the following:

Subject	Code
Azimuth	AZ
Latitude	LA
Time	TI
Longitude	LO

Type code of your choice and press Enter? **AZ**

Sub-menu
Select from:

Method	Code
Azimuth by altitude	A1
Azimuth by H.A. using *SALS* tables	A2

Type code of your choice and press Enter? **A2**

To compute azimuth from hour angle of sun or star.

If a simple stop-watch, started on object contact and stopped on clock reading, was used, you will be asked to enter, for each pointing, the clock time and then the stop-watch time of contact. If, however, a stop-watch was not used, or a stop-watch with lap-time facility was used, you will be asked to enter only the final clock time of contact for each pointing.

Was a simple stop-watch used, type Y or N? **Y**
Enter clock time, hit enter, and then enter stop-watch time and hit enter—
i.e. 2 entries for each pointing.
Clock time when s/w stopped, H,M,S? **8,6,5**
Stop-watch time, seconds? **10.5**
Clock difference from Zone time (+ always); H,M,S? **1,1,26.1**
Clock fast or slow on Zone time, type F or S? **F**
DUT1 (secs)? **.6**
If looking up *SALS* tables for "E", "R" or Dec. note the following carefully:

UT1 of observation (decimal hours) = 19.075
Look up tables for day before local date of observation.
Look up *SALS* table for "E" at 6-hour epoch before UT1 of observation.
"E" from tables, H,M,S? **11,46,5.7**
6-hour epoch used, type 0,6,12 or 18? **18**
Change in "E" in 6 hours, +/− seconds? **1.6**
UT of obs. = 19.075
Look up *SALS* table for Dec. at 6-hour epoch before UT of observation
Sun's Dec. (+ always) at correct epoch on correct UT date, D,M, (no secs)? **11,18.5**
Be careful with the sign of the next entry!
If the Dec. is N. and increasing numerically, the change is +, and if decreasing, it is −.
If the Dec. is S. and increasing numerically, the change is −, and if decreasing, it is +.
6-hourly change in Dec., +/− arc-mins? **6.0**
6-hour epoch used to look up Dec.; type 0,6,12 or 18? **18**
If you have used the single limb observing method described earlier, you will need to apply the semi-diameter (S.D.) correction.
Do you wish to apply the S.D. corrn, Y or N? **Y**
SUN'S S.D., arc-minutes? **16.2**
Which true limb of sun did you observe, L or R? **R**

** IMPORTANT **

When entering the deg, min, sec of hor. circle readings on R.M. and object, enter the deg on FR as though they had been read on FL—i.e. 180 deg different, e.g. FL = 68, FR = 247; enter FR as 67; clear? (as mud?).

No. of pointings on R.M.? **1**
Enter the hor. circle readings on the R.M., 1 pointing at a time.

Hor. circle on R.M. (careful of degrees!), D,M,S? **348,11,3**
Hor. circle on object observed (careful of degress!), D,M,S? **173,3,56**
Sum of plate level divs. left, ditto right? **2.7,2.6**
Value of 1 div. of plate level (secs)? **20**
Approx. altitude of obj., D,M? **11,0**
Azimuth of object observed = 95 41 10
From stn. Arts Block F to R.M., stn. red light 1 Cobden Street
Azimuth of R.M. = 271 4 43
READY

Appendix II

```
2 CLS
4 PRINT"1.01 GRIFFIN SOFTWARE BY J.B. MAC
KIE":PRINT"ASTRONOMY FOR SURVEYORS":PRIN
T
5 PRINT"THIS IS A COMPREHENSIVE PROGRAM
FOR PROCESSING THE"
6 PRINT"DATA FROM FIELD ASTRONOMY OBSERV
ATIONS TO YIELD"
7 PRINT"AZIMUTH, LATITUDE, TIME AND LONG
ITUDE BY METHODS"
8 PRINT"EXPLAINED IN 'ASTRONOMY FOR SURV
EYORS' BY J.B. MACKIE,"
9 PRINT"PUBLISHED BY CHARLES GRIFFIN & C
O., CRENDON STREET,"
10 PRINT"HIGH WYCOMBE, ENGLAND.":PRINT
12 PRINT"THE PROGRAM IS COPYRIGHT.":PRIN
T
13 PRINT"PLEASE READ THE INFORMATION IN
THE BOOK BEFORE USING"
14 PRINT"                    THE PROGRAM.":
PRINT
15 CLEAR500
17 DEFDBLD,F,G,P,Q
19 A=206265:B=3600:C=60:PI=3.14159265:D=
PI/2:F=5.729577951D+01
20 P=3.6D+03:Q=6.0D+01:FF=1.002737909
24 INPUT"NAME OF STATION";D$:INPUT"NAME
OF R.M. (IF NONE, TYPE X)";DD$
26 PRINT"WHEN PROMPTED, ENTER THE LATITU
DE AND LONGITUDE OF"
28 PRINT"THE OBSERVING STATION (BOTH AS
POSITIVE VALUES)."
30 PRINT"IF EITHER OF THESE IS NOT REQUI
RED, ENTER 0,0,0"
32 PRINT"FOR D (OR H),M,S."
33 INPUT"LATITUDE OF STN., D,M,S, (ALWAYS
+)";BD,BE,BF
34 BV=D-(BD*B+BE*C+BF)/A
36 INPUT"LONGITUDE OF STN.,H,M,S, (ALWAYS
+)";DD,DE,DF
38 DG=DD+DE/Q+DF/P
40 INPUT"IS LONGITUDE EAST OR WEST; TYPE
E OR W";X$
42 IFX$="W"THENDG=-DG
44 INPUT"LONGITUDE OF ZONE MERIDIAN (TYP
E 0 IF NOT REQD.),+/-H,E.G. -7.5";QJ
45 PRINT"ANSWER NEXT QUESTION WITH X IF
CIRCUM-MERID. LATITUDE."
46 INPUT"WAS OBJECT OBSERVED E OR W OF M
ERIDIAN; TYPE E OR W";F$
48 INPUT"WAS SUN OR STAR OBSERVED; TYPE
SUN OR STAR";C$
50 INPUT"WAS DECLINATION OF SUN OR STAR
NORTH OR SOUTH, TYPE N OR S";N$
52 INPUT"IS OBSERVING STATION IN N OR S
HEMISPHERE; ENTER N OR S";E$
53 INPUT"NUMBER OF POINTINGS ON SUN/STAR
";GY:GZ=GY
55 CLS
59 PRINT"MENU":PRINT
60 PRINT"SELECT FROM THE FOLLOWING:"
70 PRINTTAB(15)"SUBJECT";TAB(31)"CODE":P
RINT
80 PRINTTAB(15)"AZIMUTH";TAB(31)"AZ"
90 PRINTTAB(15)"LATITUDE";TAB(31)"LA"
100 PRINTTAB(15)"TIME";TAB(31)"TI"
110 PRINTTAB(15)"LONGITUDE";TAB(31)"LO"
160 INPUT"TYPE CODE OF YOUR CHOICE & PRE
SS ENTER";A$
165 CLS
170 IFA$="AZ"THEN260
180 IFA$="LA"THEN900
190 IFA$="TI"THEN2500
200 IFA$="LO"THEN4000
220 PRINT"YOU DID NOT TYPE A VALID CODE;
TRY AGAIN AFTER PAUSE"
230 FORI=1TO200:I=I+0:NEXTI:CLS
240 GOTO55
250 CLS
260 PRINT"SUB-MENU"
270 PRINT"SELECT FROM:":PRINT
280 PRINTTAB(15)"METHOD";TAB(44)"CODE":P
RINT
290 PRINTTAB(5)"AZIMUTH BY ALTITUDE";TAB
(45)"A1"
300 PRINTTAB(5)"AZIMUTH BY H.A. USING SA
LS TABLES";TAB(45)"A2"
330 INPUT"TYPE CODE OF YOUR CHOICE AND P
RESS ENTER";B$
340 CLS
350 IFB$="A1"THEN500
360 IFB$="A2"THEN700
390 PRINT"YOU DID NOT TYPE A VALID CODE;
TRY AGAIN AFTER PAUSE"
400 FORI=1TO200:I=I+0:NEXTI
410 GOTO260
470 END
500 PRINT"TO COMPUTE AZIMUTH BY ALTITUDE
.":PRINT
510 GOSUB10000
515 IFC$="SUN"THEN517ELSE540
517 PRINT"TO ANSWER THE NEXT QUESTION PL
EASE MAKE A MENTAL":PRINT"CALCULATION FO
R THE APPROX. UT OF OBSERVATION AND THE"
:PRINT"CORRECT DATE TO ENTER SALS.":PRIN
T
520 INPUT"APPROX. UT OF OBSERVATION, H,M
";DL,DM
530 DR=DL+DM/Q
535 PRINT"APPROX. UT OF OBS. = ";INT(DR*
1000+.5)/1000
540 IFC$="SUN"THENGOSUB10250ELSEGOSUB105
00
550 GOSUB10750:GOSUB11000
560 CJ=CC-BR:IFCJ<0THENCJ=CJ+2*PI
570 IFCJ>2*PITHENCJ=CJ-2*PI
580 CK=CJ*F:CLS:GOSUB10900:PRINT:GOSUB10
950
590 END
700 PRINT"TO COMPUTE AZIMUTH FROM HOUR A
NGLE OF SUN OR STAR."
701 PRINT
710 GOSUB11500:GOSUB11700
720 IFC$="SUN"THEN730ELSE850
730 GOSUB11800:GOSUB11900
740 FP=DR+FM+DG:GOSUB12200
```

381

```
750 PRINT"UT OF OBS. = ";INT(DR*1000+.5)
/1000
760 GOSUB10250:GOSUB12300
761 PRINT"IF YOU HAVE USED THE SINGLE LI
MB OBSERVING METHOD":PRINT"DESCRIBED EAR
LIER, YOU WILL NEED TO APPLY THE SEMI-":
PRINT"DIAMETER (S.D.) CORRECTION."
764 INPUT"DO YOU WISH TO APPLY THE S.D.
CORRN., Y OR N";OA$
766 IF OA$="Y" THEN GOSUB 13600 ELSE 770
770 GOSUB11000
780 CJ=CC-BR:IFCJ<0THENCJ=CJ+2*PI
790 IFCJ>2*PITHENCJ=CJ-2*PI
800 CK=CJ*F:CLS:GOSUB10900:GOSUB10950
810 END
850 GOSUB11800:GOSUB12000:GOSUB12100
860 FP=DR*FF+FQ-FW+DG:GOSUB12200
870 PRINT"UT OF OBS. = ";INT(DR*1000+.5)
/1000
880 GOSUB10500:GOSUB12300:GOSUB11000
890 GOTO780
900 PRINT"TO DETERMINE LATITUDE BY CIRCU
M-MERIDIAN"
901 PRINT"OBSERVATIONS OF SUN OR STAR.":
PRINT
910 IFC$="SUN"THEN920ELSE1800
920 FP=0:GOTO2520
1270 INPUT"CLOCK DIFFERENCE (+ ALWAYS) F
ROM ZONE TIME, H,M,S";DW,DX,DY
1280 DZ=DW+DX/Q+DY/P
1290 INPUT"CLOCK FAST OR SLOW ON ZONE TI
ME, TYPE F OR S";Z$
1300 IFZ$="F"THENDZ=-DZ
1310 GX=DR+QJ-DZ:IFGX<0THENGX=GX+24
1315 IFGX>24THENGX=GX-24
1320 GOSUB11500
1350 IFY$="Y"THEN1360ELSE1410
1360 PRINT"ENTER CLOCK TIME, HIT ENTER,
AND THEN ENTER STOP-"
1361 PRINT"WATCH TIME AND HIT ENTER -- I
.E. 2 ENTRIES FOR "
1362 PRINT"EACH POINTING."
1370 INPUT"CLOCK TIME WHEN S/W WAS STOPP
ED, H,M,S";DH,DI,DJ
1380 DK=DH+DI/Q+DJ/P
1390 INPUT"STOP-WATCH TIME, SECONDS";DQ:
DQ=DQ/P
1400 DK=DK-DQ:GOTO1430
1410 INPUT"FINAL CLOCK TIME OF EACH POIN
TING, H,M,S";DH,DI,DJ
1420 DK=DH+DI/Q+DJ/P
1430 DL=(((GX-DK)*P)[2)*5.454D-04
1440 BN=BN+DL:GY=GY-1:IFGY=0THEN1460
1450 IFGY>0ANDY$="Y"THEN1370ELSE1455
1455 IFGY>0ANDY$="N"THEN1410
1460 PP=BN/GZ
1470 GOSUB10000:IFC$="STAR"THEN1510
1480 PRINT"TO ANSWER THE NEXT QUESTION Y
OU WILL NEED TO TAKE":PRINT"MENTALLY THE
 MEAN OF THE 2 HIGHEST (LOWEST) ALTITUDE
S, ONE":PRINT"ON EACH FACE (ONE ON EACH
LIMB, TOO, FOR SUN).":PRINT
1510 INPUT"MEAN OF 2 HIGHEST/LOWEST ALTI
TUDES, 1 ON EA. FACE, D,M,S";AG,HH,AI
1520 BJ=(AG*B+AH*C+AI)/A:IFC$="STAR"THEN
 2110
1530 BJ=BJ+(BS+BT)/A:K=1:DP=0
1550 IFE$="N"THEN1560ELSE1630
1560 INPUT"WAS OBJ. N. OR S. OF ZENITH,
N OR S";J$
1570 IFJ$="N"THEN1580ELSE1620
1580 INPUT"WAS OBJ. AT UPPER OR LOWER TR
ANSIT, U OR L";K$
1590 IFK$="U"THEN1600ELSE1610
1600 P1=BJ+DP-(D-QU):GOTO1700
1610 P1=BJ-DP+(D-QU):GOTO1700
1620 P1=D-(BJ+DP-QU):GOTO1700
1630 INPUT"WAS OBJ. N. OR S. OF ZENITH,
N OR S";J$
1640 IFJ$="S"THEN1650ELSE1690
1650 INPUT"WAS OBJ. AT UPPER OR LOWER TR
ANSIT, U OR L";K$
1660 IFK$="U"THEN1670ELSE1680
1670 P1=BJ+DP-(D+QU):GOTO1700
1680 P1=BJ-DP+(D+QU):GOTO1700
1690 P1=D-(BJ+DP+QU)
1700 IFK>1THEN1750
1710 B9=(COS(P1)*COS(QU))/COS(BJ)
1720 DP=(PP*B9)/A
1730 IFC$="STAR"THENDP=DP*1.0055
1740 BJ=BU:K=K+1
1743 PRINT"THERE IS REPETITION HERE BECA
USE APPROX. LAT. WAS"
1744 PRINT"REQUIRED FOR SOLUTION."
1746 GOTO1550
1750 CK=P1*F:CL=INT(CK):CM=(CK-CL)*C:CN=
INT(CM):CO=(CM-CN)*C
1755 PRINTTAB(1)"STATION   ";D$
1760 IFE$="N"THEN1770ELSE1780
1770 PRINTTAB(1)"LATITUDE NORTH = ";TAB(
20)CL;TAB(25)CN;TAB(30)INT(CO):GOTO2120
1780 PRINTTAB(1)"LATITUDE SOUTH = ";TAB(
20)CL;TAB(25)CN;TAB(30)INT(CO):GOTO2120
1800 GOSUB12100
1830 INPUT"WAS OBJECT AT UPPER OR LOWER
TRANSIT, U OR L";G$
1840 IFG$="L"THENFW=FW+12
1850 GOSUB10500
1890 INPUT"DUT1, +/-SECS";FA:PRINT
1895 FP=0:GOTO2910
2035 GV=FB
2040 INPUT"CLOCK DIFFERENCE FROM LOCAL Z
ONE TIME (+ ALWAYS); H,M,S";DW,DX,DY
2050 INPUT"CLOCK FAST OR SLOW ON ZONE TI
ME; TYPE F OR S";Z$
2060 DZ=DW+DX/Q+DY/P:IFZ$="F"THENDZ=-DZ
2070 GX=GV+QJ-DZ-FA/P:IFGX<0THENGX=GX+24
2080 IFGX>24THENGX=GX-24
2090 GOTO1320
2110 BJ=BJ+BT/A:K=1:GOTO1550
2120 END
2500 PRINT"TO COMPUTE ZONE TIME FROM E-W
 OBSERVATION OF SUN OR"
2501 PRINTTAB(30)"STARS.":PRINT
2510 GOSUB10000
2515 IFC$="STAR"THEN2900
2520 PRINT"USING APPROXIMATIONS FOR SUN'
S DEC. OR 'E' WE SHALL"
```

Appendix II

```
2521 PRINT"FIND THE CORRECT UT DATE AND
A SUFFICIENTLY CLOSE"
2522 PRINT"VALUE OF UT FOR THEIR INTERPO
LATION. "
2523 PRINT"FROM A KNOWLEDGE OF YOUR APPR
OX. ZONE TIME YOU CAN"
2524 PRINT"DEDUCE THE APPROX. UT AND DAT
E TO ENTER SALS FOR"
2525 PRINT"SUITABLE VALUES OF DEC. OR 'E
'; OR YOU CAN, LESS"
2526 PRINT"ACCURATELY, TAKE THEM FOR UT
12 HRS. ON YOUR LOCAL"
2527 PRINTTAB(26)"DATE OF OBS. "
2528 IFA$="LA"THEN2580
2530 INPUT"APPROX. SUN'S DEC. (ALWAYS +)
, D, M";BA, BB
2540 BA=(BH*B+BB*C)/A: IFN$="S"THENBA=-BA
2550 IFE$="N"THENBW=(D-BA)ELSEBW=(D+BA)
2560 GOSUB13500
2580 INPUT"APPROX. VALUE OF 'E', H,M,S";
FG,FH,FI
2590 FJ=FG+FH/Q+FI/P
2600 GG=FP+24-FJ-DG: GI=GG
2610 IFGI<0THENGI=GI+24: IFGI>24THENGI=GI
-24
2615 DR=GI
2630 PRINT"APPROX. UT OF OBS. = ";INT(GI
*1000+.5)/1000
2640 IFGG<0THEN2650ELSE2660
2650 PRINT"TAKE 'E' ON DAY BEFORE LOCAL
DATE": GOTO2690
2660 IFGG>24THEN2670ELSE2680
2670 PRINT"TAKE 'E' ON DAY AFTER LOCAL D
ATE": GOTO2690
2680 PRINT"TAKE 'E' ON SAME DATE AS LOCA
L DATE"
2690 GOSUB11900
2700 PRINT"NOW USING UT = ";INT(GI*1000
+.5)/1000;"ON SAME"
2701 PRINT"DATE AS FOR 'E': "
2710 GOSUB10250
2715 IFA$="LA"THEN1270
2720 GOSUB13500
2730 INPUT"DUT1 (SECS)";FA: FA=FH/P
2740 GG=FP+24-FM-DG+QJ-FA: IFGG<0THENGG=G
G+24
2750 IFGG>24THENGG=GG-24
2760 GOSUB11500: G2=DP-GG
2820 IFG2<0THENT$="SLOW"ELSET$="FAST"
2830 G2=ABS(G2): G3=INT(G2): G4=(G2-G3)*Q:
G5=INT(G4): G6=(G4-G5)*Q
2840 IFG3<0THENG3=G3+24: IFG3>24THENG3=G3
-24
2841 PRINT: PRINT
2850 PRINTTAB(1)"CLOCK IS ";T$;" BY";TAB
(18)G3;TAB(21)"HRS";TAB(25)G5;TAB(29)"MI
NS";TAB(34)INT(G6*100+.5)/100;TAB(38)"SE
CS"
2860 END
2900 GOSUB12100: GOSUB10500: GOSUB13500
2910 INPUT"'R' AT 0 H UT ON LOCAL DATE O
F OBS. , H,M,S";FQ,FR,FS
2920 FQ=FQ+FR/Q+FS/P: K=1
2930 FA=FP+FW-FQ: IFFA<0THENFA=FA+24
2940 IFFA>24THENFA=FA-24
2950 FB=FA-DG
2955 IFK>1THEN2990
2960 IFFB<0THEN2970ELSE2980
2970 FQ=FQ-24*(FF-1): GOTO2985
2980 IFFB>24THEN2981ELSE3010
2981 FQ=FQ+24*(FF-1)
2985 K=K+1: GOTO2930
2990 IFFB<0THENFB=FB+24
3000 IFFB>24THENFB=FB-24
3010 FB=FB/FF: FC=FB+QJ
3015 IFA$="LA"THEN2035
3020 INPUT"DUT1 (SECS)";FD: FD=FD/P
3030 FC=FC-FD: IFFC<0THENFC=FC+24
3040 IFFC>24THENFC=FC-24
3045 GOSUB11500
3050 G2=DP-FC: GOTO2820
3060 END
4000 PRINT"TO DETERMINE LONGITUDE FROM E
-W OBSERVATION OF SUN"
4001 PRINTTAB(25)"OR STARS. "
4010 GOSUB10000: GOSUB11500: GOSUB11700: GO
SUB11800
4020 PRINT: PRINT"NOTE DOWN UT1 AND DATE"
: PRINT
4030 IFC$="SUN"THEN4040ELSE4200
4040 GOSUB10250: GOSUB11900: GOSUB13500
4060 IFFP<0THENFP=FP+24
4070 FD=DR+FM: IFFD>24THENFD=FD-24
4080 DG=FP-FD
4090 IFDG<-12THENDG=DG+24
4100 IFDG>12THENDG=DG-24
4110 IFDG<0THENU$="WEST"ELSEU$="EAST"
4120 G2=ABS(DG): G3=INT(G2): G4=(G2-G3)*Q:
G5=INT(G4): G6=(G4-G5)*Q
4122 H2=ABS(DG)*1.5D+01: H3=INT(H2): H4=(H
2-H3)*Q: H5=INT(H4): H6=(H4-H5)*Q
4130 PRINTTAB(1)"LONGITUDE STN. ";D$;" =
";G3;" HRS. ";" ";G5;" MINS. ";INT(G6*1D+
02+.5)/1D+02;" SECS. ";" ";U$;", OR ";H3;
" DEG. ";" ";H5;" MINS. ";INT(H6);" ";"SE
CS. ";U$
4140 END
4200 GOSUB10500: GOSUB12100: GOSUB13500
4210 GOSUB12000: FA=FP+FW: IFFA<0THENFA=FA
+24
4220 IFFA>24THENFA=FA-24
4230 FB=DR*FF+FQ: IFFB>24THENFB=FB-24
4240 DG=FA-FB: GOTO4090
4250 END
10000 PRINT"CONVERT ALL VERT. CIRCLE REA
DINGS TO TRUE ALTITUDES"
10020 PRINT"AND ENTER THEM ONE AT A TIME
"
10025 BN=0: GY=GZ
10030 INPUT"ALTITUDE, D,M,S";AQ,AR,AS
10040 BM=(AQ*B+AR*C+AS)/A: BN=BN+BM: GY=GY
-1
10050 IFGY>0THEN10030ELSE10060
10060 BO=BN/GZ
10070 INPUT"TEMPERATURE(C), PRESSURE(MB)
";AW,AX
10080 BS=8.85*COS(BO)
10090 LA=TAN(D-BO): LB=LA*LA: LC=LA*LB: LD=
```

```
AW+273.16
10100 BT=-16.271*LA*(1+3.94E-05*LB*(AX/L
D))*(AX/LD)+.0749*(LC+LA)*(AX/1000)
10110 IFC$="SUN"THENBU=B0+(BS+BT)/A
10120 IFC$="STAR"THENBU=B0+BT/A
10130 CLS
10140 BX=D-BU
10150 RETURN
10250 PRINT"LOOK UP SALS TABLE FOR DEC.
AT 6-HOUR EPOCH"
10270 PRINT"BEFORE UT OF OBSERVATION"
10280 INPUT"SUN'S DEC. (+ ALWAYS) AT COR
RECT EPOCH ON CORRECT UT DATE, D,M, (NO
SECS)";BA,BB
10290 BW=(BA*B+BB*C)/A
10300 IFN$="S"THENBW=-BW
10310 PRINT"BE CAREFUL WITH THE SIGN OF
THE NEXT ENTRY!"
10320 PRINT"IF THE DEC. IS N. AND INCREA
SING NUMERICALLY, THE"
10330 PRINT"CHANGE IS +, AND IF DECREASI
NG, IT IS -"
10340 PRINT"IF THE DEC. IS S. AND INCREA
SING NUMERICALLY, THE"
10350 PRINT"CHANGE IS -, AND IF DECREASI
NG, IT IS +"
10360 INPUT"6-HOURLY CHANGE IN DEC., +/-
ARCMINS";PK
10370 PK=(PK*Q)/A
10380 INPUT"6-HOUR EPOCH USED TO LOOK UP
DEC.; TYPE 0,6,12, OR 18";FK
10390 GW=DR-FK: QU=BW+PK*(GW/6)
10400 IFE$="N"THENBW=(D-QU)ELSEBW=(D+QU)
10410 RETURN
10500 INPUT"DEC. OF STAR AFTER USER INTE
RPOLATION FROM SALS, (+ALWAYS), D,M,S";B
A,BB,BC
10520 QU=(BA*B+BB*C+BC)/A
10530 IFN$="S"THENQU=-QU
10540 IFE$="N"THENBW=(D-QU)ELSEBW=(D+QU)
10550 RETURN
10750 BY=(COS(BW)-COS(BV)*COS(BX))/(SIN(
BV)*SIN(BX))
10760 BZ=SQR(1-BY*BY): CA=BZ/BY: CB=ATN(CA
)
10770 IFCB<0THENCB=PI+CB
10780 IFE$="N"THEN10790ELSE10820
10790 IFF$="E"THEN10800ELSE10810
10800 CC=CB: GOTO10850
10810 CC=2*PI-CB: GOTO10850
10820 IFF$="E"THEN10830ELSE10840
10830 CC=PI-CB: GOTO10850
10840 CC=PI+CB
10850 CD=CC*F
10860 RETURN
10900 CE=INT(CD): CF=(CD-CE)*C: CG=INT(CF)
: CH=(CF-CG)*C
10910 PRINTTAB(1)"AZIMUTH OF OBJECT OBSE
RVED = ";TAB(26)CE; TAB(31)CG; TAB(35)INT(
CH)
10920 RETURN
10950 CL=INT(CK): CM=(CK-CL)*C: CN=INT(CM)
: CO=(CM-CN)*C
10960 PRINTTAB(1)"FROM STN. ";D$;" TO R.
M., STN. ";DD$
10970 PRINTTAB(1)"AZIMUTH OF R.M. = ";TA
B(30)CL; TAB(35)CN; TAB(39)INT(CO)
10980 RETURN
11000 CLS
11010 PRINT@400,"** IMPORTANT **"
11020 PRINT@448,"WHEN ENTERING THE DEG,M
IN, SEC OF HOR. CIRCLE READINGS"
11030 PRINT@512,"ON R.M. AND OBJECT, ENT
ER THE DEG ON FR AS THOUGH"
11040 PRINT@576,"THEY HAD BEEN READ ON F
L -- I.E. 180 DEG DIFFERENT"
11050 PRINT@640,"E.G. FL=68, FR=247: ENT
ER FR AS 67: CLEAR? (AS MUD?)"
11055 PRINT
11060 INPUT"NO. OF POINTINGS ON R.M.";RM
11070 PRINT"ENTER THE HOR. CIRCLE READIN
GS ON THE R.M., 1 POINTING AT A TIME"
11075 BN=0: RN=RM
11080 INPUT"HOR. CIRCLE ON R.M. (CAREFUL
OF DEGREES!), D,M,S";AA,AB,AC
11090 BG=(AA*B+AB*C+AC)/A
11100 BN=BN+BG: RM=RM-1
11110 IFRM>0THEN11080ELSE11120
11120 BI=BN/RN: BN=0: GY=GZ
11130 INPUT"HOR. CIRCLE ON OBJECT OBSD.
(CAREFUL OF DEGREES!), D,M,S";AG,AH,AI
11140 BJ=(AG*B+AH*C+AI)/A
11150 BN=BN+BJ: GY=GY-1
11160 IFGY>0THEN11130ELSE11170
11170 BL=BN/GZ: BN=0: GY=GZ
11180 INPUT"SUM OF PLATE LEVEL DIVS. LEF
T, DITTO RIGHT";AM,AO
11190 INPUT"VALUE OF 1 DIV OF PLATE LEVE
L (SECS)";AY
11200 BH=(((AM-AO)/(2*GZ))*AY)/A
11210 IFBU=0THEN11220ELSE11250
11220 INPUT"APPROX. ALTITUDE OF OBJ., D,
M";RO,RP
11230 RQ=(RO*B+RP*C)/A
11240 BP=BH*TAN(RQ): GOTO11260
11250 BP=BH*TAN(BU)
11260 BQ=BL+BP: BR=BQ-BI
11270 RETURN
11500 BN=0: GY=GZ
11510 PRINT
11520 PRINT"IF A SIMPLE STOP-WATCH, STAR
TED ON OBJECT CONTACT"
11521 PRINT"AND STOPPED ON CLOCK READING
, WAS USED, YOU WILL"
11522 PRINT"BE ASKED TO ENTER, FOR EACH
POINTING, THE CLOCK"
11523 PRINT"TIME AND THEN THE STOP-WATCH
TIME OF CONTACT."
11524 PRINT"IF, HOWEVER, A STOP-WATCH WA
S NOT USED, OR A STOP-"
11525 PRINT"WATCH WITH LAP-TIME FACILITY
WAS USED, YOU WILL BE"
11526 PRINT"ASKED TO ENTER ONLY THE FINA
L CLOCK TIME OF CON-"
11527 PRINT"TACT FOR EACH POINTING."
```

Appendix II

```
11528 PRINT
11530 INPUT"WAS A SIMPLE STOP-WATCH USED
, TYPE Y OR N";Y$
11535 IFA$="LA"THENRETURN
11540 IFY$="Y"THEN11550ELSE11600
11550 PRINT"ENTER CLOCK TIME, HIT ENTER,
 AND THEN ENTER STOP-"
11551 PRINT"WATCH TIME AND HIT ENTER --
I.E. 2 ENTRIES FOR"
11552 PRINT"EACH POINTING. "
11560 INPUT"CLOCK TIME WHEN S/W STOPPED,
 H,M,S";DH,DI,DJ
11570 DK=DH+DI/Q+DJ/P
11580 INPUT"STOP-WATCH TIME, SECONDS";DQ
:DQ=DQ/P
11590 DK=DK-DQ:GOTO11620
11600 INPUT"FINAL CLOCK TIME OF POINTING
, H,M,S";DH,DI,DJ
11610 DK=DH+DI/Q+DJ/P
11620 BN=BN+DK:GY=GY-1:IFGY=0THEN11650
11630 IFGY>0ANDY$="Y"THEN11560ELSE11640
11640 IFGY>0ANDY$="N"THEN11600
11650 DP=BN/GZ:IFDP>24THENDP=DP-24
11660 RETURN
11700 INPUT"CLOCK DIFFERENCE FROM ZONE T
IME (+ ALWAYS); H,M,S";DW,DX,DY
11720 DZ=DW+DX/Q+DY/P
11730 INPUT"CLOCK FAST OR SLOW ON ZONE T
IME, TYPE F OR S";Z$
11740 IFZ$="F"THENDZ=-DZ
11750 INPUT"DUT1 (SECS)";FA:FB=FA/P
11760 FE=DP+DZ-QJ+FB:DR=FE
11770 IFDR<0THENDR=DR+24
11780 IFDR>24THENDR=DR-24
11790 RETURN
11800 PRINT"IF LOOKING UP SALS TABLES FO
R 'E', 'R' OR DEC. "
11810 PRINT"NOTE THE FOLLOWING CAREFULLY
:":PRINT
11820 PRINT"UT1 OF OBSERVATION (DECIMAL
HOURS) = ";INT(DR*1000+.5)/1000
11830 IFFE<0THEN11840ELSE11850
11840 PRINT"LOOK UP TABLES FOR DAY BEFOR
E LOCAL DATE OF OBS.":GOTO11880
11850 IFFE>24THEN11860ELSE11870
11860 PRINT"LOOK UP TABLES FOR DAY AFTER
 LOCAL DATE OF OBS.":GOTO11880
11870 PRINT"LOOK UP TABLES FOR LOCAL DAT
E OF OBS. "
11880 RETURN
11900 PRINT"LOOK UP SALS TABLE FOR 'E' A
T 6-HOUR EPOCH"
11911 PRINT"BEFORE UT1 OF OBSERVATION. "
11920 INPUT"'E' FROM TABLES, H,M,S";FG,F
H,FI
11930 FJ=FG+FH/Q+FI/P
11940 INPUT"6-HOUR EPOCH USED, TYPE 0,6,
12, OR 18";FK
11950 INPUT"CHANGE IN 'E' IN 6 HOURS, +/
-SECONDS";FL
11960 GW=DR-FK
11970 FM=FJ+((GW/6)*FL)/P
11980 RETURN
12000 PRINT"LOOK UP TABLE FOR 'R' AT 0 H
RS UT ON CORRECT DATE"
12011 PRINT"OF OBS. AS INDICTED EARLIER
."
12020 INPUT"'R' AT 0 HRS UT FROM TABLES,
 H,M,S";FQ,FR,FS
12030 FW=FQ+FR/Q+FS/P
12040 RETURN
12100 PRINT"USER -- PLEASE INTERPOLATE R
.A. OF STAR FROM SALS"
12111 PRINT"FOR DATE OF OBSERVATION"
12120 INPUT"R.A. OF STAR, H,M,S";FT,FU,F
V
12130 FW=FT+FU/Q+FV/P
12140 RETURN
12200 IFFP>24THENFP=FP-24:GOTO12230
12220 IFFP<0THENFP=FP+24
12230 IFFP<12THEN12250
12240 IFFP>12THENFP=24-FP
12250 GA=FP*1.5D+01
12260 RETURN
12300 GA=GA/F
12320 GB=((1/TAN(BW))*SIN(BV))/SIN(GA)-C
OS(BV)/TAN(GA)
12330 CB=ATN(1/GB)
12340 IFCB<0THENCB=PI+CB
12350 IFE$="N"THEN12360ELSE12390
12360 IFF$="E"THEN12370ELSE12380
12370 CC=CB:GOTO12420
12380 CC=2*PI-CB:GOTO12420
12390 IFF$="E"THEN12400ELSE12410
12400 CC=PI-CB:GOTO12420
12410 CC=PI+CB
12420 CD=CC*F
12430 RETURN
13500 GA=(COS(BX)-COS(BV)*COS(BW))/(SIN(
BV)*SIN(BW))
13520 GB=SQR(1-GA*GA):GC=GB/GA:GD=ATN(GC
)
13530 IFGD<0THENGD=GD+PI
13540 FP=GD*F/15:IFF$="E"THENFP=-FP
13550 RETURN
13600 INPUT"SUN'S S.D., ARC-MINUTES";Q1
13610 INPUT"WHICH TRUE LIMB OF SUN DID Y
OU OBSERVE, L OR R";OB$
13620 Q2=(Q1*ABS(SIN(CC))/(COS(QU)*ABS(S
IN(GA))))*2.90888D-04
13630 IF E$="N" AND OB$="R" THEN CC=CC+Q
2:RETURN
13640 IF E$="N" AND OB$="L" THEN CC=CC-Q
2:RETURN
13650 IF E$="S" AND OB$="L" THEN CC=CC-Q
2:RETURN
13660 IF E$="S" AND OB$="R" THEN CC=CC+Q
2:RETURN
```

AII.03.02 A program for finding latitude and longitude using the position line method

The program follows closely the method used in the example in Chapter 14, section **14.07**, except that no graphical solution is supplied for viewing on the computer video display unit. Instead, the solution for true position is by least squares as in **14.08**. An observation equation is found for each star, normal equations in the three unknowns are then set up, and these are solved by matrix inversion. Provision is made to view, if desired, the equations, the inverted matrix, and, if a plot is to be made, the intercepts and the azimuths of the stars.

A minimum of 4 stars should be observed, as in **14.07**, and it is always a good procedure to observe an additional pair of stars in such a group, one east and the other west near the prime vertical, to add strength to the longitude determination. If less than 3 stars are observed, the program given here returns to its starting point.

The size of the intercept is monitored (see lines 842 ff.) and if it is greater than 100 arc-seconds the user is asked if he or she wishes to include that particular observation or pass on to the next star. The value 100 in lines 842 and 844 can easily be changed by editing those lines. If an observational blunder gave rise to an oversize intercept, it would vitiate the final result unless some test were applied to draw attention to it.

The program, 9093 bytes long, is listed in the print-out at the end of the following example which uses the data from **14.07** and **14.08** in *Astronomy for Surveyors*. As in **AII.03.01**, the sequence of questions and answers is followed to show the user what to expect. As before, the answers are in bold-face type.

EXAMPLE showing the use of the program:

Do you wish to by-pass the preamble, Y or N? **N**

2.01 Charles Griffin Software by J. B. Mackie, "Astronomy for Surveyors" —comprehensive program for computing latitude and longitude from Position Lines

This is a complete program for computing latitude and longitude simultaneously using stars and the position line method. It finds the intercepts and star azimuths, and then solves for the position by using least squares. Up to 8 stars are catered for, but the number may be increased by changing the dimensions of the arrays C1 and C2 to (x,8) and (x,4) respectively, where x is the number of stars required.

The method is explained in "Astronomy for Surveyors", 9th edition (1985) by J. B. Mackie, published by Charles Griffin & Co. Ltd, Crendon Street, High Wycombe, England.

Please read the information in the book before using the program.

This program is copyright by Charles Griffin & Co. Ltd.

To commence press E, then ENTER? **E**

To find latitude and longitude simultaneously from position line (astrofix) observations of stars

Appendix II

Name of station? **ARTS BLOCK F**
Assumed latitude of stn., D,M,S (always +)? **45,52,0**
Assumed longitude of stn., H,M,S (always +)? **11,22,5**
Is longitude East or West; type E or W? **E**
Longitude of zone meridian, +/−HRS only, e.g. −7.5? **12**
Is observing stn. in N or S hemisphere; type N or S? **S**
How many stars for this astrofix? **4**

Now for star No.1

Was star observed E or W of meridian; type E or W? **W**
Was declination of star N or S; type N or S? **S**
How many pointings on this star? **4**
Convert all vert. circle readings to true altitudes and enter them one at a time.
Altitude, D,M,S? **35,24,35**
Altitude, D,M,S? **35,15,28**
Altitude, D,M,S? **35,6,40**
Altitude, D,M,S? **34,59,9**
Temperature (C), Pressure (MB)? **5,1020**

If a simple stop-watch, started on object contact and stopped on clock reading, was used, you will be asked to enter, for each pointing, the clock time and then the stop-watch time of contact. If, however, a stop-watch was not used, or a stop-watch with lap-time facility was used, you will be asked to enter only the final clock time of contact for each pointing.

Was a simple stop-watch used, type Y or N? **N**
Final clock time of pointing, H,M,S? **22,24,59.4**
Final clock time of pointing, H,M,S? **22,26,11**
Final clock time of pointing, H,M,S? **22,27,20.2**
Final clock time of pointing, H,M,S? **22,28,19.6**
Clock difference from zone time (+ always); HRS, MIN, SEC? **0,0,12.4**
Clock fast or slow on zone time, type F or S? **F**
DUT1 (secs)? **.1**

In looking up *SALS* tables for E, R or Dec. note the following carefully:
UT1 of observation (decimal hours) = 10.442
Look up tables for local date of obs.
Look up table for R at 0 HRS UT on correct date of obs. as indicated earlier.
R at 0 HRS UT from tables, H,M,S? **21,47,51.3**
User—please interpolate R.A. of star from *SALS* for date of observation
R.A. of star, H,M,S? **13,38,16.6**
Dec. of star after interpolation from *SALS*, (+ always), D,M,S? **53,20,32**

Now for star No.2

(Follow the same procedure as for Star No.1, answering the questions with the appropriate data; then follow in the same way for Stars No. 3 and 4. After the data for Star No. 4 have been entered, the following appears:-)
Do you wish to see the observation eqns., Y or N? **Y**
Observation equations:-

Astronomy for surveyors

1	−.730351	−.683072	3.33609
1	−.596674	.802824	19.3969
1	.73728	.675588	42.0713
1	.755616	−.655015	16.1096

Do you wish to see the normal eqns., Y or N? **Y**
Normal equations:-

4	.16587	.139985	80.9138
.16587	2.00397	.0232188	29.1808
.139985	.0232188	1.99603	31.1577

Do you wish to see the inverted matrix, Y or N? **Y**

.251464	−.020613	−.017396
−.020613	.500767	−.004380
−.017396	−.004380	.502265

Solution of normal equations:-

$V = -19.2034 \quad X = -12.8085 \quad Y = -14.1141$

True latitude = −45 52 14 (S)

Station   Arts Block F

True longitude = 11 22   3.8 (E)
            or 170 30 56.6 (E)

Do you wish to see intercepts and azimuths, Y or N? **Y**

Star	Intercept Arc-secs	Azimuth Deg Min
1	−3.33609	226 55
2	−19.3969	323 22.1
3	−42.0713	47 30
4	−16.1096	130 55.2

READY

The reader will note that the above solution differs slightly from the one in **14.08** of the main text. The difference is due to the rounding off to 1 arc-second in the book example, to other rounding differences in the arithmetic and to the method of deriving the refraction correction.

The curvature correction in altitude is applied only when the number of pointings on a star is 2 or more.

Appendix II

```
10 CLS
20 INPUT"DO YOU WISH TO BY-PASS THE PREAMBLE, Y OR N";Q$
30 CLS
40 IF Q$="Y" THEN 340 ELSE 50: CLS
50 PRINT"2.01 GRIFFIN SOFTWARE BY J.B. MACKIE"
60 PRINT
70 PRINT"ASTRONOMY FOR SURVEYORS -- COMPREHENSIVE PROGRAM FOR"
80 PRINT"COMPUTING LATITUDE AND LONGITUDE FROM POSITION LINES."
90 PRINT
100 PRINT"THIS IS A COMPLETE PROGRAM FOR COMPUTING LATITUDE AND"
110 PRINT"LONGITUDE SIMULTANEOUSLY USING STARS AND THE POSITION"
120 PRINT"LINE METHOD.    IT FINDS THE INTERCEPTS AND STAR AZIMUTHS,"
130 PRINT"AND THEN SOLVES FOR THE POSITION BY USING LEAST SQUARES."
140 PRINT"UP TO 8 STARS ARE CATERED FOR, BUT THE NUMBER MAY BE"
150 PRINT"INCREASED BY CHANGING THE DIMENSIONS OF THE ARRAYS C1"
160 PRINT"AND C2 TO (X,8) AND (X,4) RESPECTIVELY, WHERE X IS THE"
170 PRINT"NUMBER OF STARS REQUIRED."
180 FOR I=1 TO 5000
190 I=I+1
200 NEXT I
210 CLS
220 PRINT"THE METHOD IS EXPLAINED IN 'ASTRONOMY FOR SURVEYORS',"
230 PRINT"9TH. EDITION (1984) BY J.B. MACKIE, PUBLISHED BY"
240 PRINT"CHARLES GRIFFIN & CO..LTD., CRENDON STREET, HIGH WYCOMBE,"
250 PRINT"ENGLAND."
260 PRINT
270 PRINT"PLEASE READ THE INFORMATION IN THE BOOK BEFORE USING"
280 PRINT"THE PROGRAM."
290 PRINT
300 PRINT"THIS PROGRAM IS COPYRIGHT BY CHARLES GRIFFIN & CO. LTD."
310 PRINT
320 INPUT"TO COMMENCE PRESS E, THEN ENTER";P$
330 IF P$="E" THEN 340 ELSE 2220
340 CLS
350 PRINT"TO FIND LATITUDE AND LONGITUDE SIMULTANEOUSLY FROM"
360 PRINT"POSITION LINE (ASTRO-FIX) OBSERVATIONS OF STARS."
370 CLEAR 250
380 DEFDBL D, F, G, P, Q
390 DEFINT I, J, K, M, O, R
400 DIM C1(8, 8), C2(8, 4), C3(3, 4), C4(3, 3), C5(3)
405 DIM L(6), L1(6)
410 A=206265
420 B=3600
430 C=60
440 PI=3.14159265
450 D=PI/2
460 F=5.729577951D+01
470 P=3.6D+03
480 Q=6.0D+01
490 FF=1.002737909
500 C$="STAR"
510 INPUT"NAME OF STATION";D$
520 INPUT"ASSUMED LATITUDE OF STN., D,M,S (ALWAYS +)";BD, BE, BF
530 BD=(BD*B+BE*C+BF)/A
540 BV=D-BD
550 INPUT"ASSUMED LONGITUDE OF STN., H,M,S (ALWAYS +)";DD, DE, DF
560 DG=DD+DE/Q+DF/P
570 INPUT"IS LONGITUDE EAST OR WEST; TYPE E OR W";X$
580 IF X$="W" THEN DG=-DG
590 INPUT"LONGITUDE OF ZONE MERIDIAN, +/-HRS ONLY, E.G. -7.5";QJ
600 INPUT"IS OBSERVING STN. IN N OR S HEMISPHERE; TYPE N OR S";E$
610 INPUT"HOW MANY STARS FOR THIS ASTRO-FIX";M
615 IF M<3 THEN 10
620 K=1
630 PRINT"NOW FOR STAR NO. ";K
640 INPUT"WAS STAR OBSERVED E OR W OF MERIDIAN; TYPE E OR W";F$
650 INPUT"WAS DECLINATION OF STAR N OR S; TYPE N OR S";N$
660 INPUT"HOW MANY POINTINGS ON THIS STAR";GY
670 GZ=GY:K9=GY
680 GOSUB 2230
690 GOSUB 2700
700 GOSUB 2980
710 GOSUB 3080
720 GOSUB 3180
730 GOSUB 3240
740 GOSUB 2640
750 FZ=DR*FF+FQ
760 IF FZ>24 THEN FZ=FZ-24
770 FP=FZ+DG-FW
780 GOSUB 3300
790 GA=GH/F
800 B1=COS(BV)*COS(BW)+SIN(BV)*SIN(BW)*COS(GA)
810 B2=SQR(1-B1*B1)
820 B3=B1/B2
830 B4=ATN(B3)
840 CP=(BU-B4)*F*B
842 IF ABS(CP)>100 THEN 844 ELSE 850
844 PRINT"YOU HAVE AN INTERCEPT >100 ARC SECS (ACTUALLY";CP" ARC SECS)"
845 INPUT"DO YOU WISH TO INCLUDE THIS OBSN., YY FOR YES, NN FOR NO";RR$
```

```
846 IF RR$="YY" THEN 850 ELSE 847
847 K=K+1: IF K<=M THEN 630 ELSE 960
850 GA=GA*F
860 GOSUB 3370
862 IF K9<2 GOTO 870
863 GQ=COS(BD)*COS(CB)
864 GR=GQ*(GQ*TAN(BU)-SIN(BD))*SUM
865 CP=CP+GR
870 C1(K, 1)=1
880 C1(K, 2)=SIN(CC)
890 C1(K, 3)=COS(CC)
900 C1(K, 4)=-CP
910 C1(K, 5)=CD
920 K=K+1
930 CLS
940 IF K<=M THEN PRINT"NOW FOR STAR NO. ";K: PRINT
950 IF K<=M THEN 640
960 CLS
970 INPUT"DO YOU WISH TO SEE THE OBSERVATION EQNS., Y OR N";U$
980 IF U$="N" THEN 1080
990 PRINT"OBSERVATION EQUATIONS:-"
1000 PRINT
1010 FOR I=1 TO M
1020 FOR J=1 TO 4
1030 PRINT C1(I, J);" ";
1040 NEXT J
1050 PRINT
1060 NEXT I
1070 PRINT
1080 SUM=0
1090 FOR K=1 TO 3
1100 FOR I=1 TO M
1110 FOR J=1 TO 4
1120 C2(I, J)=C1(I, K)*C1(I, J)
1130 NEXT J, I
1140 FOR J=1 TO 4
1150 FOR I=1 TO M
1160 SUM=SUM+C2(I, J)
1170 NEXT I
1180 C3(K, J)=SUM
1190 SUM=0
1200 NEXT J, K
1210 INPUT"DO YOU WISH TO SEE THE NORMAL EQNS., Y OR N";V$
1220 IF V$="N" THEN 1310
1230 PRINT"NORMAL EQUATIONS:-"
1240 FOR I=1 TO 3
1250 FOR J=1 TO 4
1260 PRINT C3(I, J);" ";
1270 NEXT J
1280 PRINT
1290 NEXT I
1300 PRINT
1310 FOR J=1 TO 3
1320 C4(J, J)=1
1330 NEXT J
1340 FOR J=1 TO 3
1350 FOR I=J TO 3
1360 IF C3(I, J)<>0 THEN 1400
1370 NEXT I
1380 PRINT"SINGULAR MATRIX"
1390 GOTO 2220
1400 FOR K=1 TO 3
1410 S=C3(J, K)
1420 C3(J, K)=C3(I, K)
1430 C3(I, K)=S
1440 S=C4(J, K)
1450 C4(J, K)=C4(I, K)
1460 C4(I, K)=S
1470 NEXT K
1480 T=1/C3(J, J)
1490 FOR K=1 TO 3
1500 C3(J, K)=T*C3(J, K)
1510 C4(J, K)=T*C4(J, K)
1520 NEXT K
1530 FOR L=1 TO 3
1540 IF L=J THEN 1600
1550 T=-C3(L, J)
1560 FOR K=1 TO 3
1570 C3(L, K)=C3(L, K)+T*C3(J, K)
1580 C4(L, K)=C4(L, K)+T*C4(J, K)
1590 NEXT K
1600 NEXT L
1610 NEXT J
1630 INPUT"DO YOU WISH TO SEE THE INVERTED MATRIX, Y OR N";W$
1640 IF W$="N" THEN 1720
1650 PRINT"INVERTED MATRIX OF NORMAL EQN. COEFFICIENTS:-"
1660 FOR I=1 TO 3
1670 FOR J=1 TO 3
1680 PRINT C4(I, J);" ";
1690 NEXT J
1700 PRINT
1710 NEXT I
1720 SUM=0
1730 FOR I=1 TO 3
1740 FOR J=1 TO 3
1750 SUM=SUM+C4(I, J)*(-C3(J, 4))
1760 NEXT J
1770 C5(I)=SUM
1780 SUM=0
1790 NEXT I
1800 PRINT
1810 PRINT"SOLUTION OF NORMAL EQUATIONS:-"
1820 PRINT"V= ";C5(1);" ";"X= ";C5(2);" ";"Y= ";C5(3)
1830 IF E$="S" THEN BD=-BD
1840 BD=BD+C5(3)/A
1850 DG=DG+(C5(2)/COS(BD))/5.4D+04
1860 BD=BD*F
1870 T1=ABS(BD)
1880 T2=INT(T1)
1890 T3=(T1-T2)*C
1900 T4=INT(T3)
1910 T5=(T3-T4)*C
1920 IF E$="S" THEN T2=-T2
```

Appendix II

```
1930 PRINT"TRUE LATITUDE = ";T2;" ";T4;"
"; INT(T5*10+.5)/10;" ";"(";E$;")"
1940 T6=ABS(DG)
1950 T7=INT(T6)
1960 T8=(T6-T7)*C
1970 T9=INT(T8)
1980 T0=(T8-T9)*C
1990 TA=T6*15
2000 TB=INT(TA)
2010 TC=(TA-TB)*C
2020 TD=INT(TC)
2030 TE=(TC-TD)*C
2040 IF X$="W" THEN T7=-T7: IF X$="W" THEN TB=-TB
2050 PRINT
2060 PRINT"STATION  ";D$
2070 PRINT
2080 PRINT"TRUE LONGITUDE = ";T7;" ";T9;
" "; INT(T0*10+.5)/10;" ";"(";X$;")"
2090 PRINT TAB(12)"OR";TAB(17)TB;" ";TD;
" "; INT(TE*10+.5)/10;" ";"(";X$;")"
2100 INPUT"DO YOU WISH TO SEE INTERCEPTS
 & AZIMUTHS, Y OR N";S$
2110 IF S$="N" THEN 2220
2120 PRINT TAB(5)"STAR";TAB(15)"INTERCEPT";TAB(30)"AZIMUTH"
2130 PRINT TAB(16)"ARCSECS";TAB(30)"DEG"
;TAB(35)"MIN"
2140 FOR K=1 TO M
2150 C1(K, 6)=INT(C1(K, 5))
2160 C1(K, 7)=(C1(K, 5)-C1(K, 6))*C
2170 C1(K, 8)=INT(C1(K, 7)*10+.5)/10
2180 NEXT K
2190 FOR K=1 TO M
2200 PRINT TAB(6)K; TAB(15)(-C1(K, 4));
TAB(29)C1(K, 6); TAB(34)C1(K, 8)
2210 NEXT K
2220 END
2230 REM**TO COMPUTE CORRECTED ALTITUDE
AND ZENITH DISTANCE**
2240 PRINT"CONVERT ALL VERT. CIRCLE READ
INGS TO TRUE ALTITUDES"
2250 PRINT"AND ENTER THEM ONE AT A TIME"
2260 BN=0
2270 GY=GZ
2280 INPUT"ALTITUDE, D,M,S";AQ, AR, AS
2290 BM=(AQ*B+AR*C+AS)/A
2300 BN=BN+BM
2310 GY=GY-1
2320 IF GY>0 THEN 2280 ELSE 2330
2330 BO=BN/GZ
2340 INPUT"TEMPERATURE(C), PRESSURE(MB)"
;AW, AX
2350 BS=8.85*COS(BO)
2360 LA=TAN(D-BO)
2370 LB=LA*LA
2380 LC=LA*LB
2390 LD=AW+273.16
2400 BT=-16.271*LA*(1+3.94E-05*LB*(AX/LD
))*(AX/LD)+.0749*(LC+LA)*(AX/1000)
2410 IF C$="SUN" THEN BU=B0+(BS+BT)/A
2420 IF C$="STAR" THEN BU=B0+BT/A
2430 CLS
2440 BX=D-BU
2450 RETURN
2640 REM**TO FIND THE DEC. & POLAR DIST.
 OF A STAR**
2650 INPUT"DEC. OF STAR AFTER USER INTER
POLATION FROM SALS, (+ ALWAYS), D,M,S";B
A, BB, BC
2660 QU=(BA*B+BB*C+BC)/A
2670 IF N$="S" THEN QU=-QU
2680 IF E$="N" THEN BW=(D-QU) ELSE BW=(D
+QU)
2690 RETURN
2700 REM**TO FIND THE MEAN CLOCK TIME OF
 AN OBSERVATION**
2710 BN=0: GY=GZ
2715 II=1
2720 PRINT"IF A SIMPLE STOP-WATCH, START
ED ON OBJECT CONTACT"
2730 PRINT"AND STOPPED ON CLOCK READING,
 WAS USED, YOU WILL"
2740 PRINT"BE ASKED TO ENTER, FOR EACH P
OINTING, THE CLOCK"
2750 PRINT"TIME AND THEN THE STOP-WATCH
TIME OF CONTACT."
2760 PRINT"IF, HOWEVER, A STOP-WATCH WAS
 NOT USED, OR A STOP-"
2770 PRINT"WATCH WITH LAP-TIME FACILITY
WAS USED, YOU WILL BE"
2780 PRINT"ASKED TO ENTER ONLY THE FINAL
 CLOCK TIME OF CON-"
2790 PRINT"TACT FOR EACH POINTING."
2800 PRINT:PRINT
2810 INPUT"WAS A SIMPLE STOP-WATCH USED,
 TYPE Y OR N";Y$
2820 IFA$="LA"THENRETURN
2830 IFY$="Y"THEN2840ELSE2910
2840 PRINT"ENTER CLOCK TIME, HIT ENTER,
AND THEN ENTER STOP-"
2850 PRINT"WATCH TIME AND HIT ENTER -- I
.E. 2 ENTRIES FOR"
2860 PRINT"EACH POINTING."
2870 INPUT"CLOCK TIME WHEN S/W STOPPED,
H,M,S";DH, DI, DJ
2880 DK=DH+DI/Q+DJ/P
2890 INPUT"STOP-WATCH TIME, SECONDS";DQ:
DQ=DQ/P
2900 DK=DK-DQ: GOTO2930
2910 INPUT"FINAL CLOCK TIME OF POINTING,
 H,M,S";DH, DI, DJ
2920 DK=DH+DI/Q+DJ/P: L(II)=DK
2930 BN=BN+DK: II=II+1: GY=GY-1: IFGY=0THEN
2960
2940 IFGY>0ANDY$="Y"THEN2870ELSE2950
2950 IFGY>0ANDY$="N"THEN2910
2960 DP=BN/GZ
2961 IF K9<2 THEN 2970
```

```
2962 FORIJ=1TOK9:L1(IJ)=(L(IJ)-DP)*3.6D+
03:NEXTIJ
2964 FORIJ=1TOK9:L1(IJ)=L1(IJ)*L1(IJ)*5.
454D-04:NEXTIJ
2965 SUM=0:FOR IK=1TOK9:SUM=SUM+L1(IK):N
EXTIK
2966 SUM=SUM/GZ
2970 RETURN
2980 REM**TO FIND THE UT1 OF AN OBS. KNO
WING ITS MEAN CLOCK TIME**
2990 INPUT"CLOCK DIFFERENCE FROM ZONE TI
ME (+ ALWAYS); HRS,MIN,SEC";DW,DX,DY
3000 DZ=DW+DX/Q+DY/P
3010 INPUT"CLOCK FAST OR SLOW ON ZONE TI
ME, TYPE F OR S";Z$
3020 IFZ$="F"THENDZ=-DZ
3030 INPUT"DUT1 (SECS)";FA:FB=FA/P
3040 FE=DP+DZ-QJ+FB:DR=FE
3050 IFDR<0THENDR=DR+24
3060 IFDR>24THENDR=DR-24
3070 RETURN
3080 REM**SORTING OUT DATE FOR ENTERING
SALS TABLES**
3090 PRINT"IF LOOKING UP SALS TABLES FOR
'E', 'R' OR DEC"
3100 PRINT"NOTE THE FOLLOWING CAREFULLY:
":PRINT
3110 PRINT"UT1 OF OBSERVATION (DECIMAL H
OURS) = ";INT(DR*1000+.5)/1000
3120 IFFE<0THEN3130ELSE3140
3130 PRINT"LOOK UP TABLES FOR DAY BEFORE
LOCAL DATE OF OBS.":GOTO3170
3140 IFFE>24THEN3150ELSE3160
3150 PRINT"LOOK UP TABLES FOR DAY AFTER
LOCAL DATE OF OBS.":GOTO3170
3160 PRINT"LOOK UP TABLES FOR LOCAL DATE
OF OBS."
3170 RETURN
3180 REM**TO FIND 'R' AT 0 HRS UT ON UT
DATE OF OBSERVATION**
3190 PRINT"LOOK UP TABLE FOR 'R' AT 0 HR
S UT ON CORRECT DATE"
3200 PRINT"OF OBS. AS INDICATED EARLIER.
"
3210 INPUT"'R' AT 0 HRS UT FROM TABLES,
H,M,S";FQ,FR,FS
3220 FQ=FQ+FR/Q+FS/P
3230 RETURN
3240 REM**TO FIND STAR'S R.A. AT TIME OF
OBSERVATION**
3250 PRINT"USER -- PLEASE INTERPOLATE R.
A. OF STAR FROM SALS"
3260 PRINT"FOR DATE OF OBSERVATION"
3270 INPUT"R.A. OF STAR, H,M,S";FT,FU,FV
3280 FW=FT+FU/Q+FV/P
3290 RETURN
3300 REM**ROUTINE FOR GETTING H.A. LESS
THAN 12 HRS.**
3310 IFFP>24THENFP=FP-24:GOTO3330
3320 IFFP<0THENFP=FP+24
3330 IFFP<12THEN3350
3340 IFFP>12THENFP=24-FP
3350 GA=FP*1.5D+01
3360 RETURN
3370 REM**ROUTINE FOR COMPUTING ZENITH A
NGLE & AZIMUTH OF OBJECT**
3380 GA=GA/F
3390 GB=((1/TAN(BW))*SIN(BV))/SIN(GA)-CO
S(BV)/TAN(GA)
3400 CB=ATN(1/GB)
3410 IFCB<0THENCB=PI+CB
3420 IFE$="N"THEN3430ELSE3460
3430 IFF$="E"THEN3440ELSE3450
3440 CC=CB:GOTO3490
3450 CC=2*PI-CB:GOTO3490
3460 IFF$="E"THEN3470ELSE3480
3470 CC=PI-CB:GOTO3490
3480 CC=PI+CB
3490 CD=CC*F
3500 RETURN
```

AII.03.03 A program for computing the solar ephemeris

This program is based upon the algorithm given by G. G. Bennett in "A Solar Ephemeris for use with Programmable Calculators" in *The Australian Surveyor*, **30**(3), September 1980.

The data required are the year, the month, the day and the U.T. (hours, minutes and seconds); all three of the latter are required, e.g. if the U.T. is 6h, it must be entered as 6,0,0. The computation produces the GHA sun, the quantity E, the sun's declination and its semi-diameter.

As an example, suppose the information is required for U.T. 18^h on 1978 May 22. On entering the program, typing RUN and pressing the ENTER key, the following appears (note: the bold-faced type shows the answers to the questions generated by the computer program):

3.01   Charles Griffin Software by J. B. Mackie, "Astronomy for Surveyors"

This program is copyright by Charles Griffin & Co. Ltd, Crendon Street, High Wycombe, England.

Appendix II

This is a program for computing the ephemeris of the sun. The only data required are the year, the month, the day and the U.T. at which the information is required.

The calculation gives the sun's GHA, the quantity E, the sun's declination, and its semi-diameter.

UT(H,M,S)? **18,0,0**
YEAR, MONTH, DAY? **1978,5,22**
After about 65 seconds have elapsed, the following appears:
GHA OF SUN = 90 DEG 51 MIN 25.9 SEC
          OR = 6 HRS 3 MIN 25.7 SEC
E = 12 HR 3 MIN 25.7 SEC
DECLINATION = 20 DEG 24 MIN 57.2 SEC
SEMI-DIAMETER = 15 MIN 49.2 SEC

READY

From the *Star Almanac* for 1978 May 22, 18^h:
Dec. = N. 20° 25.0'
$E = 12^h\ 03^m\ 25.7^s$
Sun's S.D. = 15.8'

The program is about 4.4K bytes long.

Bennett (1980) states that the algorithm used is an approximate one and that it is intended to be used primarily for the reduction of observations made from 1980 to the end of this century, although the reduction of those made at epochs even a century removed from the present will not be seriously affected by the approximations taken. The algorithm itself has been adapted from *Meeus' Tables of Moon and Sun* (1962) which are based on *Newcomb's Tables of the Sun* (1895).

Random checks made by the author against *Star Almanac* values give differences of up to 0.2^s in the sun's GHA, which are quite tolerable for the solar observations used in field astronomy.

Program listing:

Astronomy for surveyors

```
2 CLS:PRINT"3.01 GRIFFIN SOFTWARE BY J.B
. MACKIE":PRINT"ASTRONOMY FOR SURVEYORS"
:PRINT
4 PRINT"THIS PROGRAM IS COPYRIGHT BY CHA
RLES GRIFFIN & CO. LTD.":PRINT"CRENDON S
TREET, HIGH WYCOMBE, ENGLAND"
5 PRINT
6 PRINT"THIS IS A PROGRAM FOR COMPUTING
THE EPHEMERIS OF THE SUN.":PRINT"THE ONL
Y DATA REQUIRED ARE THE YEAR, THE MONTH,
 THE DAY":PRINT"AND THE U.T. AT WHICH TH
E INFORMATION IS REQUIRED.":PRINT
7 PRINT"THE CALCULATION GIVES THE SUN'S
GHA, THE QUANTIITY 'E',":PRINT"THE SUN'S
DECLINATION, AND ITS SEMI-DIAMETER."
8 PRINT
10 DEFDBL A-F,H,P-Y,Z:DEFINT I-N:I=1:J=2
:PI=3.1415926535897932:Z9=PI/2D+00:Q9=6.
0D+01
15 FZ=PI/1.8D+02
20 INPUT"UT (H,M,S)";A1,A2,A3
25 H3=(A1+A2/6.0D+01+A3/3.6D+03)/2.4D+01
28 INPUT"YEAR, MONTH, DAY";H0,H1,H2
30 H4=(3.67D+02*H0-INT(7.0D+00*(H0+INT((
H1+9D+00)/1.2D+01))/4D+00)+INT(2.75D+02*
H1/9D+00)+H2-6.940065D+05+H3)/3.6525D+04
40 H5=3.58475D+02+3.599905D+04*H4
42 HZ=3.6D+02
44 HL=H5/HZ:HM=INT(HL):H5=(HL-HM)*HZ
50 H6=6.3D+01+2.2518D+04*H4
55 HL=H6/HZ:HM=INT(HL):H6=(HL-HM)*HZ
60 H7=3.32D+02+3.3718D+04*H4
65 HL=H7/HZ:HM=INT(HL):H7=(HL-HM)*HZ
70 H8=2.22D+02+3.2964D+04*H4
75 HL=H8/HZ:HM=INT(HL):H8=(HL-HM)*HZ
80 H9=1.01D+02+1.934D+03*H4
85 HL=H9/HZ:HM=INT(HL):H9=(HL-HM)*HZ
100 X=H5*FZ:GOSUB9803:B1=S
110 X=2.0D+00*H5*FZ:GOSUB9803:B2=S
120 X=3D+00*H5*FZ:GOSUB9803:B3=S
130 X=(2.61D+02+4.4526D+05*H4)*FZ:GOSUB
9823:B4=Q
140 X=(9D+01+H6)*FZ:GOSUB9823:B5=Q
150 X=(9D+01+2D+00*H6)*FZ:GOSUB9823:B6=Q
160 X=(2.58D+02+2D+00*H6-H5)*FZ:GOSUB982
3:B7=Q
170 X=(7.8D+01+3D+00*H6-H5)*FZ:GOSUB9823
:B8=Q
180 X=(5.1D+01+3D+00*H6-2D+00*H5)*FZ:GOS
UB9823:B9=Q
190 X=(9D+01+H7)*FZ:GOSUB9823:B0=Q
200 X=(3.06D+02+H7-H5)*FZ:GOSUB9823:BA=Q
210 X=(9.1D+01+H8)*FZ:GOSUB9823:BB=Q
220 X=(2.7D+02+2D+00*H8)*FZ:GOSUB9823:BC
=Q
230 X=(1.75D+02+H8-H5)*FZ:GOSUB9823:BD=Q
240 X=(2.93D+02+2D+00*H8-H5)*FZ:GOSUB982
3:BE=Q
250 X=(9D+01-H9)*FZ:GOSUB9823:BF=Q
260 X=(2.95D+02+2D+00*H5)*FZ:GOSUB9823:B
G=Q
270 X=H9*FZ:GOSUB9823:BN=Q
280 BH=2.7969019D+02+3.600076892D+04*H4
282 HJ=BH/HZ:HK=INT(HJ):BH=(HJ-HK)*HZ
285 HD=(1.91945D+00-4.79D-03*H4)*B1+2D-0
2*B2+2.9D-04*B3
290 HE=1.79D-03*B4
295 HF=1.34D-03*B5+1.54D-03*B6+6.9D-04*B
7+4.3D-04*B8+2.8D-04*B9
300 HG=5.7D-04*B0+4.9D-04*BA
305 HH=2D-03*BB+7.6D-04*BC+7.2D-04*BD+4.
5D-04*BE
310 BK=4.79D-03*BF+3.5D-04*BG
320 BL=BH+HD+HE+HF+HG+HH+BK
325 HI=BL*FZ
330 BO=(2.345229D+01-1.301D-02*H4+2.56D-
03*BN)*FZ
340 X=HI:GOSUB9803:BP=S
350 X=HI:GOSUB9823:BM=Q
360 X=BO:GOSUB9823:BQ=Q
370 X=BO:GOSUB9803:BU=S
380 BR=BP*BQ:BS=BR/BM
390 X=BS:GOSUB9803:BT=R
392 IFBR>0ANDBM<0THENBT=BT+PI:GOTO400
394 IFBR<0ANDBM<0THENBT=BT+PI:GOTO400
396 IFBR<0ANDBM>0THENBT=BT+2D+00*PI
400 BV=BP*BU:BW=1-BV*BV:X=BW:GOSUB9893:B
X=S
405 HA=BV/BX:X=HA:GOSUB9883:HB=R:HC=HB*1
.8D+02/PI
410 BY=H3*2.4D+01*1.5D+01
420 BZ=BY+9.96913D+01+3.600076892D+04*H4
+9.17D-01*BK-BT/FZ
430 HL=BZ/HZ:HM=INT(HL):BZ=(HL-HM)*HZ
431 L1=INT(BZ):Q1=(BZ-L1)*Q9:L2=INT(Q1):
Q2=(Q1-L2)*Q9
432 PRINT"GHA OF SUN= ";L1"DEG";L2"MIN
";INT(Q2*1.0D+01+.5)/1.0D+01"SEC"
433 Q3=BZ/1.5D+01:L3=INT(Q3):Q4=(Q3-L3)*
Q9:L4=INT(Q4):Q5=(Q4-L4)*Q9
434 PRINT"         OR= ";L3"HRS";L4"MIN
";INT(Q5*1.0D+01+.5)/1.0D+01"SEC"
435 CA=BZ-BY:CA=CA/1.5D+01
436 IF CA<0 THEN CA=24+CA
440 O1=INT(CA):CD=(CA-O1)*6.0D+01:O2=INT
(CD):CE=(CD-O2)*6.0D+01
441 PRINT"E= ";O1"HR";O2"MIN";INT(CE*10+
.5)/10;"SEC"
442 X=H5*FZ:GOSUB9823:CB=Q
443 CC=2.6696D-01+4.47D-03*CB
444 CM=HC: IF HC<0 THEN HC=-HC
445 O3=INT(HC):CG=(HC-O3)*6.0D+01:O4=INT
(CG):CH=(CG-O4)*6.0D+01
446 IF CM<0 THEN O3=-O3
447 PRINT"DECLINATION= ";O3"DEG";O4"MIN"
;INT(CH*10+.5)/10;"SEC"
448 CN=CC*6.0D+01:O5=INT(CN):CO=(CN-O5)*
6.0D+01
449 PRINT"SEMI-DIAMETER= ";O5"MIN";INT(C
O*10+.5)/10"SEC"
450 END
```

Appendix II

```
9803 Y=X:R=0: IF ABS(Y)>PI THEN R=FIX(Y/P
I):Y=Y-R*PI
9804 M=0:N=1:S=Y:T=Y:U=-Y*Y
9805 V=S:M=M+J:N=N+J:T=T*U/(M*N):S=S+T:I
F S<>V GOTO 9805
9806 IF INT(R/J)<>R/J THEN S=-S
9807 RETURN
9823 Y=X:R=0: IF ABS(Y)>PI THEN R=FIX(Y/P
I):Y=Y-R*PI
9824 M=-1:N=0:Q=1:T=1:U=-Y*Y
9825 V=Q:M=M+J:N=N+J:T=T*U/(M*N):Q=Q+T:I
F Q<>V GOTO 9825
9826 IF INT(R/J)<>R/J THEN Q=-Q
9827 RETURN
9883 U=X*X:Y=ABS(X):N=I
9884 IF Y>.77 AND Y<1.18 THEN P=U+1:S=SQ
R(P): GOTO 9887
9885 IF U<I THEN R=Y:T=Y:S=-U ELSE T=-I/
Y:R=PI/J+T:S=-T*T
9886 V=R:N=N+J:T=T*S:R=R+T/N: IF R<>V GO
TO 9886 ELSE 9890
9887 V=S:S=(P/S+S)/J: IF S<>V GOTO 9887
9888 N=0:R=Y/S:T=R:U=R*R
9889 V=R:N=N+J:T=T*U*(N-I)/N:R=R+T/(N+I)
:IF R<>V GOTO 9889
9890 R=R*SGN(X)
9891 RETURN
9893 S=X
9894 IF X<0 PRINT"****ERROR, SQ. ROOT NE
G. NO. ":GOTO 450
9895 S=SQR(X)
9896 V=S:S=(X/S+S)/J: IF S<>V GOTO 9896
9897 RETURN
```

APPENDIX III
THE CONVERGENCE OF MERIDIANS

The line of sight of the telescope of a theodolite in accurate adjustment traces out a vertical plane as the telescope is turned about its horizontal axis. This, if we regard the earth as spherical, we may consider to be a plane passing through the centre of the earth. Therefore the "straight" line that is set out by a theodolite is in reality always the arc of a great circle on the earth's surface. Now, unless it happens to coincide with the equator or with a meridian of longitude, any great circle will cut different meridians at different angles. In other words, its bearing will vary from point to point. Thus as we proceed along a straight line set out by a theodolite on the earth's surface, the bearing of the line will not remain constant but will gradually alter. The line the bearing of which was everywhere the same would not be a straight line. A parallel of latitude, for instance, is such a line, but if the telescope of a theodolite is set out truly east and west at any place, its direction would not mark out the parallel of latitude, which is a small circle, but a great circle that would ultimately intersect the equator.

This alteration in the bearing of a straight line is an important matter in surveys of any magnitude, as in latitudes in the neighbourhood of 60° it amounts to almost a minute of arc in a line one kilometre long, and in higher latitudes the alteration is still greater.

In Fig. A.III.1, let N and S denote the north and south terrestrial poles, $ELMQ$ the equator, and A and B any two points between which the great circle arc AB has been set out.

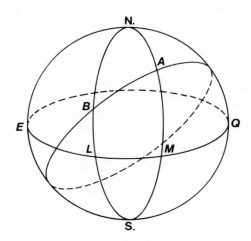

Fig. A.III.1

Appendix III

Let *NAMS* and *NBLS* be the meridians through A and B. Then the bearing of the line BA at B is the angle NBA, and the bearing of the same line at A is $180° - NAB$.

The difference between the bearings of the line AB at the points A and B is known as the *convergence* of the meridians between A and B.

If AB is plotted as a straight line on a plane, then the meridians through A and B will not be drawn as parallel lines, but as lines making an angle with each other equal to the convergence. Denote the convergence by c. Then,

$$c = 180° - NAB - NBA.$$

Let ϕ = latitude of A, and ϕ' = the latitude of B.

$$NA = 90° - \phi, \quad NB = 90° - \phi'.$$

Denote the difference of longitude between A and B by m, so that m = angle BNA. Then in the spherical triangle NBA, given two sides and the included angle,

$$\tan\tfrac{1}{2}(NBA + NAB) = \frac{\cos\tfrac{1}{2}(NB - NA).\cot\tfrac{1}{2}m}{\cos\tfrac{1}{2}(NB + NA)}.$$

Therefore

$$\cot\tfrac{1}{2}(180° - NBA - NAB) = \frac{\cos\tfrac{1}{2}(\phi - \phi').\cot\tfrac{1}{2}m}{\cos\tfrac{1}{2}(180° - \phi - \phi')}$$

so that

$$\cot\tfrac{1}{2}c = \frac{\cos\tfrac{1}{2}(\phi - \phi')}{\sin\tfrac{1}{2}(\phi + \phi')}\cot\tfrac{1}{2}m$$

or, inverting,

$$\tan\tfrac{1}{2}c = \frac{\sin\tfrac{1}{2}(\phi + \phi')}{\cos\tfrac{1}{2}(\phi - \phi')}\tan\tfrac{1}{2}m. \quad (A.III.1)$$

In any ordinary survey, the length of the line AB will be very small compared with the earth's radius, and the angles c and m will be so small that $\tan\tfrac{1}{2}c$ and $\tan\tfrac{1}{2}m$ may be replaced by $\tfrac{1}{2}c$ and $\tfrac{1}{2}m$ respectively without appreciable error. Therefore,

$$c \text{ (in circular measure)} = \frac{\sin\tfrac{1}{2}(\phi + \phi')}{\cos\tfrac{1}{2}(\phi - \phi')}.m \text{ (in circular measure)},$$

and

$$c \text{ (in seconds of arc)} = \frac{\sin\tfrac{1}{2}(\phi + \phi')}{\cos\tfrac{1}{2}(\phi - \phi')}.m \text{ (in seconds of arc)}.$$

Again, unless the line joining A and B is a very long one, $\cos\tfrac{1}{2}(\phi - \phi')$ differs from unity by but a very small quantity, so that for short lines,

Convergence in seconds
= sin (mid-latitude) × difference of longitude in seconds. $\quad (A.III.2)$

Astronomy for surveyors

Formula (A.III.2) may be used where the distance from observing station to Circuit Initial is up to about 16 km. For distances between 16 and 40 km formula (A.III.1) should be used. For distances over 40 km more exact formulae are necessary to compute the convergence on the spheroid; these may be found in textbooks on geodesy.

It will be obvious that the convergence increases very rapidly in high latitudes; for instance, in latitude 60° the bearing of a straight line 5 km long running approximately east and west will at one extremity be different by 04' 40" from what it is at the other, whereas the same line in latitude 20° has only 01' of convergence between its ends.

In most work carried on by the land surveyor in New Zealand, traverses for control are tied to minor triangulation points or to standard survey points which lie within a particular meridional circuit (a limited area covered by a plane survey projection). All the bearings between these points in one circuit are in terms of the meridian of the initial station of the circuit. Thus, if a surveyor wishes to check a traverse bearing by means of an astronomical observation, he must make a correction for convergence between the meridian through the initial and that through his observing station. The amount of the correction is obtained as described in the foregoing part of this Appendix. The *sense* in which it must be applied is found as follows.

For places in the southern hemisphere

(a) where the observing station lies to the east of the initial (Fig. A.III.2(a)):
   NS is the true meridian through the initial.
   PQ is the true meridian through observing station *A*.
   N'S' is the direction of the true meridian through the initial transferred to station *A* (i.e. *N'S'* is parallel to *NS*).

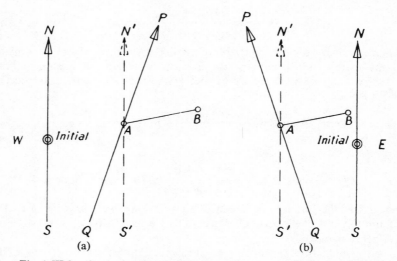

Fig. A.III.2 Correction for convergence of meridians: southern hemisphere

Appendix III

The traverse bearing of AB, obtained by the surveyor, will be in terms of $N'S'$, and will be measured by the angle $N'AB$.

The true bearing of AB, obtained astronomically, will be in terms of PQ, and will be measured by the angle PAB.

The convergence between the meridians is measured by the angle $N'AP$.

Thus, to convert the true bearing of AB into the bearing in terms of the initial, the convergence ($N'AP$) must be added.

(b) where the observing station lies to the west of the initial (Fig. A.III.2(b)):

From the diagram it will be obvious that the convergence must be subtracted from the bearing obtained astronomically.

For places in the northern hemisphere

(a) where the observing station lies to the east of the initial:

Fig. A.III.3(a) is self-explanatory. The convergence must be subtracted from the bearing obtained by astronomical observation to bring it into terms of the meridian through the initial.

(b) where the station lies to the west of the initial (Fig. A.III.3(b)):

It is clear from the diagram that the convergence must be added to the astronomical azimuth to convert this into the circuit bearing.

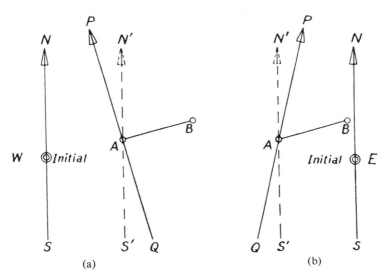

Fig. A.III.3 Correction for convergence of meridians: northern hemisphere

EXAMPLE. We shall find the azimuth of the R.M. in the example in **10.13** in terms of the initial (Trig. A, Taieri, New Zealand).

Initial: Lat., 45° 51' 40" S., Long. 170° 16' 57" E.
Observing Station: Lat., 45° 52' 15" S., Long. 170° 30' 59" E.
Convergence in seconds = sin (mid-latitude) × difference of longitude in seconds
= sin (45° 51' 58") × 842"

Astronomy for surveyors

Convergence		604″ = 10′ 04′
Astronomical azimuth of R.M. (see **10.13**)	271° 04′ 45″	
Convergence (*add*, since station is in southern hemisphere and lies to east of initial)	+10 04	
Bearing of R.M., in terms of initial	271° 14′ 49″	

Deviation of the vertical

The above correction for convergence takes no account of deviation of the vertical (see **3.09**). If the latter is known at the locality of the observing station, then Laplace's equation may be used to get the geodetic azimuth from the astronomic azimuth:

$$\text{Geodetic azimuth} = \text{Astronomic azimuth} - \eta \tan \phi$$

where η is the component of the deviation in Prime Vertical in seconds of arc, reckoned positive when the downward vertical is deviated to the west of the inward spheroidal normal.

In the above example η is +6.5″ and the correction to the astronomic azimuth to bring it to geodetic azimuth is:

$$\eta \tan \phi = 6.5'' \tan(-45° \, 52' \, 15'') = -6.7'', \quad \text{say} \; -7''.$$

Then:

Astro azimuth of R.M.	271° 04′ 45″
Deviation ($-\eta \tan \phi$)	+07
Geodetic azimuth of R.M.	271 04 52
Convergence	+10 04
Bearing in terms of initial	271° 14′ 56″

INDEX

References are to page numbers

Aberration, annual and diurnal, 49
Accuracy, astronomical azimuths, 228
—, — latitudes, 264, 293, 313
—, — longitudes, 293, 301, 313
— of position-line fixes, 293, 313
— of time from astronomical observations, 298
Admiralty List of Radio Signals, 115
Alidade level correction to altitudes, 133, 145
Almanac, Star, for Land Surveyors, see Star Almanac for Land Surveyors
Almucantar observations for time, 288
Altitude, azimuth by, 171–84, 190–204, 330, 333
— and azimuth tables, 315
— of celestial pole, 31
—, change in short period, 159
—, correction for curvature of path, 127, 145
—, prediction of, 157, 325, 328
— of star, 21
—, vertical collimation correction, 132, 145
Altitudes, corrections to observed, 119–34
—, equal, azimuth by, 171
—, —, latitude by, 260
—, —, time by, 288
American Ephemeris, 53
Angle, hour, 24
Annual aberration, 49
Antarctic Circle, 30
Antarctica, observation of daylight stars, 170
Apparent motion of stars, 18
— place of star, 49
— — — —, reduction to, 51
Apparent Places of Fundamental Stars, 49
APFS Tables, extract from, 58
Arctic Circle, 30
Aries, First Point of, 22, 69
—, — — —, time of transit, 87
Assistant, observer's, 150
Astrofix, 306
—, accuracy of, 293, 313
Astrolabe attachment, 293–4
Astrolabes, 291–4, 313–14
—, observations for time and latitude by, 291–4

Astronomical coordinates, 21
— corrections to observations, 119–46
— *Ephemeris,* 53, 68
— terms, synopsis of, 25
— Triangle, 2, 25, 26
Atmospheric refraction, 122
— — tables in *SALS,* 124
Atomic second defined, 65
— Time, 65, 68
Attachment, astrolabe, 293–4
—, Roelofs solar prism, 126, 145, 195
Australia, PMG Research Laboratories, time-signals, 116
Autumnal equinox, 47
Azimuth by altitude, 171–84, 190–204, 330, 333
BASIC programs for, 376–85
change in short period, 159
by circum-polar star near elongation, 174
corrections to, 134–44
correction for curvature of path, 136, 145
— for horizontal collimation error, 141, 146
— for trunnion axis off-level, 142, 146
defined, 167
determination, 167–228, 330–40
by elongating stars, 174
by equal altitudes of star, 171
exercises, 230
by ex-meridian altitudes, 171–84, 190–204, 330, 333
HP-67/97 programs for, 368–76
by hour angles, 184–90, 204–27, 331, 334
methods, choice of, 228
—, summary, 227
need for, 167
observations, corrections to, 134–44
using Polaris at any hour angle and Pole Star Tables, 336
using Pole Star Tables, 223, 336–40
prediction of, 157, 325, 328
of sun, semi-diameter correction, 134, 145
of star, 21
by time observations of a close circum-polar star, 221–7

401

Barometer, aneroid, 150
BASIC computer language, 11
  microcomputer programs, 376–95
  program for azimuth by altitudes, 376–85
  – – – by hour angles, 376–85
  – – circum-meridian latitudes, 376–85
  – – time by ex-meridian altitudes, 376–85
  – – longitude, 376–85
  – – position-line fixes, 386–92
  – – the solar ephemeris, 392–5
Battery, electric, 148
Bearing, 167
Bennett, G. G., planisphere for predicting positions of σ and B Octantis, 222
–, –, solar observations by hour angle, 209, 371
Besselian Day Numbers, 53
Booker, duties of, 150–1
Booking forms, 151
Boss Star Catalogue, 50
Bureau International de l'Heure, 116

Caesium atom, 65
Calculator, HP, 16
–, pocket, with electronic stop-watch, 109
Cancer, Tropic of, 30
Capricorn, Tropic of, 30
Cassinis, 35
Catalogue, star, *APFS*, 49, 58
–, –, *Boss*, 50
–, –, *Smithsonian Astrophysical Observatory,* 50, 59
Catalogues, star, 49
Celestial equator, 20
– horizon, 18
– poles, 19
– refraction, 122, 145
– sphere, 1
Chronocord, 112
Chronographs, 110
Chronometers, 110
Circle, great, 1
–, hour, 23
–, small, 3
–, vertical, index error, 132
Circular Parts, Napier's Rules of, 4
Circum-elongation observation of stars, 184
Circum-meridian observation for latitude, 241, 340
Circum-polar star, observation near elongation for azimuth, 174
Civil (Zone) Time, 73
Clock error, determination of, 116
– rate, determination of, 116
Close circum-polar star, for azimuth, 221–7
– – –, latitude by altitude, 255

Coefficient of refraction, 122
Collimation, horizontal, azimuth correction, 141, 146
–, vertical, correction to altitudes, 132, 145
Collimator, reference, 150
Comparison of clocks with time-signals, 116
Computer, 11, 152, 293, 319, 366 ff.
– programs, 366, 376–95
Constants, precessional, 53
Convergence of meridians, 396–400
Conversion of time, 74–96
Coordinated Universal Time, 65
Coordinates, astronomical, 21
–, terrestrial, 28–9
Correction to altitudes, alidade level, 133, 145
– – – for celestial refraction, 122, 145
– – – for curvature of path, 127, 145
– – – for parallax, 119, 144
– to azimuth observations, 134–44
– – – – for curvature of path, 136, 145
– – – – for horizontal collimation error, 141, 146
– – – –, sun, for semi-diameter, 134, 145
– – – – for trunnion axis off-level, 142, 146
–, parallax, 119, 144
–, refraction, 122, 145
–, semi-diameter of sun, 125, 134, 145
Corrections, curvature of path, 127, 136, 145
– to observations, 119–46
– – –, summary, 144–6
– – observed altitudes, 119–34
Critical table, 77, 121
Culmination, 24
– of celestial object, time of, 87, 94, 235–6, 240, 246, 272
Culminations of Moon, for longitude, 301
Curvature correction to altitude, 127, 145
– – to azimuth, 136, 145
Cyclic formulae, 3

Dark glass, 125, 148
Date, change of, 82
Datum, geodetic, 39
Day, astronomical, 71
–, civil, 71, 73
– Numbers, Besselian, 53
– –, Independent, 53
–, Sidereal, 64
–, Solar, 65
Daylight observation of stars, 169, 170, 328
– – of elongating stars, 184
Declination, 23, 48
–, parallel of, 23
– of star, 21

Index

− of sun, calculation of, 197
Deflection of the vertical, 38
Degree of longitude, terrestrial, length of, 29, 311
Determination of azimuth, 167–228, 330–40
− of clock rate and clock error, 116
− of latitude, 233–70, 306–21, 340–5
− of longitude, 300–4, 306–21, 349–50
− of time, 271–98
Deviation of the vertical, 37, 400
Diagonal eyepiece, 147, 153
Digital watches, 109
Dislevelment corrections, 133, 142
− of trunnion axis, error in azimuth, 142, 146
Diurnal aberration, 49
Dreisonstok's Tables, 315
DUT1 correction, 115

Earth, 28–39
−, climatic zones, 30
−, figure of, 34
− an oblate spheroid, 35
−, orbit, 43, 66
−, radius, 34–6
East-West stars for longitude, 303–4
− − − for time, 279–88
Eclipses of Jupiter's satellites for longitude, 301
Ecliptic, obliquity of, 47
Electronic eyepieces, 114
Elevated pole, 20
Ellipsoid, 36
Elongating stars, daylight observations of, 184
Elongation, calculation of time of, 178
− of star, 175
− of sun, 177
Ephemeris, Astronomical, 53–4
−, solar, BASIC program for computing, 392–5
− Time, 53
−, USA, extracts from, 353–63
Epoch, 50
Equal altitudes, stars, for azimuth, 171
− −, −, for latitude, 260
− −, −, for time, 288
Equation, personal, 112
− of the Equinoxes, 65
− of Time, 67, 69
Equator, celestial, 20
−, terrestrial, 28, 44
Equinox, autumnal, 47
−, vernal, 47
Equinoxes, 44
−, Equation of, 65
−, precession of, 49

Equipment for field astronomy, list of, 153
Error, clock, determination of, 116
−, horizontal collimation, 141, 146
−, index, vertical circle, 132
Errors in azimuth by hour angles, 205–8
− in direction of meridian in time determination, 273
−, effects of, in azimuth observations, 182–4
−, ex-meridian azimuths, 191
Excess, spherical, 2
Exercises, azimuth, 230
−, celestial sphere and astronomical coordinates, 27
−, the Earth, 39–41
−, latitude, 266
Exercise, location of objects on celestial sphere, 165
Exercises, longitude, 304
Exercise, position lines, 321
Exercises, sun and stars, 62
−, time, 298
Ex-meridian altitude observations for azimuth, 190–204, 330, 333
− altitudes, time by, 279–88, 346–8
Eye, protection when observing the sun, 125, 148
Eyepiece prism, 147, 153
−, Hunter shutter, 113
Eyepieces, impersonal, 113
−, electro-mechanical, 113
−, electronic, 114

Figure of Earth, 34
Filter (dark glass), for observing sun, 125, 148
First Point of Aries, 22, 47, 69
− − of Libra, 47
Fix, position-line, 50, 306–21
Fixed stars, 19
FK4 Star Catalogue, 49, 58
Focus of telescope, importance of, in observing daylight stars, 170
Forms, booking, 151
Formulae, basic, for solution of spherical triangles, 3
−, cyclic, 3
− for spherical triangles, derivation, 12–16
French Academy, 35
Frigid Zones of earth, 31

Geocentric horizon, 121
− latitutide, 36
Geodesy, International Association of, 36
Geodetic datum, world, 39
Geographic latitude, 36
− position of star, 306
Geoid, 36

Glass, dark (filter), 125, 148
Gnomon, of sundial, 295
Gravitation theory, Newton, 35
–, law of, 66
Great circles, 1
Greek alphabet, xii
Greeks, 34
Greenwich, 29, 72
– hour angle, 68, 300
– Mean Time, 72
– Sidereal Time, 70

Horizon, celestial, 18
–, geocentric, 121
–, sensible, 121
Horizontal parallax, 120
– sundial, 295
Horrebow level, 239
Hour angle, 24
– –, computation of, 178-9, 204, 279-80
– circle, 23
HP calculator, 16
HP-67/97 computer programs (RPN) for azimuth, 367-76
Hughes' Tables, 315
Hunter shutter eyepiece, 113
Hydrographic Department, Admiralty, 315

Identification of stars, 162, 326
Illuminating gear for theodolite, 148, 153
Impersonal telescope eyepieces, 113
Inclination, earth's axis to plane of orbit, 43
Independent Day Numbers, 53
Index error, vertical circle, 132
–, refractive, 122
Initial, or origin of survey circuit, 168, 398-400
Instants of time, conversion, 74-96
Intercept, position-line, 310
International Association of Geodesy, 36
Interpolation table for sun, 89
Intervals of time, conversion, 75-8

Jupiter's satellites, eclipses of, for longitude, 301

Kern DKM-3A astronomical theodolite, 112, 114

Laplace's equation, 38
Lapland, 35
Latitude, astronomic, geodetic, geographic, geocentric, 36
– by altitude of close circum-polar star, 255, 342
–, approximate, by meridian transits, 235

–, determination of, 233-70, 340-5
–, error, effect on azimuth, 182, 191, 207
–, –, effect on time, 281
–, exercises, 266
–, geodetic, 36-9
– and longitude simultaneously from position lines, 306-21, 351-2
–, methods, choice of, 264
–, –, summary, 264
–, need for, 233
–, parallel of, 29
–, using Polaris at any hour angle, 342-5
– from position lines, 306-21
– by Talcott's method, 238
–, terrestrial, 28
– from two stars at the same altitude, 260
– from zenith pairs of stars, 237
Least-squares solution of position-line fix, 319-21
Length, degree of longitude, 29, 311
Level, alidade, correction to altitudes, 133, 145
– correction, azimuth, 142-4
–, plate, 143
–, striding, 143
Libra, First Point of, 47
Lighting system for theodolite, 148, 153
Limb, of sun, 46, 125
Lines, position, 306-21, 351-2
List of Radio Signals, Admiralty, 115
Local hour angle, First Point of Aries, 69
– – –, sun, 89
– Mean Time, 71
– Sidereal Time, 24, 69, 78
– Standard Time, 73
Location of objects on the celestial sphere, 157, 325, 328
Longitude, determination of, 300-4, 306-21, 349-50
–, exercises, 304
–, geodetic, 36-9
– by older methods, 301
– by position lines, 306-21, 351-2
–, terrestrial, 28
Lower transit, 24
Lunar observations for longitude, 301

Magnitude of stars, 169
Mark, reference, 149, 153, 169
Mean Place of star, 50
– Solar Day, 67
– – Time, 67
– Sun, 67
– Time, Greenwich, 72
– –, Local, 71
Meridian, 21, 28
– altitudes for latitude, 235

Index

—, plane of, 21, 234
—, right ascension of, 25, 69
— transits, approximate latitude by, 235
— —, time by, 272, 345, 349
Meridional circuit, 168, 398
Meridians, convergence of, 396–400
Moon culminations, 301
Motion, apparent, of stars, 18
—, proper, 49
— of sun, in right ascension and declination, 44

Nadir, 18
Napier's Rules of Circular Parts, 4
— — for solution of right-angled spherical triangles, 4
Navigation Tables, 315
Newton, Isaac, 35
Noon, 71
North Temperate Zone, 30
Notation, xi–xii, 325
Numbers, Besselian Day, 53
—, Independent Day, 53
Nutation, 49

Object, reference, 149, 153, 169
Oblate spheroid, Earth, 35
Oblique spherical triangles, solution of, 6
— sundials, 297
Obliquity of Ecliptic, 47
Observation of circum-polar star near elongation for azimuth, 174
— of stars in daylight, 169, 170, 328
Observations, corrections to, summary, 144–6
—, preparations for, 147–56
—, timing of, 109–18
Observer's assistant, 150
Observing equipment, list of, 153
— programme, 151
Occultations of stars by the Moon, for longitude, 301

Parallactic angle, 26, 193, 206, 283
Parallax in altitude, table of sun's, 121
— corrections to altitude, 119, 144
—, horizontal, 120
— of near stars, 19
Parallel of declination, 23
— of latitude, 29
Personal equation, 112
Peru, 35
Photo-electric cell, 114
Photographs of the Moon, for longitude, 301
Pisces, 49
Place, Mean, of star, 50

Planisphere, 151
— for predicting positions of σ and B Octantis, 222
Plate level correction to azimuth, 142–4, 146
Plotting position lines, 310–13, 318
Polar distance, 23
— motion, 73
Polaris, *see* Pole Star
Pole, celestial, 19
—, elevated, 20
—, terrestrial, 28
Pole Star, 19, 49, 221, 256, 336, 342, 344
— —, azimuth from, 221–7, 336–40
— —, — —, at any hour angle, 336
— —, latitude from, 255, 342
— — tables, 353–5, 359–61, 364–5
— — —, azimuth using, 221, 336–40
— — — for azimuth and latitude, 223, 336, 342–4
Position, assumed, for position-line fix, 309
—, geographic, of star, 306
— lines, 306–21, 351–2
— —, exercise, 321
— —, fix, 50
— —, plotting, 310–13, 318
— —, —, for simultaneous latitude and longitude, 306–21, 351–2
Precession of the equinoxes, 49
Precessional constants, 53
Prediction of star position, 157, 325, 328
— of positions of σ and B Octantis by planisphere, 222
Preparation for observations, 147–56
Prime vertical, 25
— — sundial, 296
Prism, eyepiece, 147, 153
— attachment, Roelofs solar, 126, 145, 195
Prismatic astrolabe, 291–4
Programme for observing, 151
Programs, computer, 366
—, BASIC, comprehensive, for azimuth, latitude, time and longitude, 376–85
—, —, for sun's ephemeris, 392–5
—, —, for position-line fix, 386–92
—, HP-67/97, for azimuth, 367–76

Quartz stop-watch, 111

Radio receiver, 110
— *Signals, Admiralty List of,* 115
— time-signals, 115–16
Rate, clock, determination of, 116
Recorder, portable tape, 110

405

Reference collimator, 150
— mark or object, 149, 153, 169
Refraction, atmospheric, in altitude, 122, 145
—, coefficient of, 122
Refractive index, relative, 122
Reverse Polish Notation (RPN), 16, 366
Right-angled spherical triangles, solution of, 4 ff.
Right ascension, 23, 48
— — of star, 21
— — of sun, 45
Roelofs solar prism attachment, 126, 145, 195
RPN computer programs for azimuth, 367–76

Saastamoinen's formula for refraction, 125
SALS tables, extracts from, 56-7, 60-1, 98–105, 364–5
Satellites of Jupiter, eclipses of, for longitude, 301
Second, atomic, defined, 55
Seiko quartz stop-watch, 111
Semi-diameter correction to sun altitudes, 125, 145
— — — — azimuths, 134, 145
— of sun, 45
Shimmer, 207
Shutter eyepiece, Hunter, 113
Sidereal Day, 24, 64
— Time, 24
— —, mean and apparent, 64, 65, 68, 69
— —, conversion of instants, 74, 78–97
— —, — — intervals, 75–8
— —, Greenwich, 70
— —, Local, 24, 69, 78
Sight Reduction Tables for Air Naviagtion, 315
Signals, Radio, Admiralty List of, 115
—, radio, time, 115–16
Small circles, 3
Smithsonian Astrophysical Observatory Star Catalogue, 50, 59
— — — — —, extract from, 59
Solar Day, 65
— ephemeris, BASIC program for computing, 392–5
— prism attachment, Roelofs, 126, 145, 195
— Time, 65, 69
— —, Mean, 67
Solstitial points, solstices, 47
Solution of spherical triangles, 1–16
South Temperate Zone, 31
Southern Cross, 19
Sphere, celestial, 1, 17
Spherical excess, formula for, 2

— triangles, solution of, 1–16
— trigonometry, derivation of formulae, 12–16
Spheroid, Earth an oblate, 35
Standard Time, Local, 73
Star Almanac for Land Surveyors, 23, 45, 48, 49
— — — — —, extracts from, 56, 57, 60, 61, 97, 98–105, 364–5
Star, altitude of, 21
—, Apparent Place, 49
— atlas, 51
—, azimuth of, 21
— *Catalogue, Boss,* 50
— —, *Smithsonian Astrophysical Observatory,* 50, 59
— catalogues, extracts from, 56–61, 98–105, 353–65
— charts (from *SALS*), 60, 61
—, declination of, 21
—, elongation of, 175
— globe, 151
— identification, 162, 326
—, Mean Place, 50
—, right ascension of, 21
Stars, apparent motion of, 18
—, circum-elongation observation of, 184
—, fixed, 19
—, observation in daylight, 169, 170, 328
Stereographic projection for star predictions, 151
Stile, of sundial, 295
Stop-watch, 111, 117
—, Seiko quartz, 111
Striding level, 143
Substellar point, 306
Sun, 42–8
—, Earth's orbit around, 43
—, elongation of, 177
—, interpolation table for, 89
—, mean, 67
—, motion of, in R.A. and Dec., 44
—, semi-diameter, 45
—, —, coefficients for computing, 96
—, —, correction for, 125, 134, 145
—, time of local upper transit, 94
—, true or apparent, 89, 92, 94
Sun's declination, coefficients for computing, 96
— —, calculation of, 197
— ephemeris, BASIC program for computing, 392–5
— limb, 46, 125
— parallax in altitude, table, 121
— path on celestial sphere, 47
Sundials, 294–8
Symbols, xi–xii

Index

Synopsis of astronomical terms, 25
Système International, 65

Table, pressure conversion, xiii
–, temperature conversion, xiv
*Tables of Computed Altitude and Azimuth,
   Admiralty*, 315
–, *Hughes', for Sea and Air Navigation*, 315
–, *Navigation, for Mariners and Aviators*
   (Dreisonstok), 315
–, *Sight Reduction, for Air Navigation*, 315
–, see also *Apparent Places of Fundamental
   Stars; Ephemeris, Astronomical;
   Smithsonian Astrophysical Observatory; Star Almanac for Land
   Surveyors*
Talcott's method for latitude, 238
Tape recorder, portable, 110
Temperate Zones of Earth, 31
Terrestrial latitude and longitude, 28
Theodolites, 147
–, illuminating gear for, 148, 153
–, slow-motion screws of, 147
Thermometer, 150, 153
Time, Atomic, 65, 68
–, Civil (Zone), 73
–, determination of, 271–98
–, – methods, choice of, 298
– diagrams, 79–95
– of elongation, calculation of, 178
–, Ephemeris, 53
–, Equation of, 67, 69
–, exercises, 298
– by ex-meridian altitudes, 279–88, 347-8
–, Greenwich Mean, 72
–, Local Mean, 71
–, Local Standard, 73
–, Mean Solar, 67
–, by meridian transits, 272, 345, 349
–, – – –, both sides of zenith, 276
–, need for astronomically determined, 271
– recording, 110, 112
–, Sidereal, 24
–, –, mean and apparent, 64, 65, 68, 69
– signals, radio, 110, 115–16
–, Solar, 65, 69
–, from stars at the same altitude, 288
– transformations, 74–97
– of transit of sun or star, calculation of, 246
–, Universal, 72
–, Zone, 73
Timekeeping, 110

– equipment, 109–18, 150, 153
Timepieces, 109
Timing of observations, 109–18
– – –, accuracy of, 117
Torrid Zone of Earth, 30
Transformations, time, 74–97
Transit detector, electronic, 114
–, First Point of Aries, 87
–, lower, 24
– of sun or star, calculation of time of, 246
–, upper, 24
Transits, meridian, time by, 272, 345, 349
Triangle, astronomical, 2, 25, 26
–, spherical, 1
Trigonometry, spherical, 1–16
Tropics of Cancer and Capricorn, 30
TRS-80 microcomputer, 16, 367
Trunnion axis, dislevelment error in azimuth, 142, 146
Tycho Brahe, 42

Universal Time, 72
– –, Coordinated, 65
Upper transit, 24
USA Bureau of Land Management, *Ephemeris*, use of, 325–65
– – – – –, –, extracts from, 353–63
UT, UT0, UT1, 73
UTC, 65

Venus, 42
Vernal Equinox, 47
Vertical circle, index error correction, 132
– collimation correction to altitudes, 132, 145
–, deviation of, 37, 400
–, prime, 25
VNG time-signals, 115

Watch, stop-, 111
Watches, digital, 109
Wild T4 astronomical theodolite, 112
Wind-break, 147
World geodetic datum, 39
WWV and WWVH time-signals, 115

Zenith, 18
– distance, 22
– pairs of stars, latitude by, 237
– telescope, 239
Zodiac, 22
Zone time, 73
Zones, climatic, of Earth, 30